USING CATIA® V5

Fred Karam
Charles Kleismit

THOMSON
DELMAR LEARNING

Australia   Canada   Mexico   Singapore   Spain   United Kingdom   United States

THOMSON

∗

DELMAR LEARNING

Using CATIA® V5

Fred Karam and Charles Kleismit

Vice President, Technology and Trades SBU:
Alar Elken

Editorial Director:
Sandy Clark

Senior Acquisitions Editor:
James DeVoe

Development Editor:
Jaimie Wetzel

Marketing Director:
Cyndi Eichelman

Channel Manager:
Fair Huntoon

Marketing Coordinator:
Casey Bruno

Production Director:
Mary Ellen Black

Production Manager:
Andrew Crouth

Production Editor:
Thomas Stover

Technology Project Specialist:
Kevin Smith

Editorial Assistant:
Katherine Bevington

Freelance Editorial:
Carol Leyba, Daril Bentley

Cover Design:
Cammi Noah

Library of Congress Cataloging-in-Publication Data
Karam, Fred, 1965
Using CATIA V5 / Fred Karam
p. cm.
1. CATIA (Computer file) 2. Computer-aided engineering— Computer programs. I. Title.
TA345 .K38 2003
620.0042'0285'5369—dc21
2003020304

ISBN: 1-4018-1944-X

NOTICE TO THE READER

For my wife, Christine, and my two wonderful children, Hannah and Alexander.

—Fred Karam

For my wife, Trudy, who always encouraged and supported me, even when life seemed so unfair to her. I love you. And for my children, Conner and Julia. I truly love being your daddy.

—Charles Kleismit

Acknowledgments

We would like to take this opportunity to thank a number of people who have contributed to this book. First, our appreciation to Delmar Publishing for granting us the opportunity to write this book. Although it has been several years since the last book, we saw a need in the marketplace for this book and felt compelled to offer our experiences.

Second, we would like to thank the Delmar team in regard to all aspects of the book, including the fine team that worked with us on editing, formatting, and designing the book, Mr. Daril Bentley and Ms. Carol Leyba. Their constant feedback, encouragement, and persistence gave us the inspiration to complete this project.

Fred would like to thank the most important person in his life, his wife Christine. It was her love and support that ultimately enabled him to complete this book. She sacrificed more than he can imagine in allowing him all the late nights and weekends required to complete this project. He is deeply indebted to her for being such a wonderful, caring, loving, and beautiful wife and mother.

About the Authors

Fred Karam is founder and president of Engineering Solid Solutions, Inc., a leading engineering consulting firm in the Detroit metro area. He has twenty years of automotive engineering experience working for such companies as General Motors, Chrysler Motors, Parametric Technology Corporation, and United Technology. He holds a B.S.M.E. in Mechanical Engineering from Lawrence Technological University, and along with a vast engineering design background has experience in manufacturing operations such as castings, stampings, plastic injection molding, and rapid prototyping. Fred is very active in the CATIA and Pro/ENGINEER communities, regularly presents papers at national user group meetings, and is in addition to CATIA coauthor of books on Pro/ENGINEER.

Charles Kleismit is a CAD/CAM research and development engineer for a Daimler Chrysler. He has over 14 years of automotive-related product design experience, with proficiency in CATIA V5, V4, and VPM, and in Pro/ENGINEER. He has primary responsibility for Daimler Chrysler's interface with Dassault Systèmes in the United States, France, and Germany.

CONTENTS

Contents

INTRODUCTION

THIS BOOK IS INTENDED TO PROVIDE the beginning V5 user and migrating V4 user with a basic understanding of CATIA V5 via practical experience and a theoretical perspective regarding feature-based design and modeling. The methods and techniques covered in this book are intended to make the learning curve more manageable in the context of the entire product development process.

Approach and Content

The single most important aspect for the beginning user of CATIA V5, as reflected in the approach to this book, is to understand how robust models are developed and how the features of robust models are organized within the specification tree structure. This is critical because it is important to understand the basics of model organization before moving forward to more advanced methods and techniques.

Approach

The ability to conceptualize and build robust, manageable models within the context of the product development process is the key building block for future use of CATIA V5. The flow of chapter content in this book is intended to help develop the knowledge and skills required to begin working with CATIA V5 and toward eventual mastery.

Chapter organization is such that each chapter is a building block that leads to the next, ultimately taking you through the basics of creating robust models, assemblies, and detail drawings via a path from general to specific and from more simple concepts and techniques to more complex. In addition, the chapters follow the path of the workbench structure of CATIA V5, from those workbench environments in which you set up a working session and create geometry to the Drafting workbench in which you employ all of your skills and knowledge of the overall CATIA V5 environment to produce flexible detail drawing documents.

Content

Chapter 1 explores the structure and system features of CATIA V5, which leads to specifics of the CATIA environment (sessions, document types, the graphical user interface, and manipulating the environment) covered in Chapter 2. Chapter 2 begins an exploration of CATIA's framework of workbenches, with subsequent chapters examining the various workbench environments in a sequence that best explains and helps you develop a facility with creating and organizing flexible models.

Chapter 3 covers the Sketcher workbench, the basis from which models are created. Chapter 4 broadens your knowledge of the Sketcher workbench by exploring the Part Design workbench, in which features created in the Sketcher workbench are enhanced and manipulated. At this point, with the ability to create features and parts, you need a sense of how to organize model components toward further work within the context of design intent, which is the subject of Chapter 5.

With a practical knowledge of modeling organization, Chapter 6 moves into the larger arena of robust modeling, including what it is, design intent analysis, dealing with design change, and building in design flexibility. With the further foundation of chapters 5 and 6, Chapter 7 returns to CATIA's workbench environments to examine the practical applications of model organization, robustness, and flexibility.

Chapter 7 covers the Wireframe and Surface workbench, which complements the Part Design workbench environment of Chapter 4, examining the relationship of surface and solid modeling methods and how these can be made to work together within CATIA to maximize the utility of both the software and the models you create. Chapter 8 ties the modeling approaches, techniques, and theory of previous chapters together in an exploration of the Assembly workbench environment, in which modeling components come fully together. The chapter covers manipulation of the environment, the details of assembly features, and constraint and overall model analysis.

Chapter 9 covers in summary the Drafting workbench, which brings all of the theory and technical know-how of the book together in a picture of how detail drawing documents are created that best employ CATIA's capabilities to arrive at robust, dynamic models suitable to downstream applications with or without design modification. Along the way, the book provides methods, examples, and exercises related to specific applications and design situations in an easy-to-understand fashion.

Exercises and examples cover the options and total range of CATIA functionality. The exercises and examples within the book happen to be taken from the world of automotive design for the most part, but the functionality covered may be applied to any industry or manufacturing process, including casting, stamping, plastic injection molding, and others.

Book Mechanics and Use

The chapter content of this book, and thus the exercises and hands-on examples therein, both within chapters and from chapter to chapter is developed from basic knowledge and skills to more advanced. Thus, for the beginner and intermediate perhaps the best approach to using the book is to proceed from first chapter to last. The advanced intermediate or advanced user might follow this approach or explore chapters for their emphasis on a particular topic or workbench.

Exercises and examples are presented as numbered steps, with intervening commentary typically set apart from the step structure, so that each step represents the practical "to do" aspect of a process and the commentary explaining the reasoning behind it, finer details of its implementation, connections with other aspects of the process under discussion or with CATIA use, or how one step leads to another or what the effect of a step will be. Illustrations support exercises and examples where a visual is necessary to fully convey what a step refers to.

Throughout the book, you will find Notes, Tips, and Warnings, indicated as follows. The book also contains a detailed table of contents and a complete index, making fingertip access to specific content possible. You will also find a companion CD-ROM at the back of the book (see "About the Companion CD-ROM" following).

NOTE: *Notes highlight points of particular stress or alert you to connections you might not otherwise make or to other modeling considerations and techniques.*

TIP: *Tips convey practical information on maximizing CATIA functionality, avoiding mistakes, and making the modeling process as efficient and convenient as possible.*

WARNING: *Warnings prevent you from making mistakes that lead to unintended results or that create a situation in which you might lose data or a situation for which part of a work process must be redone or for which there is no solution or a less-than-desirable solution.*

About the Companion CD-ROM

The companion CD-ROM located at the back of the books contains the models used in exercises and examples found throughout the book. They are intended for use with CATIA V5 on Windows or UNIX platforms. Contact Fred Karam at *fkaram@essweb.com* for any comments or questions about the book or the companion CD-ROM.

CHAPTER 1

GETTING STARTED: CATIA V5

Introduction

CATIA V5 IS A BRAND NEW AND EXCITING computer-aided design (CAD) tool used by engineers and designers to develop products. With any new software product, it is important that you have a fundamental understanding of the capability and limitations of the tool. CATIA V5 was written as a platform-independent solution that offers countless scalability and customization. This platform solution is largely centered on a native Windows "look and feel." The goal of this chapter is to familiarize you with the key features, benefits, and configurations of CATIA V5 technology.

Objectives

The following are the key subjects of this chapter.

- CATIA V5 technology
- Windows integration
- Architectural software features
- Software configurations
- Platform portfolios P1, P2, and P3

CATIA V5 Technology

CATIA V5 is a "next-generation" design collaborative software that seamlessly integrates all facets of the product development process. This includes the concurrent sharing of data and geometric information from the design concept through the manufacturing process. At the heart of CATIA V5 is the integration of associative data structure that allows product life-cycle collaboration and faster evolution of products.

CATIA V5 was developed on a scalable architecture that concentrated largely on the interaction between user and software. The ease of use of CATIA V5 is centered on the new graphical user interface (GUI), which was developed utilizing Microsoft Windows and web technologies. This allows new users the comfort of familiar Windows operations, which promotes a higher level of productivity and a shorter learning curve. Although primarily developed for Windows, CATIA V5 is also ported to all major UNIX platforms.

The entire GUI is completely redesigned based on toolbars and icons and the Microsoft Windows architecture. Whether you are migrating from V4 or are a new user, the migration to V5 should be easier to make in terms of learning and use than with previous versions.

CATIA Background

CATIA was first developed by Dassault Systemes in the early 1980s for the aerospace industry. CATIA stands for Computer-aided Three-dimensional Interactive Application. Further developed with the backing of IBM, CATIA V2/V3/V4 became a powerful program. Dassault Systemes has set a new standard by introducing CATIA V5, which provides dramatic improvements in the system architecture and user interface.

CATIA is used primarily by the automotive and aerospace industries for automobile and aircraft product development. CATIA is also found in a variety of other industries, including aerospace, appliances, architecture, automotive, construction, consumer goods, electronics, medical, furniture, machinery, mold and die, and shipbuilding.

Windows Integration

The CATIA V5 interface has been optimized around a native Windows-type environment. This is a large step forward compared to V4, which was primarily a text-driven menu system that ran primarily on UNIX. The main focus of CATIA V5 was to combine the ease of use of a Windows interface with an industrial-strength solid modeling kernel.

The new user interface has been optimized to require the least number of commands in performing the greatest number of tasks. The user interface lets you work in a variety of configurations. For example, new users who require drop-down menus can work in this manner, and experienced users can make use of hotkeys. CATIA V5 offers many new Windows-like features, including the following.

- OLE integration
- Copy-and-paste functions
- Contextual and drop-down menus
- Traditional toolbars
- Drag-and-drop functionality
- Keyboard shortcuts (hotkey combinations)

Object Linking and Embedding (OLE)

OLE stands for Object Linking and Embedding. This provides the ability to integrate objects into a document either through linking or embedding, described in the following sections.

Linking

A reference, linked to a component of an outside application, is placed in the source document.

Embedding

The outside object is embedded to become a standalone application within the document. Even if the original file is deleted, the object remains integrated as an entire application.

Cut, Copy, and Paste Functions

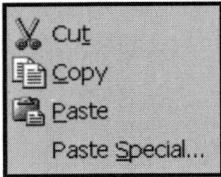

Cut, copy, and paste operations (depicted in figure 1-1) are the easiest way of sharing objects with very few operations. These commands are executed through menu picks or keyboard shortcuts. The concept is to select an object, run your mouse over either the Cut or Copy icon, and then click the first mouse button. The object is then pasted by moving the mouse to the desired location and clicking again to "paste" the object.

Fig. 1-1. Cut, copy, and paste functions.

Contextual and Drop-drown Menus

A *contextual menu* is a menu displayed when a user presses the middle mouse button while the pointer is over an object. Contextual menus, shown in figure 1-2, are one of two main types of menus, the other type being drop-down menus, which you select from a menu bar.

Contextual and drop-down menus are sometimes referred as pop-up menus if they visually pop up rather than drop down. These

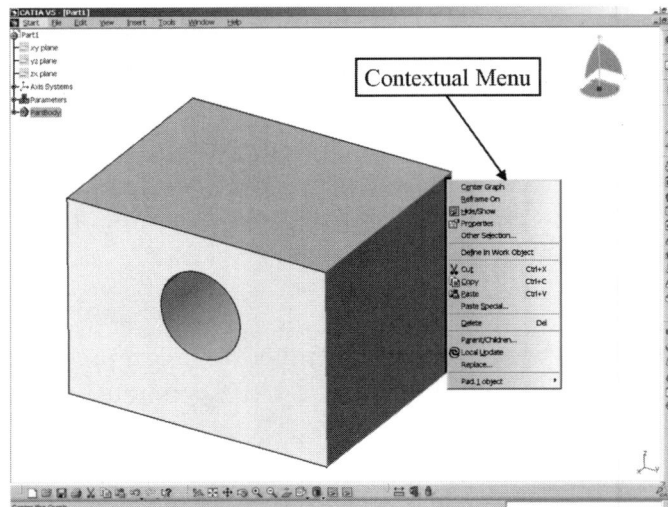

Fig. 1-2. Contextual menus.

menus provide quick access to menu items available elsewhere in an application. A contextual menu typically includes frequently used menu items. Examples of drop-down menus are shown in figure 1-3. Although contextual menus and drop-down menus are alike in most respects, contextual menus differ from drop-down menus in the following ways.

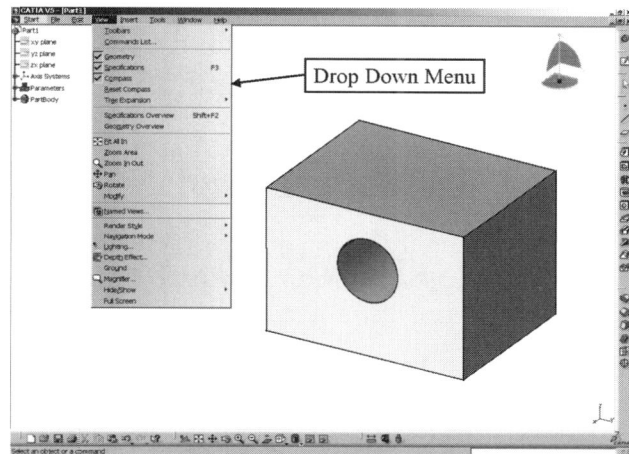

Fig. 1-3. Drop-down menus.

- Contextual menus are displayed only when a user presses the middle mouse button.
- Contextual menus consist of the following.
    — Menu items that affect the object
    — Menu items for the entire window
    — Menu items that do not require a selection

Traditional Toolbars

A toolbar is a strip (row or column) of command icons. Toolbars, which typically reside on the outer boundary of the GUI, can be relocated as independent menu bars. A typical Windows toolbar is shown in figure 1-4.

Fig. 1-4. Typical Windows toolbar.

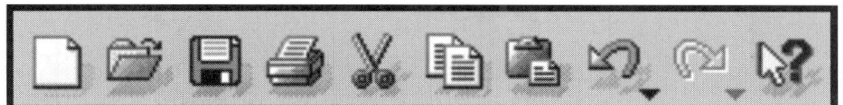

Drag-and-Drop Functionality

Drag-and-drop functionality allows you to transfer information between two entities logically associated within the GUI. You can dynamically drag and drop objects between forms using the mouse. An object might be an item you drag or a target upon which you drop an item after dragging it. Dragging is performed by clicking on an icon on the desktop, holding down the mouse button, and dragging the icon somewhere else on screen. Dropping refers to releasing the mouse button when the dragged item is at the location you want to place it.

Keyboard Shortcuts

The following are copy, cut, and paste keyboard shortcuts. In that these are frequently used functions, many users find these shortcuts convenient. Figure 1-5 shows examples of these functions.

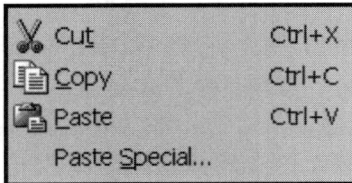

✄ Cut	Ctrl+X
📋 Copy	Ctrl+C
📋 Paste	Ctrl+V
Paste Special...	

Fig. 1-5. Copy, cut, and paste keyboard shortcuts.

- *Ctrl C:* Copies whatever was highlighted to the computer's temporary memory (clipboard).

- *Ctrl X:* Cuts whatever was highlighted to the computer's temporary memory (clipboard).

- *Ctrl V:* Pastes the previously copied or cut object at the point you choose.

NOTE: *Paste Special offers advanced part-to-part geometric and parametric linking capability.*

CATIA V5 System Features

The sections that follow describe various CATIA V5 system features. These include the sketcher, the parametric variational modeler, and associative functionality.

Sketcher

The sketcher is an essential and integral part of CATIA V5. The sketcher environment is the bridge between 2D elements and 3D geometry. The sketcher provides the functionality for creating and modifying 2D geometries used in the construction of 3D solids and surfaces.

Many solid features are started by first developing a 2D profile, called a sketch profile, in which you capture the design intent. Parameters and constraints may also be applied to the geometry. The sketcher environment, depicted in figure 1-6, is the key to a feature-based system in that it links 2D elements with 3D features, commonly referred to as sketcher-based features. Common sketcher-based features are pads (extrusions), pockets (cuts), shafts (revolves), and ribs (sweeps).

Fig. 1-6. Sketcher environment.

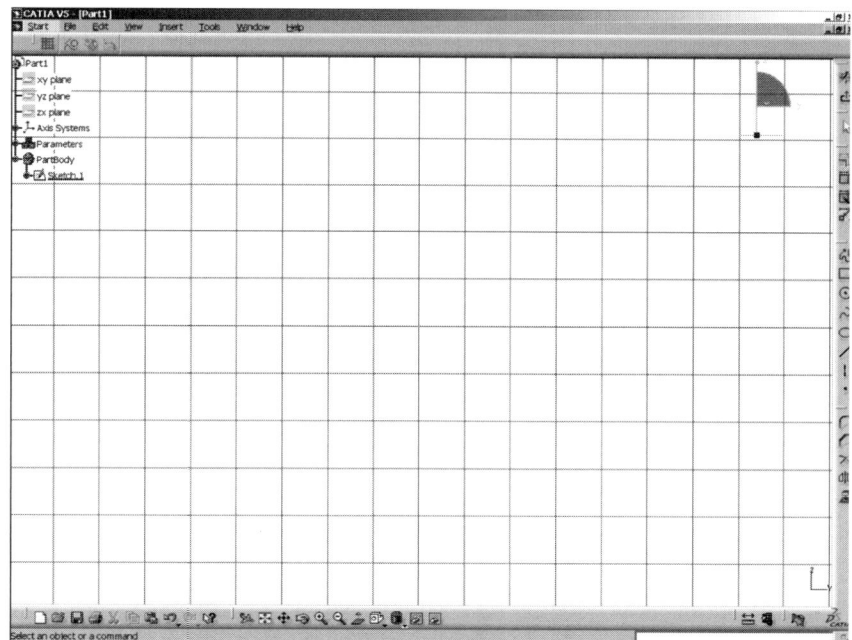

You know you are in a sketcher session when you see the grid structure used to assist in the creation of 2D geometry. Sketched elements can be manually or automatically dimensioned, based on personal preference. The ability to constrain or not constrain makes CATIA V5 very flexible in meeting the challenges of any design.

Parametric Variational Modeler

CATIA V5 is a 3D parametric variational modeling software that allows you to capture design intent by adding parameters or dimensions to drive the creation and manipulation of models. *Parameterization* adds intelligence to a part by capturing and maintaining design intent via definition of the relationships among elements, dimensions, and parameters of a model. This facilitates downstream changes by updating the model to a desired new configuration while maintaining the original design intent.

The power of CATIA V5 provides you with the ability to parameterize all geometric entities, including solids, surfaces, wireframe, and construction features. The entire design or portion of the design can be parameterized to allow for greater flexibility in developing quick design alternatives. You have the ability to add or remove dimensions at any time during the product design process. The choices you make in dimensioning and constraining a model are key in developing robust models.

Once a model has been parameterized, you can change the design's geometry by inputting of a new value for a dimension and updating the model. The update of the model will drive the geometry to its new location. In addition, parametric-based modeling allows solid modeling entities such as holes, fillets, bosses, and pockets to be associated with specific edges or faces. When the edges or faces move because of an update, dress-up features move along with it, maintaining original relationships.

Associative Functionality

Associativity is defined as the link between two different functions or objects within CATIA V5 such that when a change is made to one area all related areas are changed accordingly. For example, a change made to a CATPart model will update the CATDrawing and associated views.

Bidirectional associativity refers to two-way update, meaning that the update can occur from model to drawing or drawing to model. This means that a change made at any stage of the development process has the ability to propagate both ways from the point of origin.

CATIA V5 is an integrated associative system that facilitates changes to all downstream applications. This end-to-end capability allows you to control the propagation of design changes to all members of the product development team.

Software Configurations

CATIA V5 is an enterprise-wide software solution configured in various arrangements, as follows. These configurations are further explored in the sections that follow.

- Platform configurations
- Application portfolios
- Workbenches

Platform Configurations

CATIA V5 is organized into three different "platforms," designated P1, P2, and P3. These platforms are configured for different entry points into the CATIA V5 product development solution. Data created in one platform can be seamlessly utilized by another platform.

Platform P1

This platform provides core solid modeling capability for users who look to get introduced into high-end design solutions. This solution is ideal for entry-level users who wish to make the transition from legacy 2D systems. The P1 platform offers growth capacity in that all products are fully operational on the P2 platform.

Platform P2

The P2 platform is a process-oriented solution that supports more of an integrated environment for product life cycle development. P2 offers an end-to-end design, analysis, manufacturing, and product infrastructure solution.

The P2 platform is also based on a Windows look and feel but with a more integrated Windows user environment. It is mainly distinguishable by the user interface style, which integrates the configuration tree within the geometry creation window.

Platform P3

The P3 platform offers advanced knowledge based solutions specifically customized for certain industries, such as automotive and aerospace. The user interface style for the P3 environment is similar to the P2 environment. The only difference is that certain P3 products are optimized for the specialization of a product within a certain environment.

Application Portfolios

CATIA V5 organizes the software product offerings into categories known as *application portfolios*. These application portfolios are designed to group product applications (workbenches) to meet the needs of users at various points in the product development process.

CATIA V5 offers a wide range of applications and products well beyond the scope of this book. This book has as its focus the Mechanical Design portfolio for V5R11. Figure 1-7 shows the core application portfolios found in CATIA V5. Consult your local CATIA dealer for the latest information on current products.

Workbench "Products"

The term *workbench* is frequently used within the CATIA V5 environment. A workbench essentially means a working environment within an application portfolio that allows you unique functionality to create and manipulate geometry. The majority of workbenches are application portfolio specific. However, some workbenches (such as the Sketcher workbench) are integrated across multiple portfolios.

CATIA V5 offers a wide range of workbenches and related products well beyond the scope of this book. Consult your local CATIA dealer for the latest information on current products. This book explores the workbenches shown in figure 1-8.

Application Portfolio	
	Mechanical Design Core Solid Modeling, Drawing, and Assembly.
	Shape & Styling Free Form and Advanced Surface Applications
	Product Synthesis Advanced Digital Mockup and Simulation Applications
	Equipment & Systems Engineering Electrical, Hydraulic, and Mechanical Applications
	Analysis FEA Simulations Applications
	NC Manufacturing Manufacturing Numerical Control Applications
	Plant Factory and Plant Layout Applications
	Infrastructure Products for Integration of All Applications

Fig. 1-7. Application portfolios.

Workbench Products	
	Product Structure Structure and Organization of Products
	Sketcher Environment Used to Create 2D Geometries
	Part Design Core Mechanical Solid Modeling Capability
	Wireframe and Surface Design Basic Wireframe and Surface Creation
	Assembly Design Facilitates the Assembly of Parts
	Drafting Detail Drawing of Parts or Assemblies

Fig. 1-8. Workbenches.

Summary

The philosophy of CATIA V5 is based on the concept of the integration of digital products into the product development life cycle. The software has been completely reengineered from the ground up to offer industry leading technology.

CATIA V5 offers a new process structure developed around a unique set of software development tools, optimized around the Microsoft Windows environment. The entire user interface is completely redesigned based on toolbars, icons, and context-driven menus. This new icon-based menu structure is intended to shorten the user's learning curve.

The most impressive aspects of CATIA V5 are the integration of various workbenches and the seamless flow of the GUI. Solids are primarily created from sketcher-based and dress-up features. This provides much more capability compared with previous versions of CATIA.

Review Questions

1   What does the acronym CATIA stand for?

2   Name several ways in which CATIA V5 is integrated with the Microsoft Windows environment.

3   What is variational modeling as opposed to fully constrained modeling?

4   (True or false?): CATIA V5 requires that all created elements be constrained.

5   What is the difference between an application portfolio and a workbench?

CHAPTER 2

THE CATIA V5 ENVIRONMENT

Introduction

THIS CHAPTER IS INTENDED TO INTRODUCE you to the CATIA V5 environment. The ease of use of CATIA V5 is centered on the new graphical user interface (GUI), which was developed utilizing Microsoft Windows and web technology. This allows new users the comfort of familiar windows operations which will in return promote a higher level productivity and a shorter learning cycle.

Workbenches and toolbars, described in Chapter 1, are the key features of the interface that allow you to perform tasks and actions to quickly create and modify designs. Various methods of customization allow you to tailor the environment for maximum productivity.

Objectives

The following are the topics explored in this chapter.

- Starting a CATIA V5 session
- CATIA V5 document types
- Introduction to the GUI
- Customizing the environment
- Options for selecting and viewing

Starting CATIA V5

CATIA V5 offers various ways of initiating a user session from within the Windows environment. Most commonly this is done either from the Start menu or from a desktop icon.

Start Menu

Clicking on the Start button, located at the bottom left-hand portion of the Windows interface, will bring up a contextual menu, shown in figure 2-1. A secondary menu will automatically appear when you place the mouse cursor over the CATIA V5 option. Click on the CATIA V5 icon with the left mouse button to initiate the session. See figure 2-1.

Desktop Icon

Shortcut icons may be created and placed on the Windows desktop to quickly initiate programs such as CATIA V5. A double click on this type of desktop icon will also start a CATIA V5 session. This will initiate the icon's Start menu, show in figure 2-2.

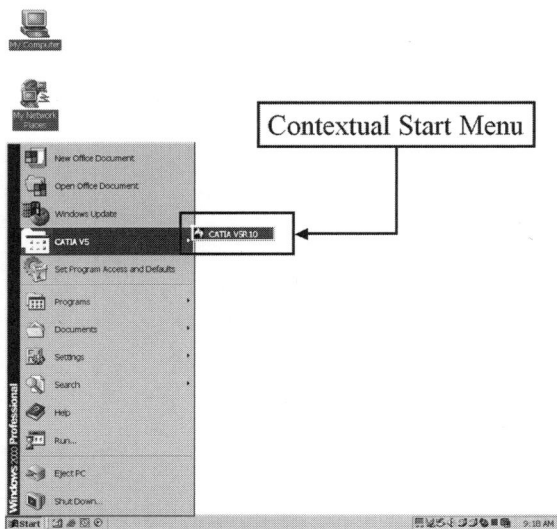

Fig. 2-1. Contextual Start menu.

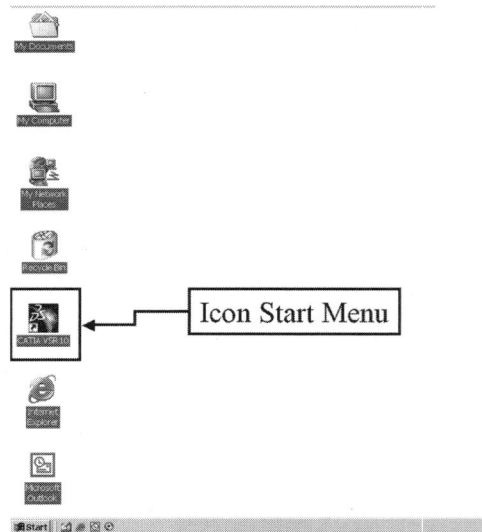

Fig. 2-2. Icon Start menu.

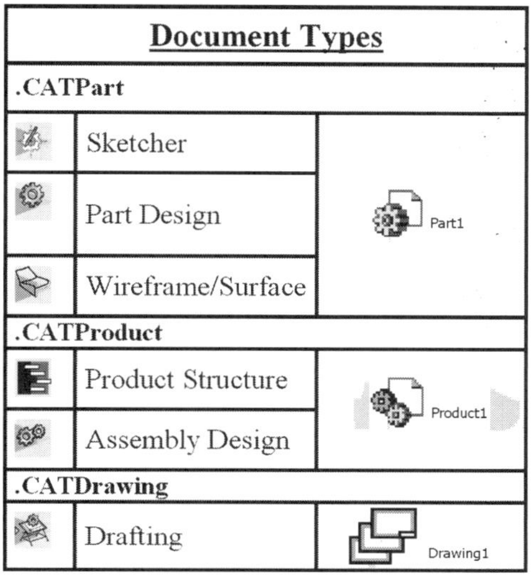

Document Types		
.CATPart		
	Sketcher	Part1
	Part Design	
	Wireframe/Surface	
.CATProduct		
	Product Structure	Product1
	Assembly Design	
.CATDrawing		
	Drafting	Drawing1

Fig. 2-3. CATIA document types.

CATIA V5 Document Types

CATIA V5 differs from V4 in that various parts of geometry are stored in different document formats. These different document formats are based on Microsoft's Multi-Document Standard. This eliminates the legacy V4 session, model, and EXP file formats. The type of information stored in these new document formats is dependant on the workbench used to create and modify the geometry. The document file format will be stored and located in the Windows directory or folder of choice. Figure 2-3 depicts the three distinct document formats found in CATIA V5.

GUI Introduction

The CATIA V5 GUI, shown in figure 2-4, is both elegant and refined. The new user interface is packed with menus and toolbars that are consistent with the traditional Windows look and feel. The attention to the ergonomic layout offers a fast and efficient way for new users to immediately be productive. Experienced Windows users should be very comfortable with the layout and location of familiar icons. The numbers in figure 2-4 refer to the numbered sections that follow.

Screen Layout

The sections that follow describe the main components of the new CATIA user interface. These include the Start menu, the Standard menu, the Active Workbench icon, geometry creation toolbars, the power input zone, the View toolbar, the feedback/prompt zone, the Standard toolbar, the model configuration tree, default planes and axes, and the Compass tool.

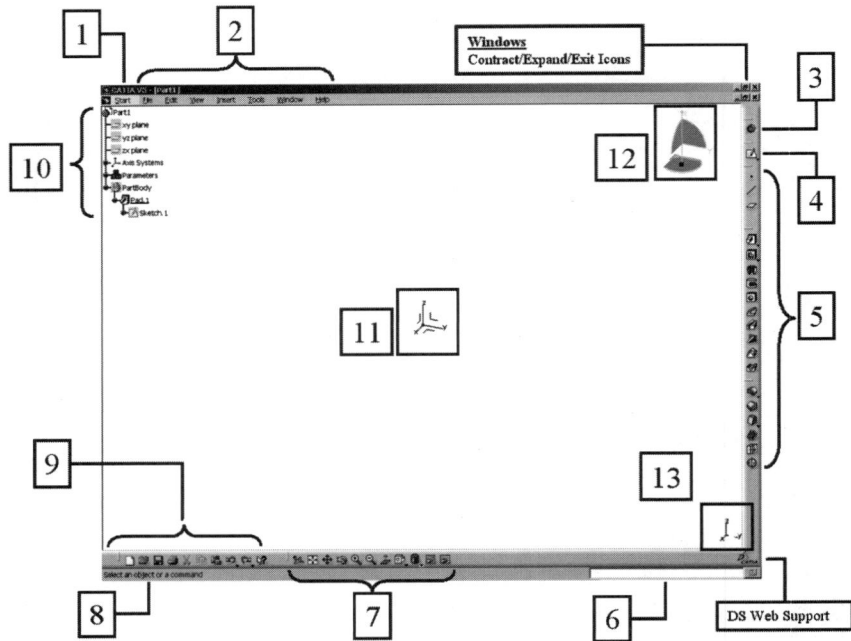

Fig. 2-4. CATIA's new GUI.

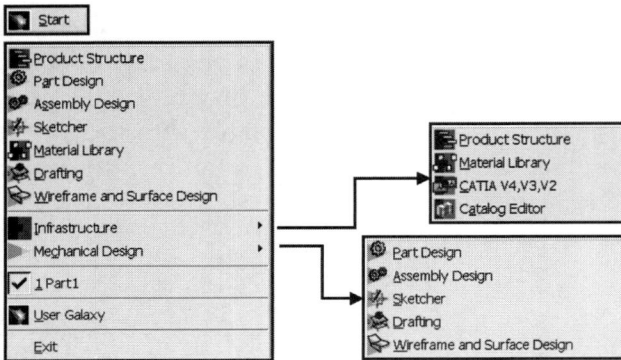

Fig. 2-5. Start menu options.

Start Menu [1]

The Start menu bar is the main navigator for switching between different environments. You have access to various workbenches and product portfolios, depending on the installed configurations and products. The Start menu also keeps track of the last models used in a session, providing quick retrieval. The CATIA V5 environment should be exited through this menu to ensure proper storage and configuration of your session. The options of the Start menu are shown in figure 2-5.

Standard Menu [2]

The Standard menu bar is consistent with the Microsoft Windows environment. The Standard menu is the primary access to initiating the majority of functions within CATIA V5. For ease of use, various shortcuts are built into the user interface, which references most of the functionality contained in the Standard menu. The sections that follow describe the options of the Standard menu.

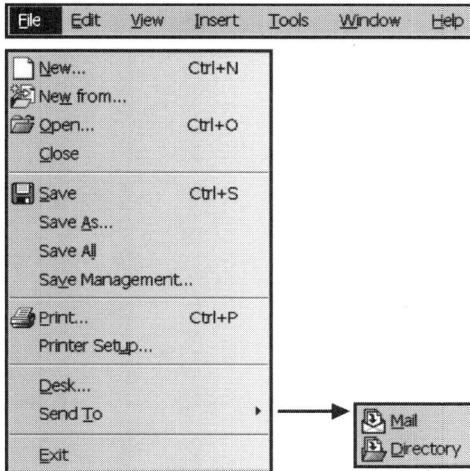

Fig. 2-6. File menu options.

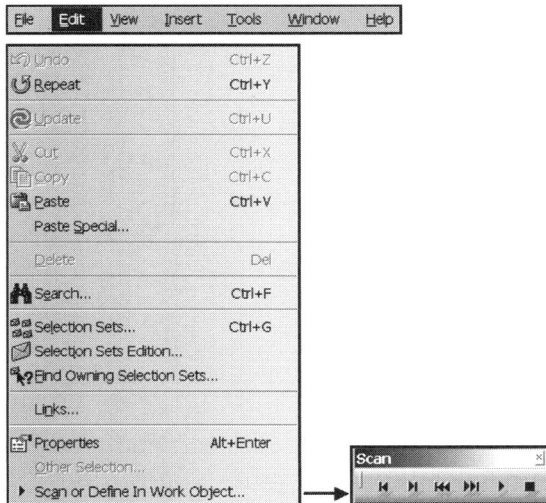

Fig. 2-7. Edit menu options.

File

The File option facilitates control of CATIA V5 documents. Opening and closing documents, and creating new documents, are the primary functions of this selection, which is consistent with most Microsoft applications. The File menu's options are shown in figure 2-6.

Edit

The Edit option allows for the manipulation of objects within the CATIA V5 environment. As mentioned in Chapter 1, common Windows functions such as cut, copy, and paste are accessed through this selection. Other functions, such as selection options and object properties, are also accessed within this menu, shown in figure 2-7.

View

The View menu, shown in figure 2-8, provides a comprehensive set of tools that allows you to manipulate the display of model geometry. Toolbar menus are also toggled on and off within this menu option for the purpose of customizing each user environment.

Insert

The Insert option allows for the creation, manipulation, and insertion of geometry elements within the model.

*Fig. 2-8. View
menu options.*

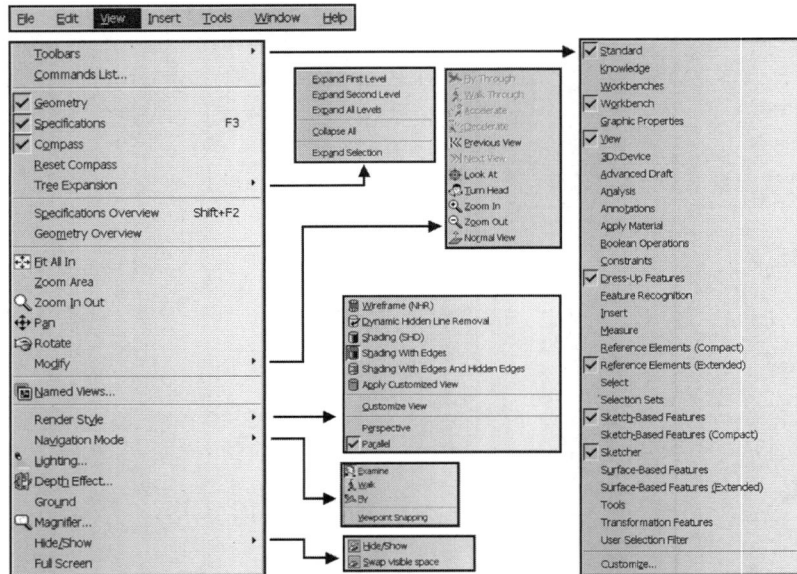

*Fig. 2-9. Insert
menu options.*

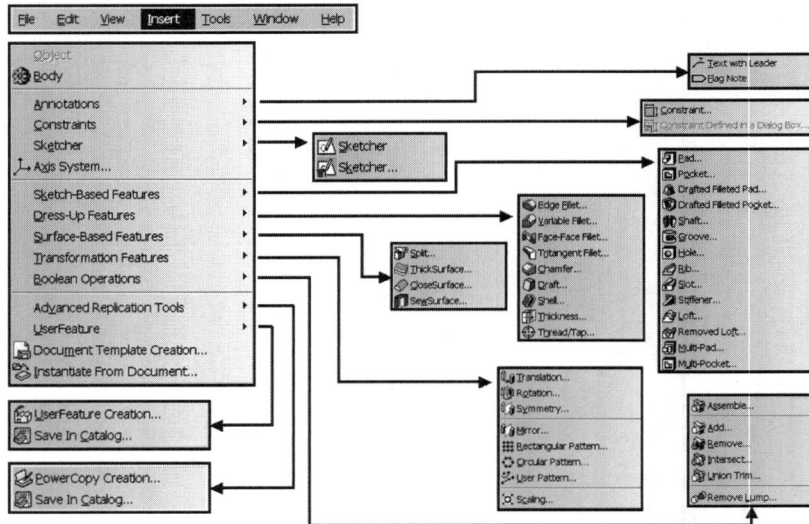

Depending on the active workbench, different geometry creation
options are accessed and introduced into the model from this
menu, shown in figure 2-9. The primary geometry creation
menus are Sketch-Based and Dress-Up (features). Other functions, such as post-feature constraints, annotations, and Boolean
operations, are also available.

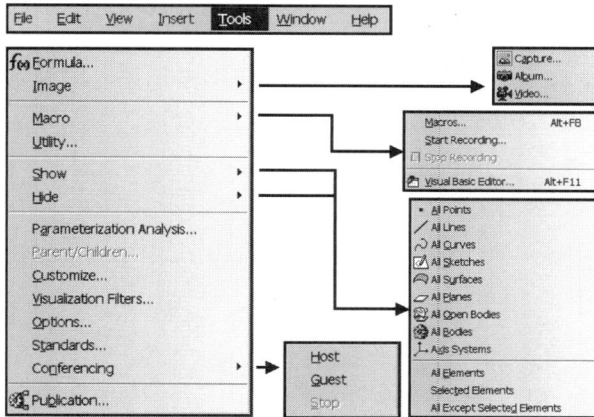

Fig. 2-10. Tools menu options.

Tools

The Tools menu, shown in figure 2-10, provides utilities for customizing the user environment, capturing images, and quickly showing or hiding common geometry element types.

Windows

CATIA V5 makes it possible to open several models independently or alternatively dependent from each other via the Windows menu, shown in figure 2-11. Common Windows tiling functionality allows each model session to be displayed vertically, horizontally, or in cascading fashion. This is useful for coordinating the display when moving geometry from one model to another.

Help

The Help menu, shown in figure 2-12, provides access to local help files or web-based learning tools.

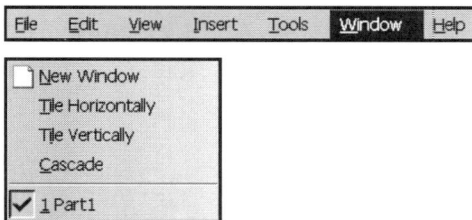

Fig. 2-11. Windows menu options.

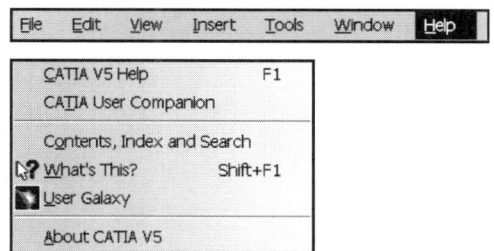

Fig. 2-12. Help menu options.

Active Workbench Icon [3]

The Active Workbench icon displays the active workbench and allows for quick access between other workbenches, such as part design, assembly, design, and draft. Selecting this icon will display a separate menu, shown in figure 2-13, which contains options for toggling between alternative workbenches. The availability of workbenches within this menu is dependant on the installed software and custom configuration.

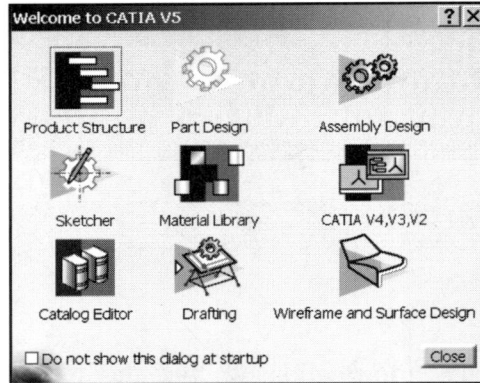

Fig. 2-13. Workbench menu options.

Sketcher Workbench Icon [4]

The Sketcher Workbench icon allows you to enter and exit the Sketcher workbench environment.

Geometry Creation Toolbars [5]

The right-hand side of the user interface is typically where the majority of the geometry creation toolbars are located. The available toolbars are dependent on the active workbench. These toolbars can be detached from the Windows interface border and dragged to any location within the Windows environment.

There are three methods of activating an icon within a given toolbar. A single click of mouse button 1 (MB1) on the icon will allow you to activate an icon once, as in creating one fillet on a part. Double click MB1 on the icon and you will be able to perform that function as many times as needed until you exit that function. To exit the function, simply click on the Cancel button, if it is available, or click once more on the icon. The third method of activating an icon is to use MB1 and drag and drop the icon onto the geometry you wish to work on. An example of this would be to drag and drop the Edge Fillet icon onto the edge of the part you want to fillet.

TIP: *If the icon you wish to use has a carrot symbol in the lower right corner of the icon this indicates that there are more options of that function available. Simply click on the carrot to display these options and click on the one that you would like to use. CATIA V5 will hold in memory the last set of toolbar configurations.*

Power Input Zone [6]

The power input zone is an advanced function that allows you to directly enter commands or queries within CATIA V5. These commands, however, are beyond the scope of this book. You can obtain more information from the CATIA software user's manual.

View Toolbar [7]

The View toolbar, shown in figure 2-14, provides options for viewing, displaying, and orientating the model within the user environment.

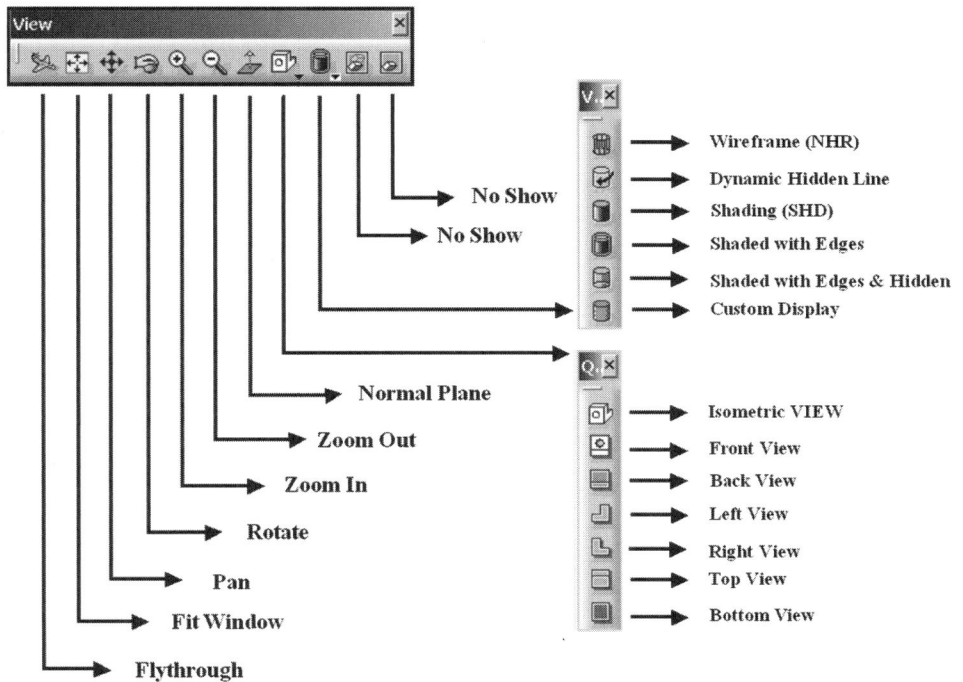

Fig. 2-14. View toolbar.

Feedback/Prompt Zone [8]

This area provides critical feedback and input prompts based on the location of the mouse or function being initiated.

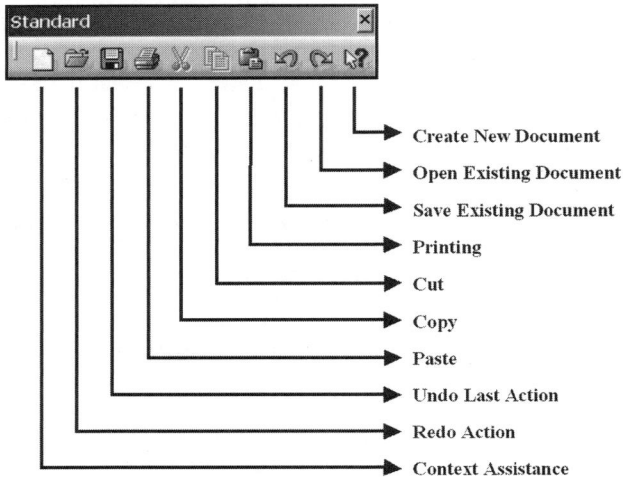

Fig. 2-15. Standard toolbar.

Standard Toolbar [9]

The Standard toolbar, shown in figure 2-15, is consistent with the Microsoft Windows architecture, displaying familiar icons that are associated with frequently used functions. These functions are derived from various menus located within the Standard menu.

Configuration Tree [10]

The configuration tree is an essential tool for manipulating, controlling, and organizing model elements and features. The branches along the tree contain the history and processes used to create the parts. These branches are organized using a multi-body principle (solid and open), covered in more detail in Chapter 3.

Default Planes and Axes [11]

CATIA V5 creates default planes and axes upon creation of a new model. These are always the first features located in the configuration tree.

Compass Tool [12]

The Compass tool allows for the modification of the location and/or orientation of a part relative to the X-Y-Z coordinate system. The Compass tool also locates parts relative to one another in the assembly environment.

Axis Display [13]

Default axis display shows the orientation of the part object.

Customizing the User Interface

The sections that follow explore issues regarding customization of the user interface.

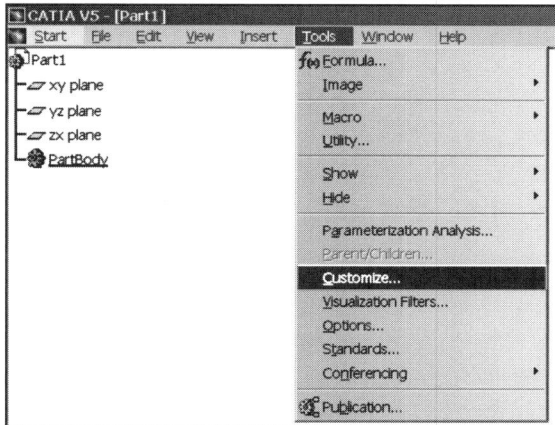

Fig. 2-16. Customize menu.

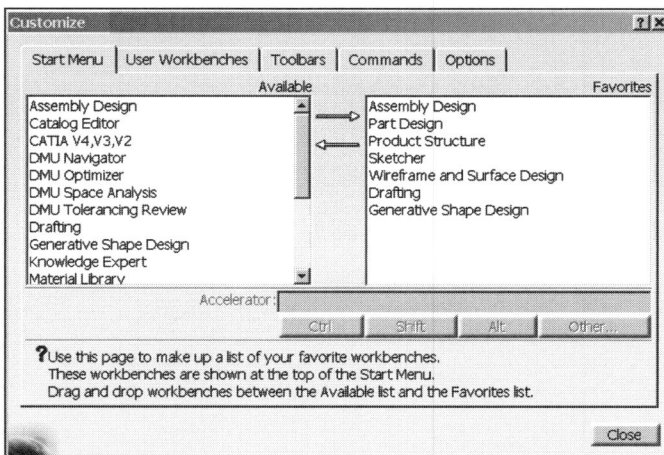

Fig. 2-17. Start menu.

User Environment

One of the most powerful features of CATIA V5 is the ability to customize the user environment. The Customize option is located under the Tools menu at the top of the user interface. Selecting this option will display a separate menu panel, shown in figure 2-16, which offers many options for customizing the GUI.

Start Menu

The Start menu has five tabs at its top. These tabs allow you to individualize the environment. The Start Menu tab, shown in figure 2-17, comes up by default. You can drag and drop workbenches between the Available list and the Favorites list. These favorites will then be displayed in the Start menu. These same favorite workbenches can also be accessed through the Active Workbench menu icon, shown in figure 2-18.

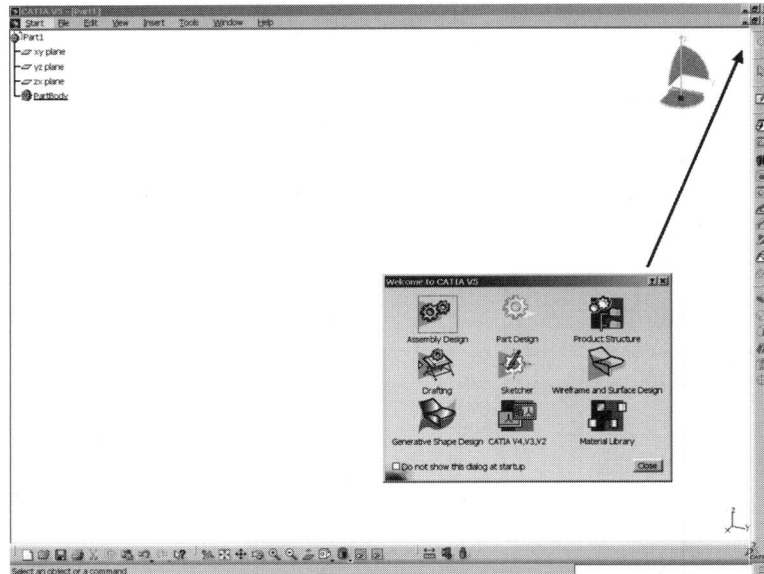

Fig. 2-18.
Workbench menu.

TIP: *An accelerator is another option that can be created for your favorite workbenches. Using MB1, click on one of your favorites. Then use the Ctrl, Shift, or Alt button plus any key found under the Other button to create an accelerator to whatever workbench you would like. Pressing these keys in the proper order will take you to that workbench without the use of your mouse. These same key combinations will now appear beside that workbench in the Start menu.*

User Workbenches

The second tab in the Customize window is User Workbenches, shown in figure 2-19. Like the standard workbenches, such as Part Design and Wireframe/Surface Design, you can create your own workbench with the toolbars from various workbenches. These toolbars can be from a combination of different standard workbenches. An example would be a toolbar from Part Design and a toolbar from Free-Style in a workbench you named Combination.

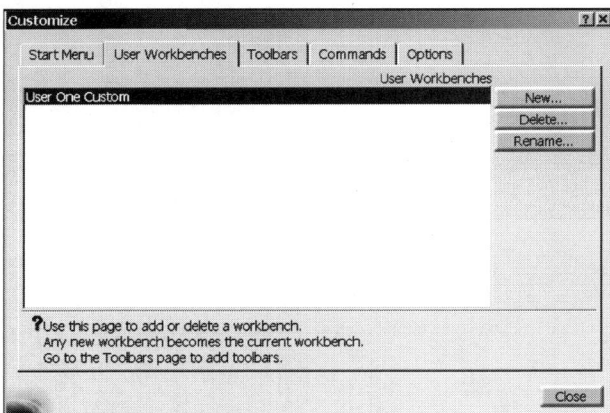

Fig. 2-19. User Workbenches tab.

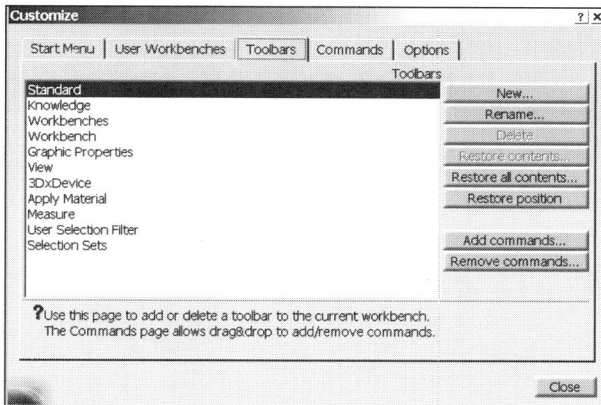

Toolbars

The third tab in the Customize widow is the Toolbars tab, shown in figure 2-20. In the New Toolbar window, you can now select and create any toolbars from any of the standard workbenches. Once created, you can then drag and drop these toolbars wherever you want.

Fig. 2-20. Toolbars tab.

Go to the Toolbars tab and select New. In the New Toolbar window you can now select toolbars from any standard workbench and add them to your newly created workbench. If the toolbar does not snap onto your workbench bar, simply drag and drop the toolbar there.

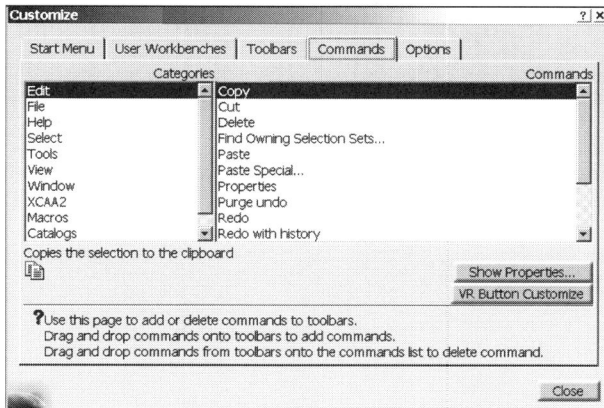

Commands

The fourth tab in the Customize window is the Commands tab, shown in figure 2-21. This tab creates command icons that can be added to any workbench or toolbar. Click on any of the command options under Categories and click on a command option, such as Select. Drag and drop this command option onto any toolbar.

Fig. 2-21. Commands tab.

NOTE: *To delete a command option (icon) from a toolbar, simply drag and drop the option back to the Commands area.*

Options

The fifth tab in the Customize window is the Options tab, shown in figure 2-22. This tab gives you the ability to toggle between large and small icons. The tab also allows you to toggle on or off tooltips. You can also choose to lock or unlock the position of tooltip toolbars and select the language of the user interface.

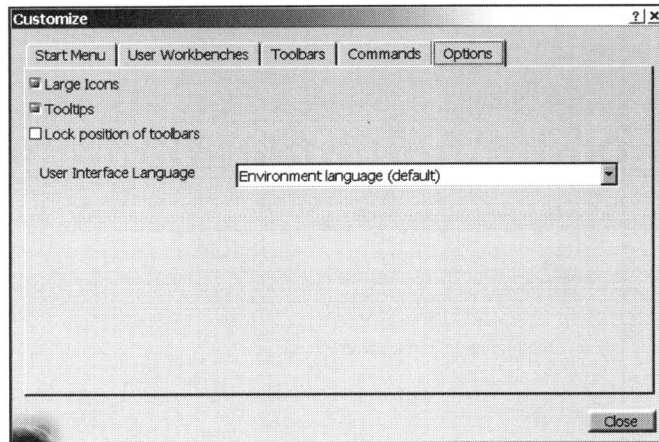

Fig. 2-22.
Options tab.

Options in the CATIA V5 Environment

This section explains how to set certain user-specified options and customized settings within the CATIA V5 environment. The Options menu selection is located at the top of the GUI under the Tools menu, as shown in figure 2-23.

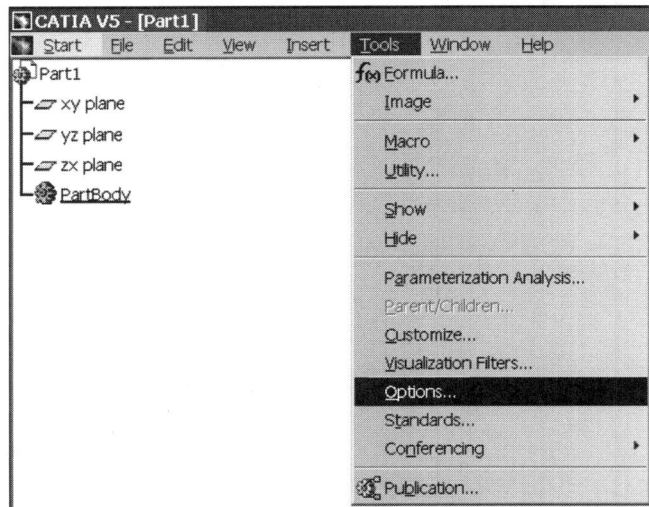

Fig. 2-23. CATIA V5
Options menu
selection.

General Panel

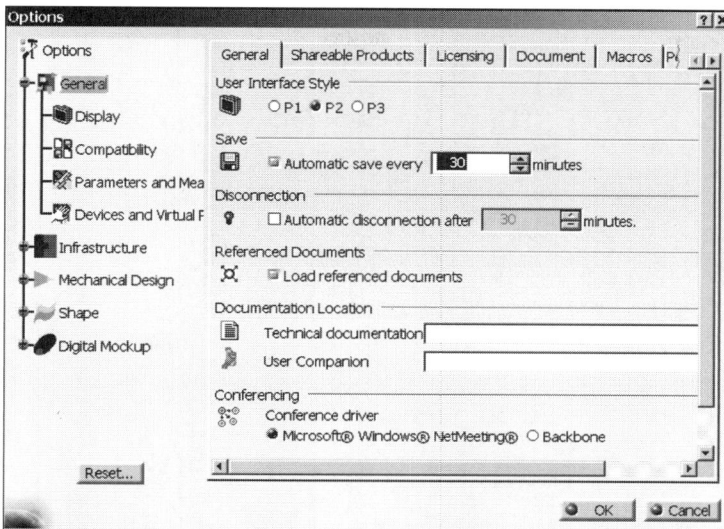

The General panel, shown in figure 2-24, is displayed when you select the Options menu. Multiple categories are present on the General panel. These categories (fields) contain numerous options for customizing the user environment. The goal of this chapter is to cover the basic options for customizing the environment.

Fig. 2-24. General panel.

User Interface Style

The User Interface Style category allows you to select among P1, P2, and P3 styles. Both P1 and P2 users can use the P1 style. The P2 style is for use by P2 users only.

Save

The second category is Save. Here you can toggle on or off automatic save mode. If you activate this mode, you can specify the duration of time between saves.

Disconnection

The third category is Disconnection. Here you can toggle on or off automatic disconnection mode. If you activate this mode, you can specify the duration of time before the session is terminated when CATIA V5 is not being used.

Reference Documents

The fourth category is Reference Documents. This setting is toggled on by default. What this means to you is that if there were any children linked to that document they would also be loaded. This option must be toggled on or off before loading the parent document, in order to achieve the results you want.

CATIA Documentation Location

The fifth category is CATIA Documentation Location. This option shows you the location of online documents. The CATIA Documentation Location field displays the path name to stored documents.

Drag & Drop

The seventh category is Drag & Drop. Here you can toggle on or off the ability to perform drag-and-drop operations. (See Chapter 1 for details on drag-and-drop functionality.)

Display

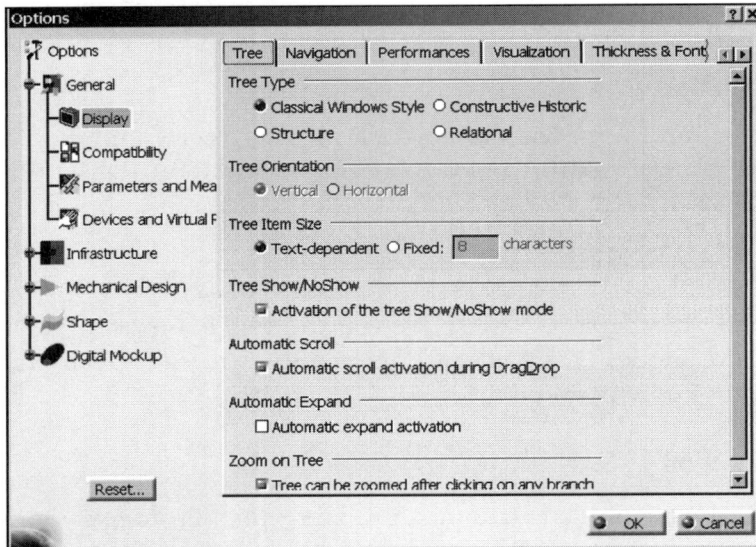

Fig. 2-25. Display panel.

The Display panel, shown in figure 2-25, offers many tabs for customizing various items within the CATIA V5 environment. These options are described in the sections that follow.

Tree

Select the Tree tab. This tab allows you to set the display of the CATIA V5 specification tree via the following category options.

Tree Type. The Tree Type category contains four styles from which to choose. For the purpose of using this book, we suggest selecting Classical Windows Style. As your proficiency level grows, experiment with the other styles to suit your needs.

Tree Orientation. There are two tree orientation options from which to choose. For the purpose of using this book, we suggest selecting Vertical.

Tree Show/No Show. With the Tree Show/No Show option toggled on, when you select something from the tree and then click on the Hide/Show icon, whatever you selected is grayed out in the tree. This is a good visualization aid when looking at the tree, allowing you to determine what will be shown on screen and what will not.

Navigation

Select the Navigation tab, shown in figure 2-26. This tab allows you to customize the display of navigation settings.

Fig. 2-26. Navigation tab.

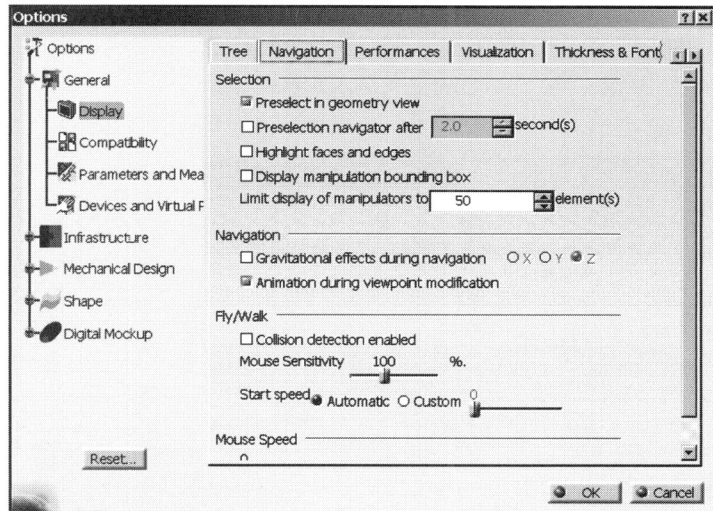

Selection. The Selection category also contains several options. With this option toggled on, as you place the cursor on different pieces of geometry they are highlighted (as are their names in the specification tree). With this option toggled on, faces and edges stay highlighted after they have been selected.

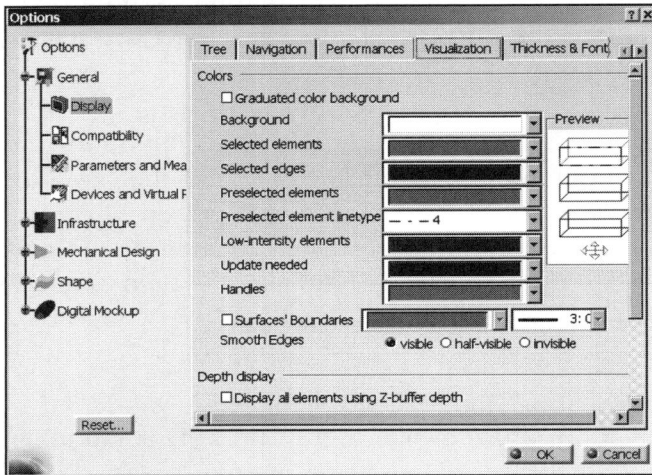

Fig. 2-27. Visualization tab.

Visualization. Select the Visualization tab, shown in figure 2-27. This tab allows you to customize the display of visualization settings.

Colors. With this option toggled on, a graduated color background appears for any open document. This is also true for the Preview window and the workbench tree on the Visualization tab. An example of a graduated color is one that goes from lighter to darker moving horizontally or vertically on screen.

TIP: *Keep the graduated color background turned off if you want better graphical response time.*

Parameters

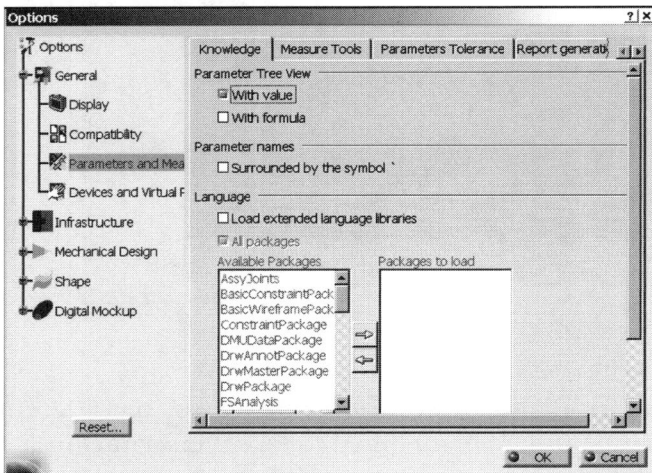

Fig. 2-28. Parameters options.

With the Options window open, click on the plus symbol to the left of General to expand that node, and then select Parameters, the options of which will be displayed, as shown in figure 2-28.

Knowledge

Select the Knowledge tab. This tab allows you to set the display of the CATIA V5 specification tree via the following options.

Parameter Tree View. These options allow you to toggle on and off the display of values or formulas within the configuration tree.

Symbols

Select the Symbols tab, shown in figure 2-29. This tab allows you to customize the display, color, and style of constraints and dimensions to your personal preferences.

Fig. 2-29. Symbols tab.

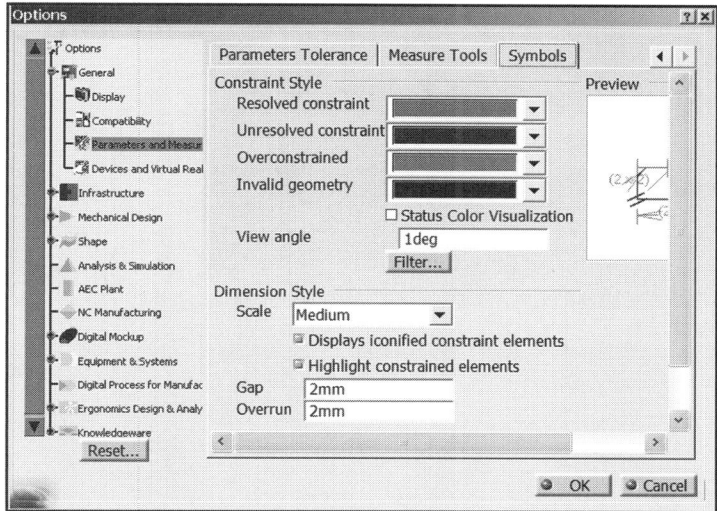

Units

Select the Units tab, shown in figure 2-30. This tab allows you to customize units.

Fig. 2-30. Units tab.

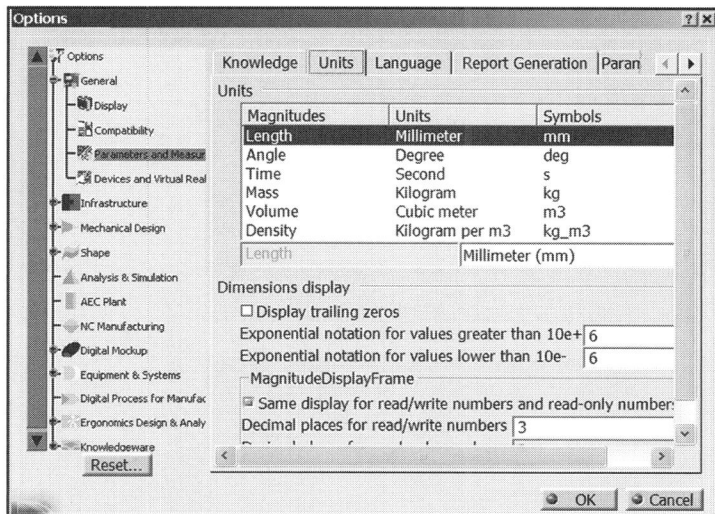

Units. These options allow you to change the system units within the model.

Dimensions Display. These options allow you to establish the number of digits and trailing zeros for dimensions.

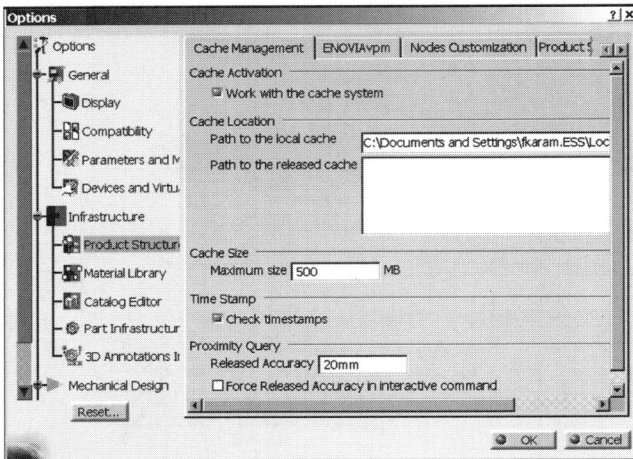

Fig. 2-31. Infrastructure panel.

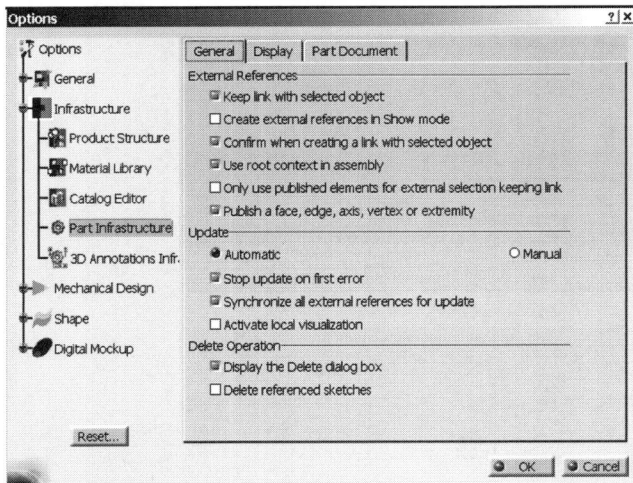

Fig. 2-32. Part Design panel.

Infrastructure

With the Options window open, click on the plus symbol to the left of Infrastructure to expand that node, and then select Product Structure. This will open the Infrastructure panel, shown in figure 2-31.

Cache Management

This option creates a lightweight CGR file for easier loading of files into a session. Refer to the CATIA V5 documentation for further details on this functionality.

Part Infrastructure

This option, shown in figure 2-32, controls the options that relate to updating models, managing external references, and using general display options.

General

External References. These options deal with linking and publishing external references among multiple models.

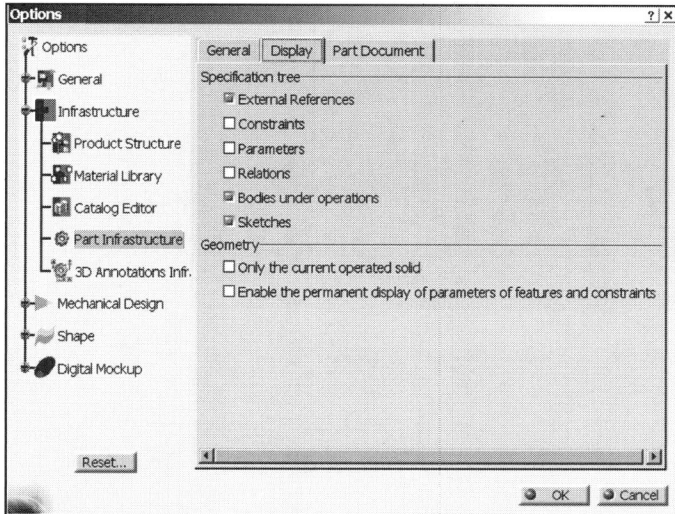

Fig. 2-33. Display settings.

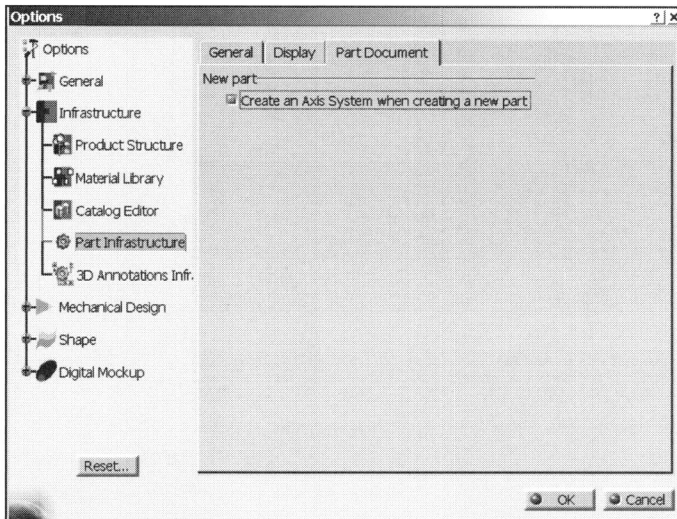

Fig. 2-34. Part Document option.

Update. This option allows you to automatically or manually update models and to synchronize multi-model links.

Display

The Display tab, shown in figure 2-33, deals primarily with display options within the Specification tree.

Part Document

This option, shown in figure 2-34, allows you to automatically create an AHS system when creating a new part.

Assembly Design

Select the Navigation tab. This tab allows you to customize the behavior of assembly designs via the Assembly Design option, shown in figure 2-35.

General

This option allows you to automatically or manually update an assembly.

Constraints

The *Paste components* option, shown in figure 2-36, provides various options for controlling and managing constraints within the assembly environment.

Fig. 2-35. Assembly Design option.

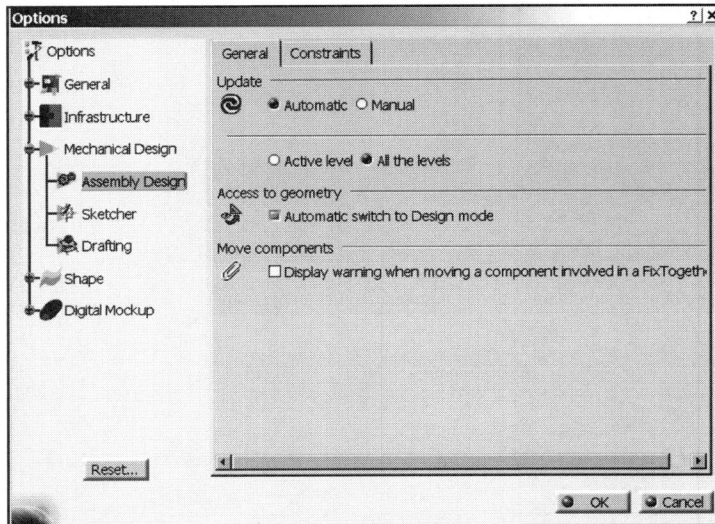

Fig. 2-36. Constraints option.

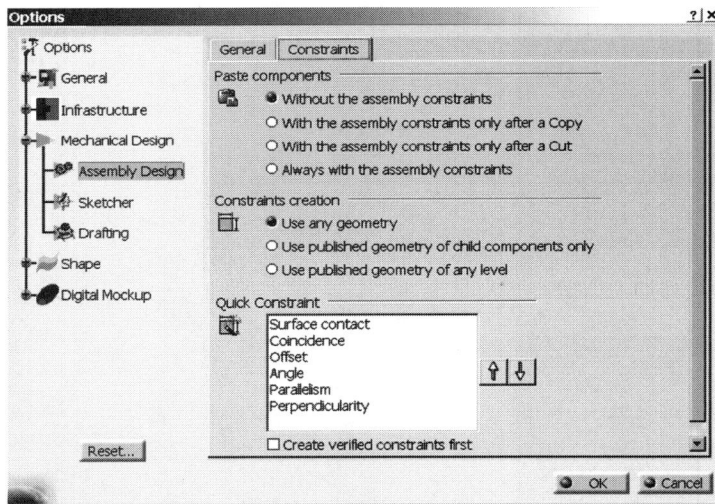

Constraints Creation

This option allows you to constrain any elements or only published elements within an assembly.

Quick Constraint

This option sets the order and priority of quick constraints within an assembly.

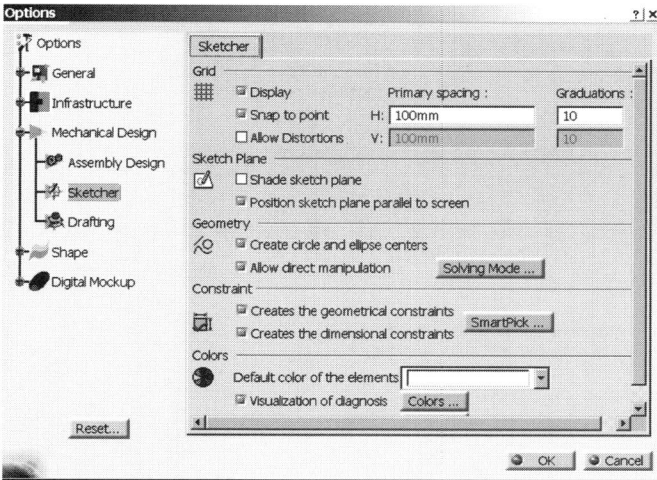

Fig. 2-37. Sketcher.

Sketcher

This tab allows you to set the default options under Sketcher, shown in figure 2-37. These options are described in the sections that follow.

Grid. This option controls grid size and display options.

Sketch Plane. This option controls the sketcher view environment and shaded sketch planes.

Constraint. Controls default options that determine whether these functions are activated automatically when you enter the sketcher.

Colors. This option controls the default color scheme for element types within the sketcher environment.

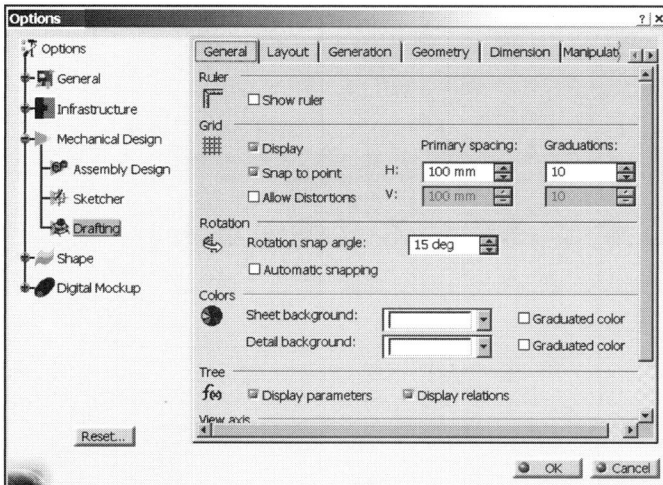

Fig. 2-38. Drafting option.

Drafting

This option, shown in figure 2-38, offers options for customizing the drafting environment.

General

This option offers general display options for grids, colors, and tree display.

Layout

This option, shown in figure 2-39, offers options for view display, sheet creation, and management of background views.

*Fig. 2-39.
Layout
option.*

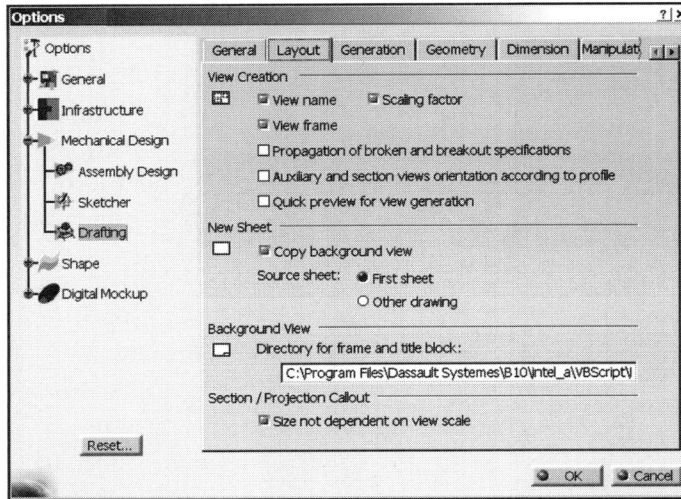

*Fig. 2-40.
Generation
options.*

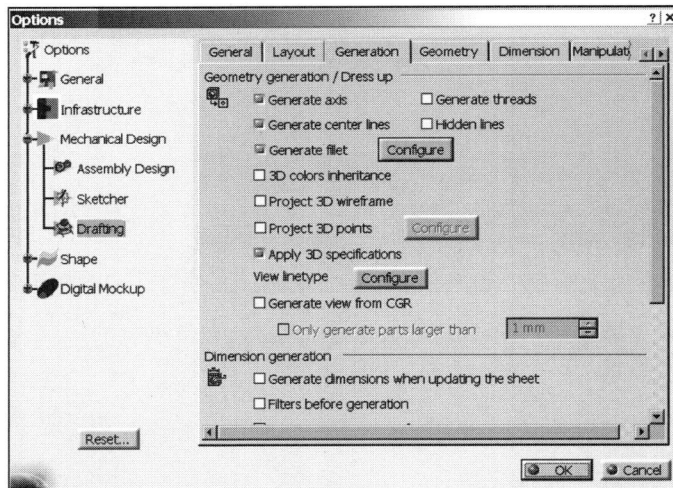

Generation

This option, shown in figure 2-40, offers options for generating and dressing up drawing elements.

Geometry

This option, shown in figure 2-41, offers options for geometry creation, constraints, display, and color schemes.

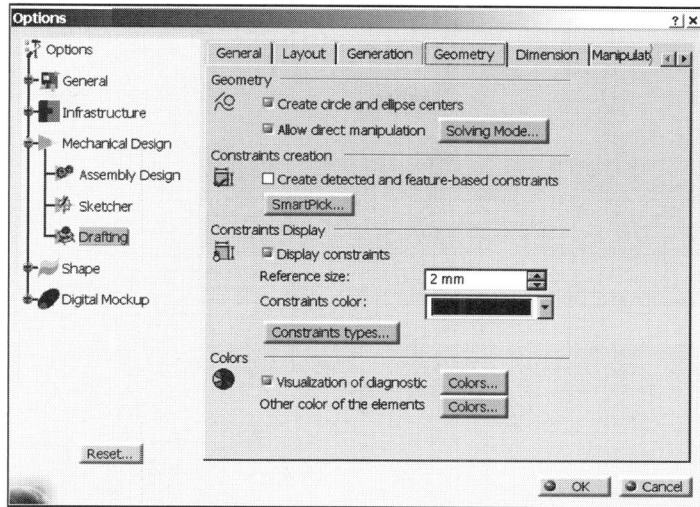

Fig. 2-41. Geometry options

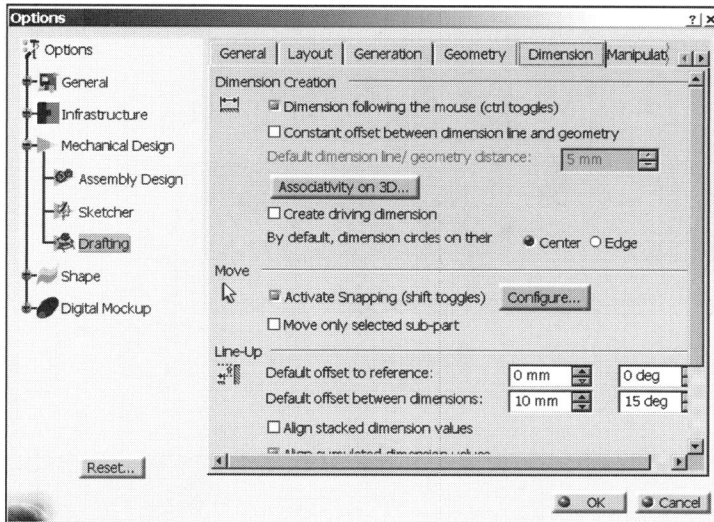

Fig. 2-42. Dimension options.

Dimension

This option, shown in figure 2-42, offers options for dimension creation, orientation, and organization within the drawing environment.

Selection, Manipulation, and View Methods

CATIA V5 provides you with many methods of selecting, manipulating, and viewing objects. Icon menu picks, keyboard shortcuts, mouse button actions, and combinations of these three are various ways of interacting with CATIA V5. The sections that follow explore these various methods.

Mouse Terminology

Figure 2-43 presents CATIA-specific terminology related to mouse operations. Before exploring the object manipulation methods previously cited, you need to understand the terminology associated with various mouse operations.

Mouse Button Terminology	
(MB1)	*Mouse Button One*
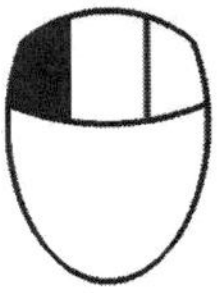	Used for selecting and editing objects.
(MB2)	*Mouse Button Two*
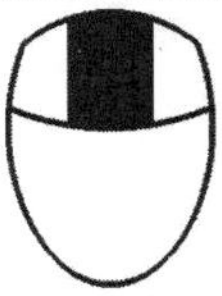	Used for manipulating objects.
(MB3)	*Mouse Button Three*
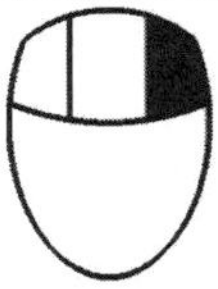	Used for displaying contextual menus.

Fig. 2-43. Mouse operations terminology.

Selecting Objects

CATIA V5 requires user interaction that is only attained by utilizing the mouse and cursor. Objects selected may be menu items, toolbar icons, history tree items, or geometry features located within the view port window. The sections that follow explore the selection of these various objects.

Menu Items

Menu items are selected by placing the mouse pointer over the desired area and pressing MB1. This mouse button action will invoke a contextual menu that displays other options, as shown in figure 2-44.

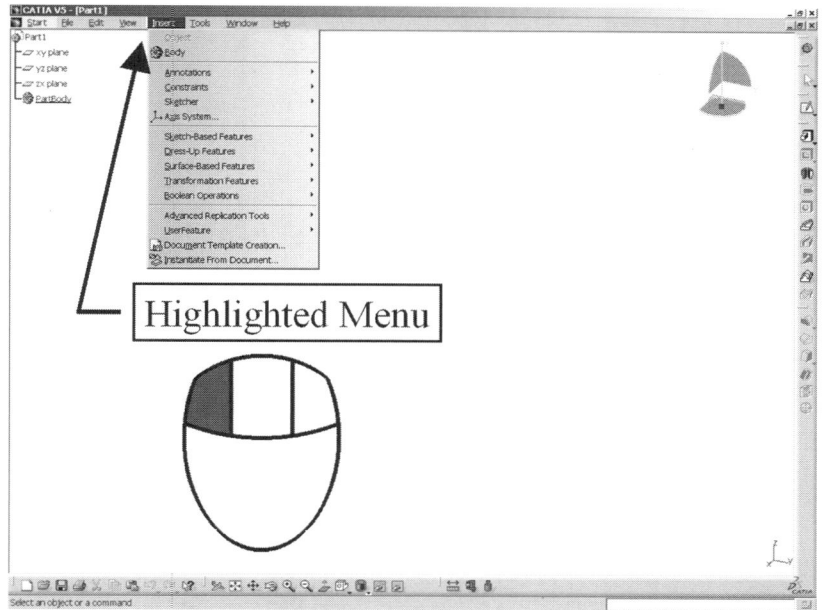

*Fig. 2-44.
Contextual menu
options.*

Highlighted Menu

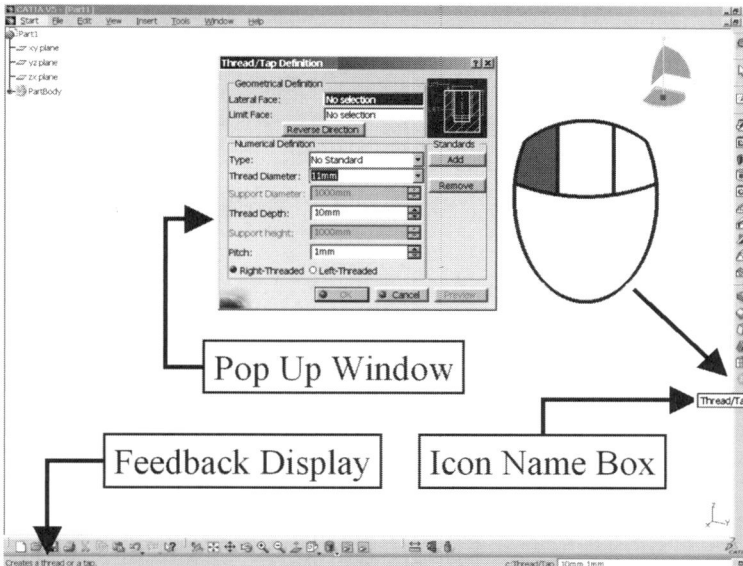

Fig. 2-45. Toolbar icon selection dialog.

Pop Up Window

Feedback Display

Icon Name Box

Toolbar Icons

Toolbar icons are selected by placing the mouse pointer over the desired icon and pressing MB1. This mouse button action will invoke a separate dialog box, shown in figure 2-45, which requests additional information for continuing with feature creation.

NOTE: Prior to selection of an icon, a small icon label will appear when the cursor is placed over the icon. The selected icon is highlighted orange until the operation is complete. The feedback display area will provide additional information associated with the selected icon.

Configuration Tree

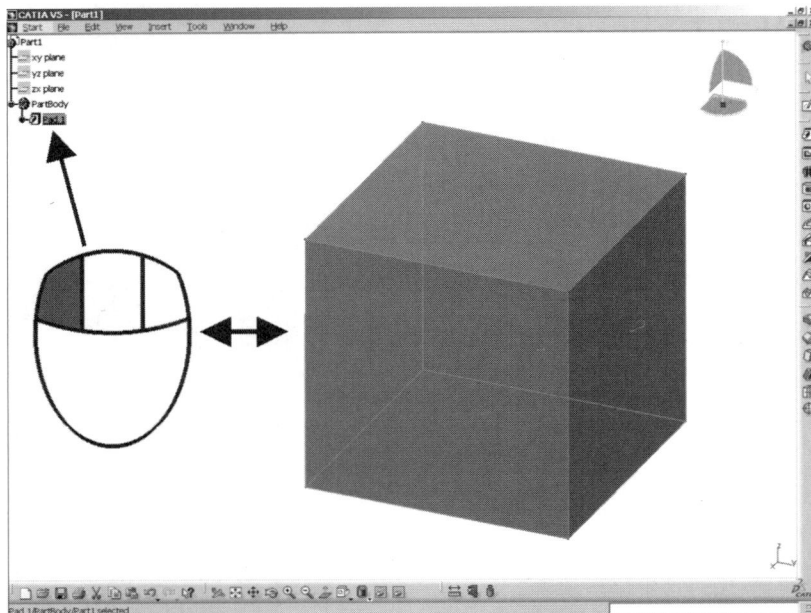

Fig. 2-46. Configuration tree.

Configuration tree items are selected also by placing the mouse pointer over the desired area and pressing MB1. Upon selection of the intended menu tree item, the desired area will become highlighted in orange, which will distinguish it from the rest of the items on the history tree, shown in figure 2-46.

NOTE: *The corresponding geometry also becomes highlighted in orange upon selection of items in the configuration tree.*

TIP: *The configuration tree is activated and deactivated by pressing F3.*

Geometry Features

CATIA V5 offers many methods of directly selecting single or multiple geometry objects. These methods are examined in the sections that follow.

Single Objects

Geometry features are selected by placing the mouse pointer over the desired element (i.e., line, surface, face, vertex, and so on) and pressing MB1. The active feature will highlight in orange when the mouse is placed over the object. The corresponding item on the configuration tree will also become highlighted orange.

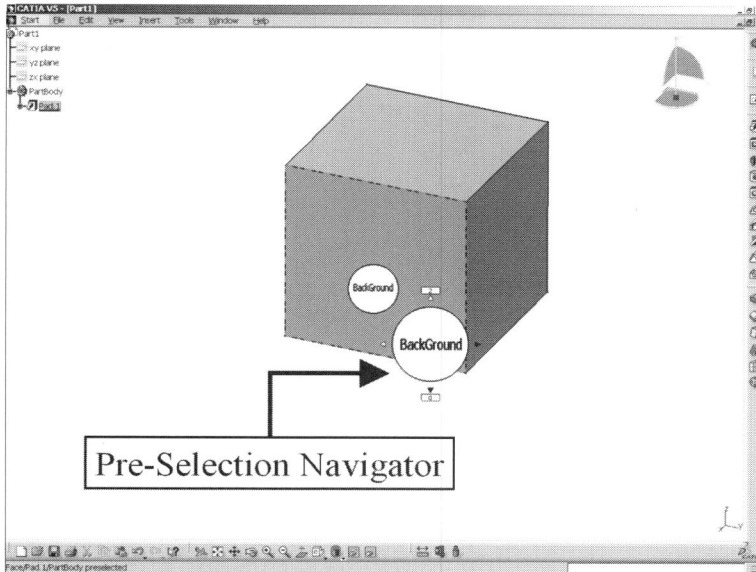

Fig. 2-47. Single object selection.

A preselection navigator is also available to help in selecting features behind the primary visible feature. This is helpful in that you do not have to rotate the model to a view of the back side in order to select the correct object. The preselection navigator is invoked by pressing one of the four keyboard arrow keys when the cursor is placed in the desired area, as shown in figure 2-47.

NOTE: *When the cursor is over an object it will change from an arrow to a pointing finger.*

Multiple Objects

This section explores the selection of multiple objects. Methods for doing so involve the keyboard and use of the Selection toolbar, described in the sections that follow.

Keyboard. Multiple features may also be selected by holding down the Ctrl key and selecting individual geometry objects.

Toolbar. The most commonly used method of selecting multiple features is to drag a selection box around the desired geometry. This is accomplished by pressing MB1 and dragging a box, as indicated in figure 2-48. The Select toolbar, shown in figure 2-49, also offers various trapping options for more control.

Criteria Search

A quick way to select like features is to use the Criteria Search option, located under the Edit menu. This option can also be

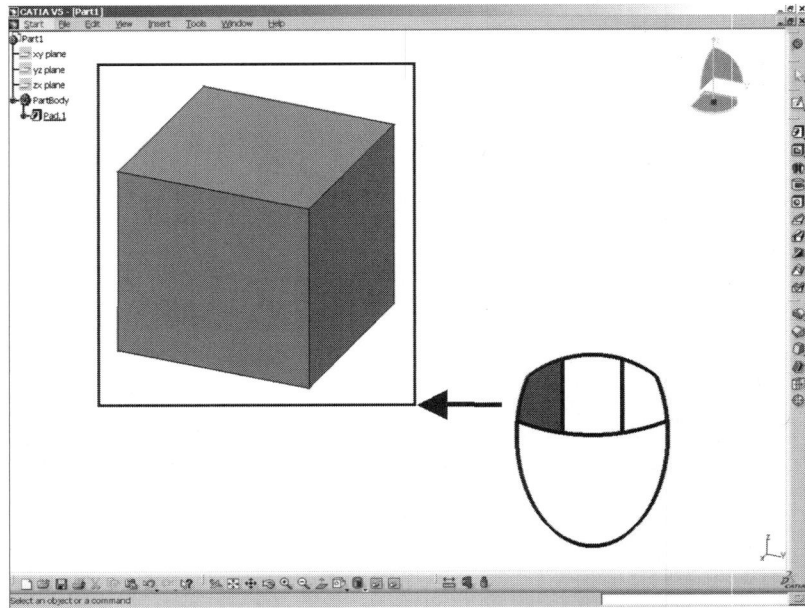

*Fig. 2-48.
Selection box.*

Fig. 2-49. Selection toolbar.

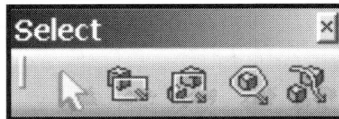

*Fig. 2-50. Defining
search criteria.*

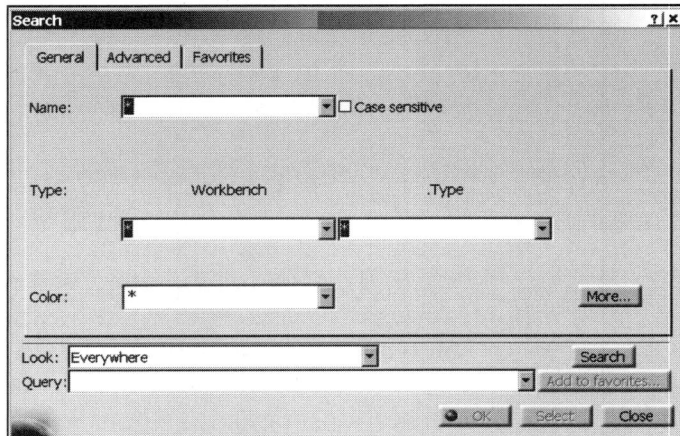

accessed by pressing Ctrl F. This allows you to define search criteria such as color, types, and so on, as indicated in figure 2-50.

Viewing and Manipulating Objects

The sections that follow explore various aspects of viewing and manipulating objects. These include pan, rotation, and zoom functionality.

Panning

Panning is the action of moving objects around the screen without rotating them or zooming in or out. CATIA V5 provides several methods for panning part geometry around the active view port window. These methods are described in the sections that follow.

Icon Selection

One method of panning is to select the Pan icon with MB1 located on the View toolbar at the bottom of the window, as shown in figure 2-51.

NOTE: *An axis appears at the location of the cursor as a point of reference for panning.*

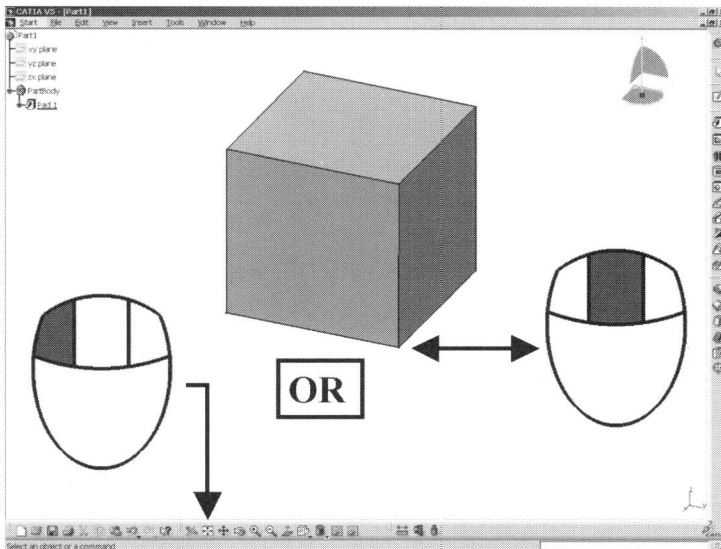

Fig. 2-51. Icon selection method of panning.

Mouse Actions

A second, more common, method of panning is to select MB2 within the view port and to drag the geometry to the desired location, as indicated in figure 2-51. Holding down MB3 in conjunction with holding the Alt key down also allows for panning of the object.

Menu Pick

A third method of panning is to select the Pan option, located under the View menu pick in the Standard menu (located at the top of the window).

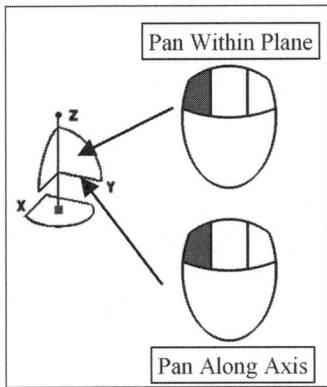

Fig. 2-52. Panning using the Compass tool.

Compass Tool

The last method of panning is to use the Compass tool, located at the top right-hand corner of the view port. Geometry objects may be dragged either along a 2D plane or along an axis, as indicated in figure 2-52.

Rotation

Rotation is the action of rotating objects around the screen without panning them or zooming in or out. CATIA V5 provides the same several methods of rotating part geometry as for panning part geometry around the active view port window. These are described in the sections that follow.

Icon Selection

The first method of rotation is to select the Rotate icon with MB1 located on the View toolbar at the bottom of the window, as indicated in figure 2-53.

Mouse Action

A second, more commonly used, method of rotation is to select and hold down MB1 and then select MB2. This activates a sphere around the part geometry used for free rotation, as shown in figure 2-53. A small cross, indicating the cursor location during free rotation, is visible on the sphere.

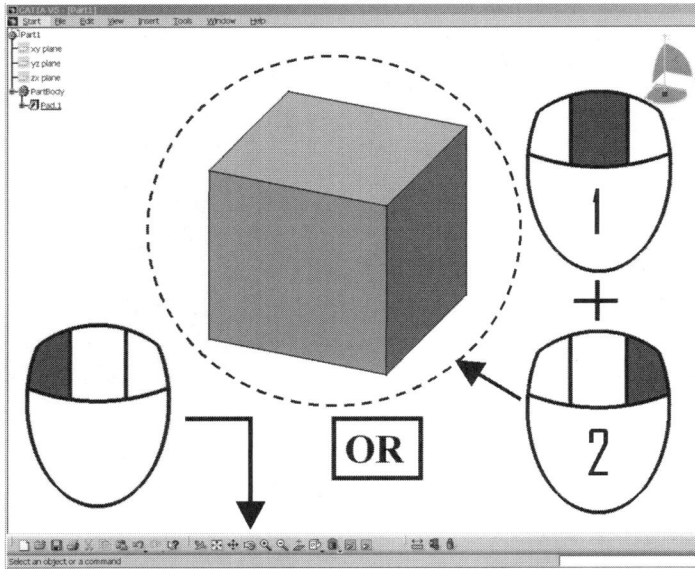

Fig. 2-53. Rotation using the Rotate icon.

Menu Pick

A third method of rotation is to select the Rotate option under the View menu pick in the Standard menu (located at the top of the window).

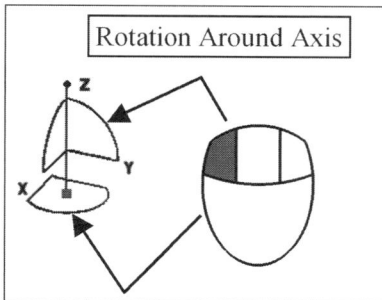

Fig. 2-54. Rotation using the Compass tool.

Compass Tool

The last method of rotation is to use the Compass tool, located at the top right-hand corner of the view port. Geometry objects may be rotated around an axis using this tool, as indicated in figure 2-54.

NOTE: *If the cursor is placed outside the sphere, rotation will occur around the Z plane.*

Zoom In/Out

Zooming in or out is the action of growing or shrinking the display of an object without panning or rotation actions. This is strictly a visual effect unrelated to the scaling or sizing of an object. CATIA V5 provides several methods of zooming in and out on part geometry in the active view port window. These methods are described in the sections that follow.

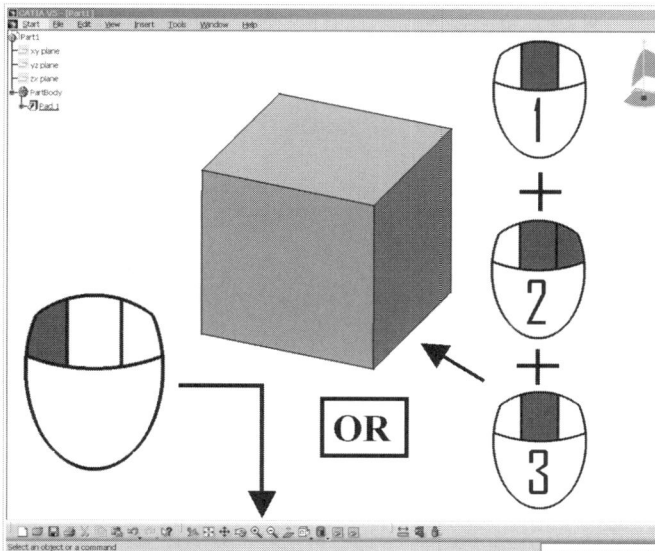

Fig. 2-55. Zooming using the Zoom In and Zoom Out icons.

Icon Selection

The first zoom method is to select either the Zoom In or Zoom Out icon with MB1 located on the View toolbar at the bottom of the window, as shown in figure 2-55.

Mouse Action

A second, more commonly used, zoom method is to select and hold down MB1 and then select and release MB2. This allows for the free zooming of active part geometry, as indicated in figure 2-55.

Menu Pick

A third zoom method is to select the Zoom In or Zoom Out option located under the View menu pick in the Standard menu (located at the top of the window).

Summary

Chapter 2 has introduced you to the GUI and basic functions of CATIA V5. User defaults, options, and customization are made easy by alternating various options within the user environment. Panning, rotating, and zooming were also covered in the context of the fundamentals of model display control. You should have a basic understanding of the following after reading this chapter.

- Starting a CATIA V5 session

- CATIA V5 document types

- Introduction to the GUI

- Customizing the environment

- Selection and view methods

Review Questions

1   List several ways in which to execute a CATIA V5 session.

2   Name the three native primary document types CATIA V5 supports.

3   By what means can the graphical user interface (GUI) be customized?

4   How is the Standard Tools toolbar similar to other Microsoft applications?

5   What operations do the three mouse buttons (MB1, MB2, and MB3) perform?

6   Name several ways to pan, zoom, and rotate the graphical display of an object within CATIA V5.

CHAPTER 3

THE SKETCHER WORKBENCH

Introduction

THE KEY ENABLER TO THE CORE FUNCTIONALITY in feature-based modeling is learning how to master the sketcher environment. The sketcher workbench provides the ability to create and edit 2D elements used for the creation of 3D features. The majority of solid features are started by first drawing a 2D profile, called a sketch. Design intent is then captured by applying parameters and constraints to the geometry.

Sketcher-based features such as pads, pockets, shafts, and ribs are created referencing 2D sketched profiles. This chapter covers in depth how to create, constrain, and develop relationships between and among 2D geometrical elements.

Objectives

The following are the main objectives of this chapter.

- Introduction to the CATIA sketcher
- Sketching simple and predefined profiles
- Sketcher tools
- Parameters and constraints
- Sketcher operations
- Sketcher tips

Sketcher Introduction

The sketcher is an integral part of CATIA V5. The sketcher environment is the bridge between 2D elements and 3D features. The sketcher provides the functionality for creating and modifying 2D elements used in the construction of 3D solids and surfaces.

Entering the Sketcher

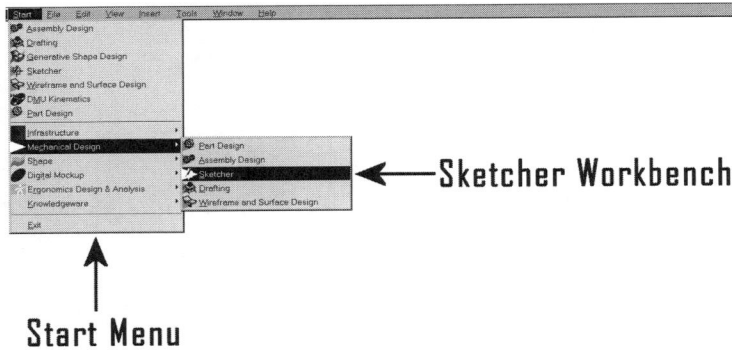

Fig. 3-1. Sketcher Start menu.

There are several ways to enter the sketcher environment. One method is to select the CATIA V5 Sketcher Start menu (shown in figure 3-1), select Mechanical Design, and then select Sketcher.

This initiates the Part Design workbench, and places you in the Sketcher workbench. However, this will probably not be the method you will use most frequently to enter the sketcher environment. Considering that we may have already started up CATIA V5 and have selected the Part Design workbench, the easiest and most frequently used method is to simply click on the Sketcher Workbench icon, shown in figure 3-2, and then select a reference plane, geometry planar face, or existing sketch. Passing the mouse over any such entity highlights it, and a single click opens the

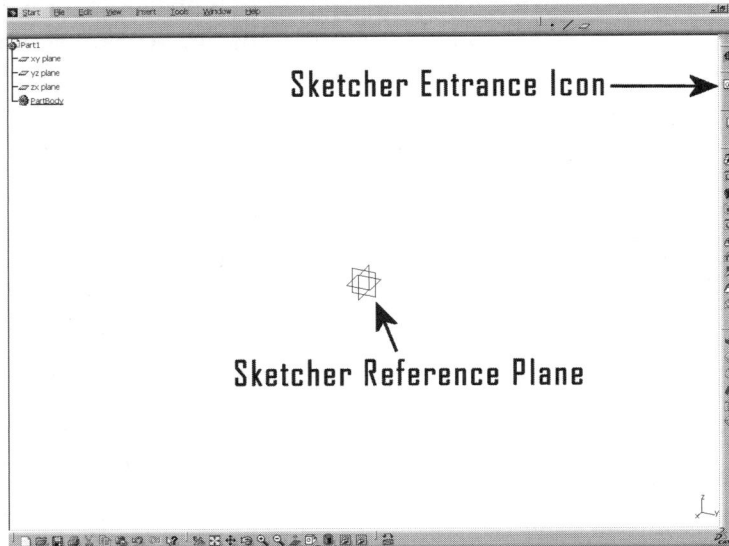

Fig. 3-2. Sketcher Workbench icon.

sketcher while displaying an orthographic view. A sketcher session is unmistakably identified by the grid structure used to assist in the creation of 2D elements, as shown in figure 3-3.

This grid structure is used to assist in sketching 2D profile elements and snapping to hard coordinate positions. Default options for the grid, such as spacing, may be controlled as discussed in Chapter two (see, particularly, figure 2-37). The following are a few notes to keep in mind.

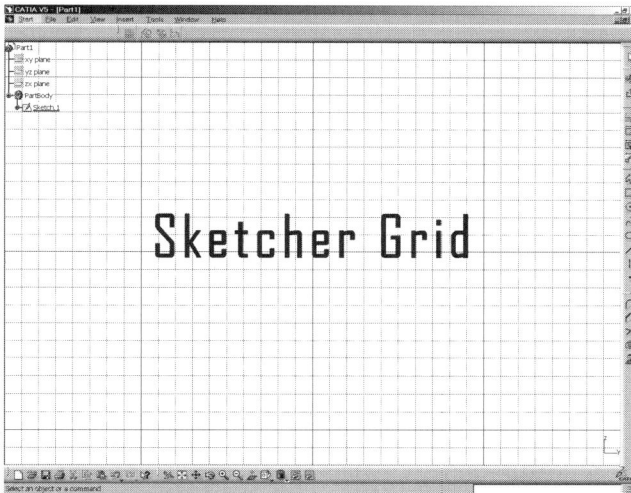

Fig. 3-3. Sketcher environment.

- Datum reference planes may also be selected from the history tree to initiate a sketched feature.

- The sketcher environment may also be initiated by multi-selecting two lines that create a plane.

- The Sketcher icon is highlighted while in the sketcher environment. To exit sketcher mode, simply click on the Sketcher icon and return to the 3D environment.

Sketcher Functionality

When within the Sketcher workbench, you will notice that there are four primary toolbars that can be used to create, edit, control, and constrain 2D elements. These toolbars are shown in figure 3-4.

Sketcher Tools

The Sketcher Tools toolbar, shown in figure 3-5, contains four icons associated with controlling the sketcher environment. These icons are discussed in the sections that follow.

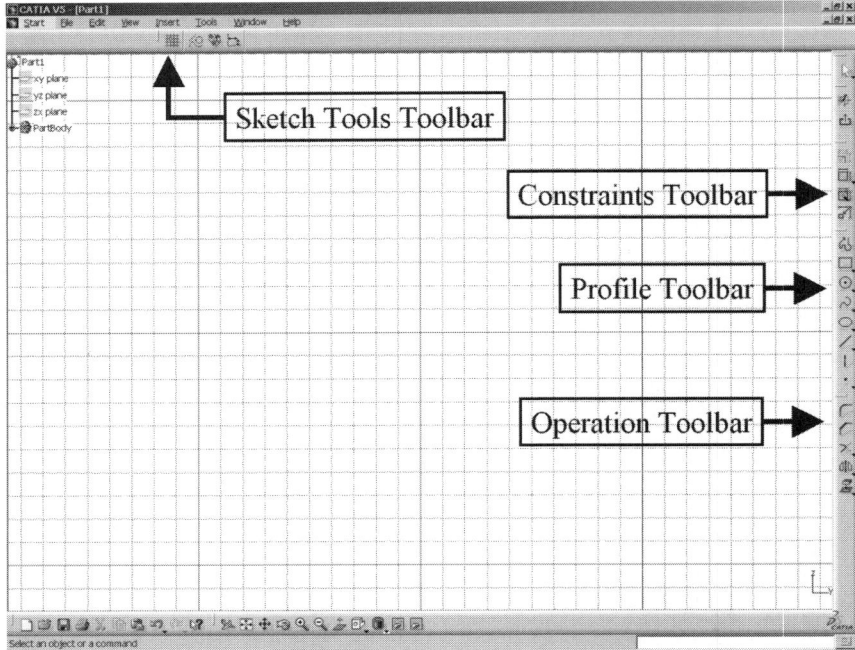

Fig. 3-4.
Primary sketcher
toolbars.

Fig. 3-5. Sketcher
Tools toolbar.

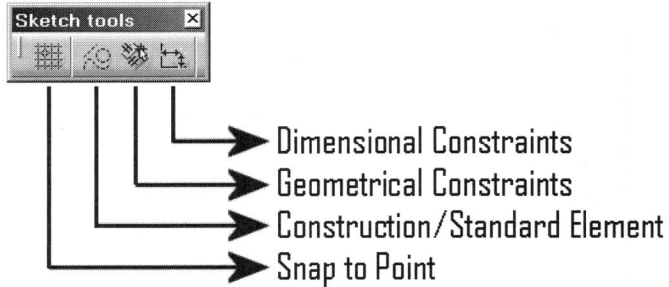

Dimensional Constraints
Geometrical Constraints
Construction/Standard Element
Snap to Point

Snap to Point

With the Snap to Point icon toggled on, during the creation of a profile the pointer will automatically snap to the closest corner point or intersection on the sketcher grid. This icon forces the sketch to begin or end on a grid point.

The horizontal and vertical coordinates are also seen on screen during this process. The horizontal number is on top, with the vertical number below it. This option, shown in figure 3-6, may be toggled on and off during the sketch operation.

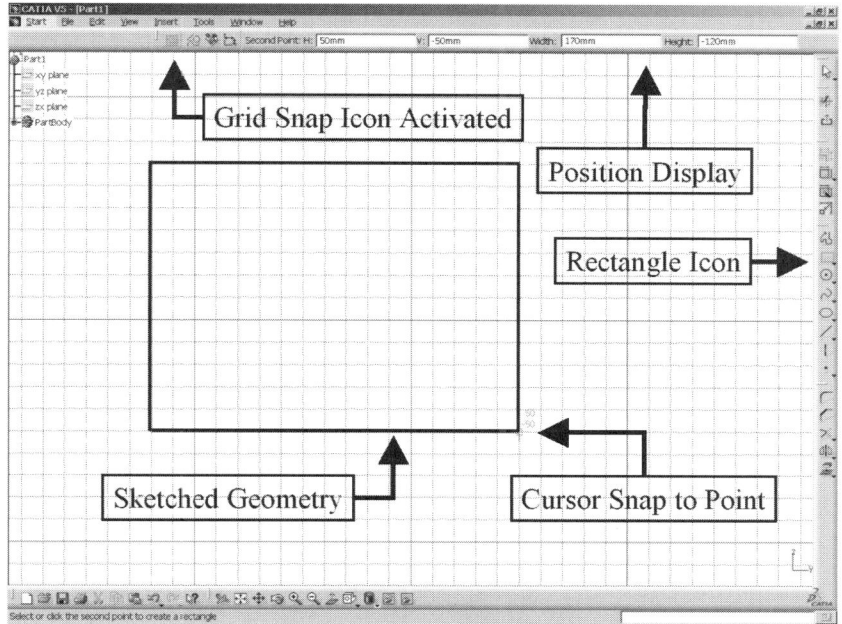

Fig. 3-6. Snap to Point icon.

Construction/Standard Element

There are two types of geometry elements within the sketcher environment: construction and standard. With the Construction/Standard Element icon toggled on, you can create and use lines, points, and curves as construction geometry that assists in the creation of 2D profiles. The appearance of these elements will be different than those of the actual profile.

Elements may be quickly changed from construction to standard by selecting or multi-selecting the element and toggling the Construction/Standard Element icon, shown in figure 3-7. These construction elements are part of your sketch, and will be there if you reenter the sketch, but will not be part of your profile when you exit the Sketcher workbench.

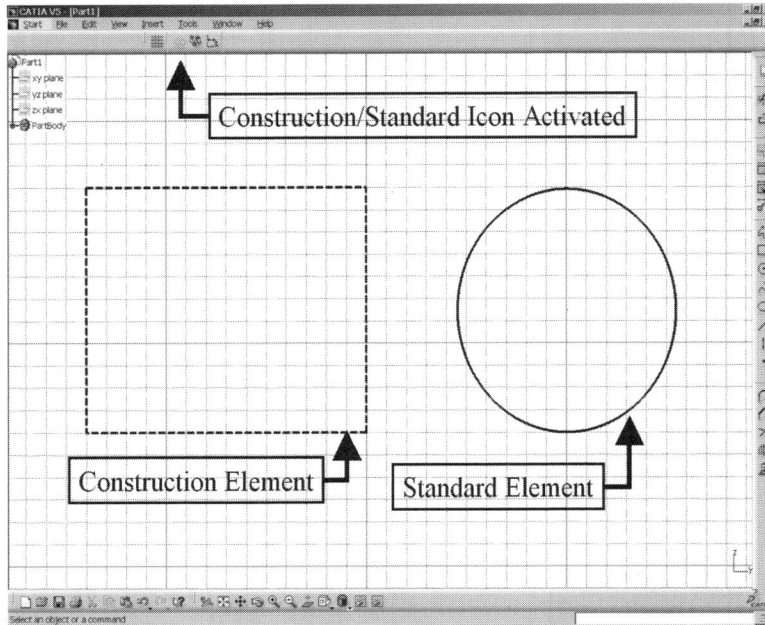

Fig. 3-7.
Construction/
Standard
Element icon.

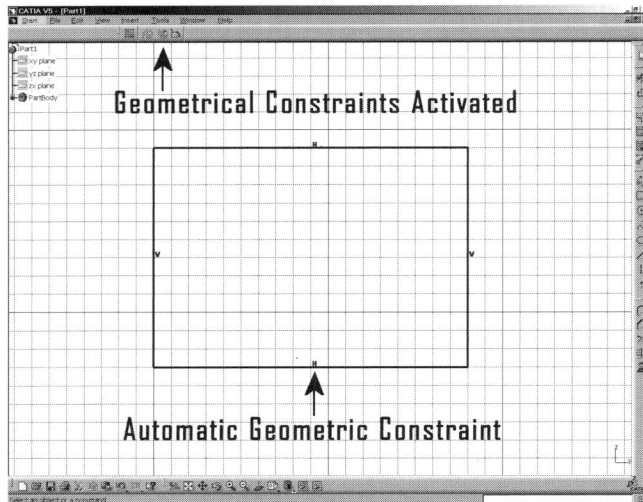

Fig. 3-8. Geometrical constraints indicated by their
symbols.

Geometrical Constraints

With the Geometrical Con-straints icon toggled on, certain geometrical constraints (such as horizontal, vertical, tangency, and coincidence) will be imposed during the profile cre-ation process. As you are creat-ing geometry for your profile, you will visually see these geo-metrical constraint symbols (shown in figure 3-8) on screen, near the cursor or near those elements where that condition exists or is being created.

TIP: *Use of the Geometrical Constraints function is effective when creating profiles under known conditions. It will save significant time in that you will not have to later redefine the sketch and impose these constraints.*

Dimensional Constraints

With the Dimensional Constraints icon toggled on, you have the ability to type in your dimensions that will create the exact geometry you want, and will place these dimensional constraints in the sketch associated with that geometry. Here, you simply type the information in the Sketcher Tools toolbar as you create that particular piece of geometry for your profile. Figure 3-9 shows an example of dimensional constraints.

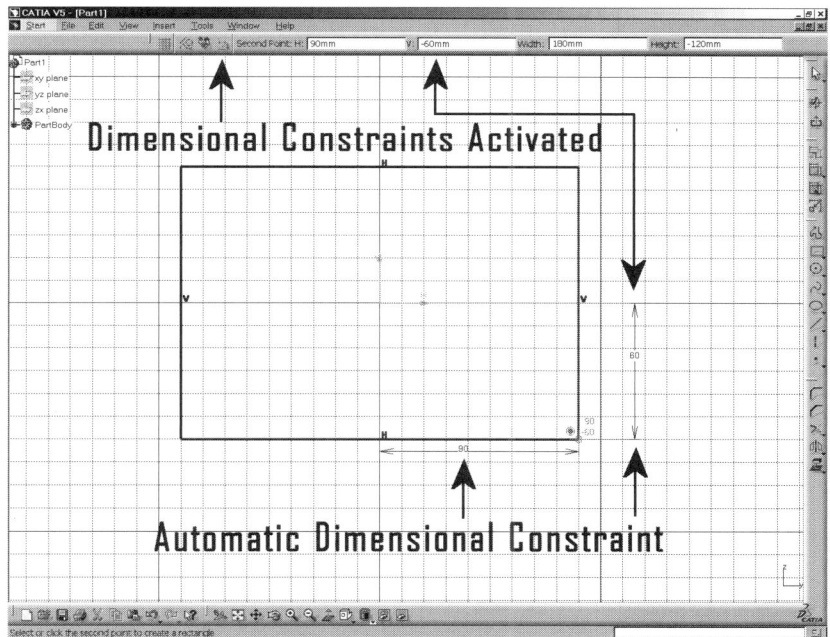

Fig. 3-9. Dimensional constraints.

NOTE: *Both the Geometrical and Dimensional icons are automatically highlighted when first entering the sketcher environment. If these icons are not highlighted during sketcher operations, geometric elements will have no relational constraints. Constraints may be later applied to selected elements by using options within the Constraint toolbar.*

Value Fields

When sketching profile elements, the values of these elements are dynamically updated as the cursor is moved within the sketcher environment. The values updated are H (horizontal), V (vertical),

L (length), and A (angle), which correlate respectively with the position of the cursor as related to the references used for sketching. These value fields are shown in figure 3-10.

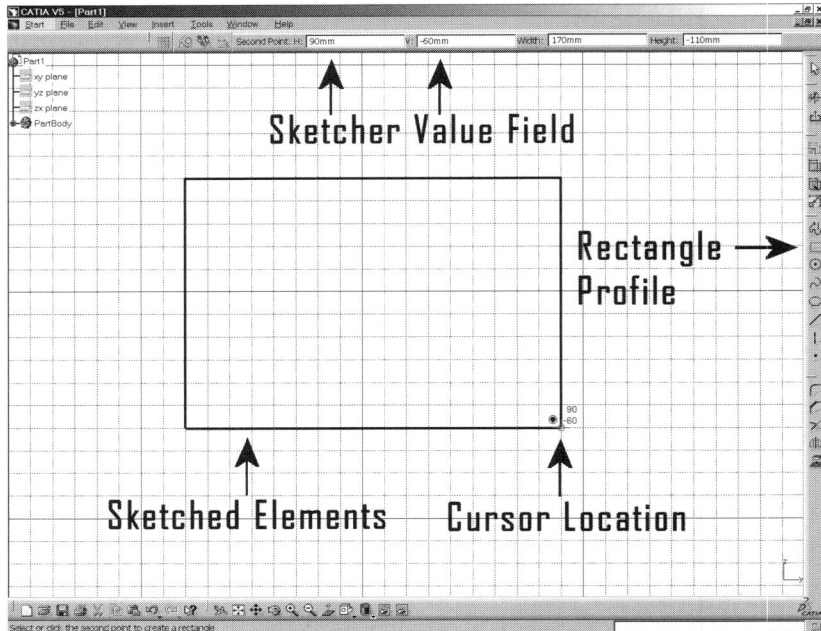

Fig. 3-10.
Value fields.

TIP: *These fields may also be used to enter hard values to locate the element position within the sketch.*

Geometry Profiles

Fig. 3-11. Sketcher Geometry toolbar.

The sketcher's Geometry (creation) toolbar, shown in figure 3-11, offers an array of functionality for creating sketched 2D geometric elements (such as lines, rectangles, circles, arcs, points, and splines). This toolbar contains eight main icons associated with creating either predefined sections or singular sketched elements.

The sections that follow describe functions performed for predetermined profile sections. The sketcher's geometry profiles are shown in figure 3-12.

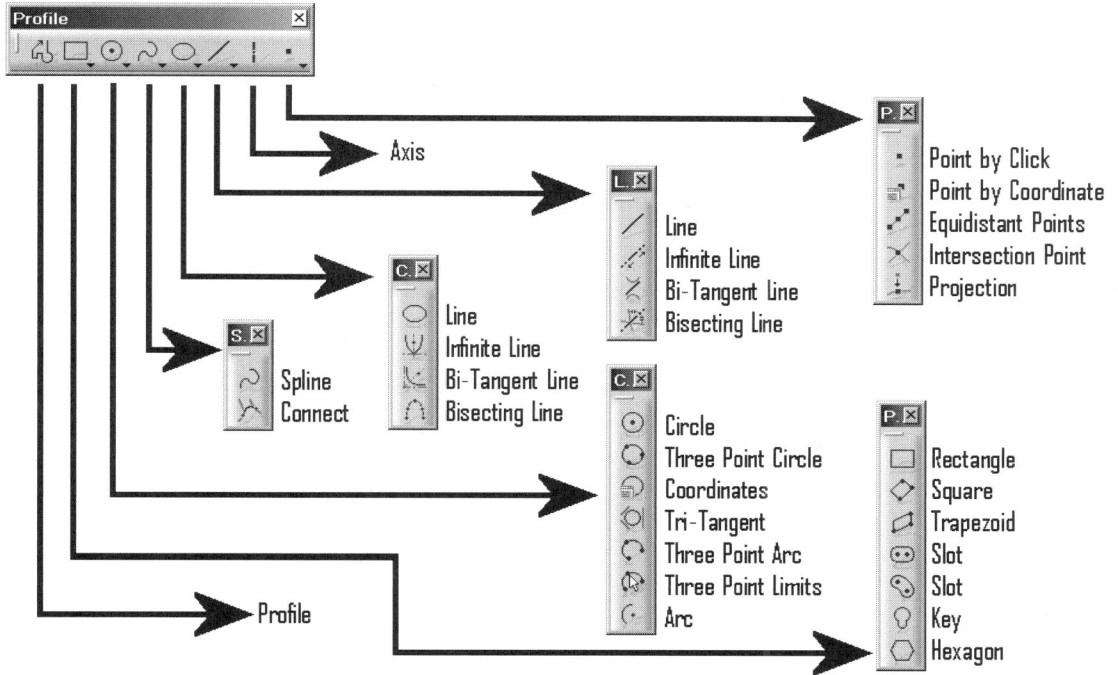

Fig. 3-12. Sketcher geometry profiles.

Creating Sketched Elements

2D sketched elements are created by first selecting the appropriate profile icon and moving the cursor within the Sketcher window while holding down the first mouse button (MB1).

The CATIA V5 sketcher offers an intelligent auto-detection sketching assistant tool called SmartPick. SmartPick is designed to assist you with creation and editing tasks within the sketcher environment.

SmartPick, shown in figure 3-13, is identified at the cursor location and dynamically detects geometric constraints and relationships. When the sketcher auto-detects a possible relationship, a graphical symbol will dynamically display the appropriate feedback. See the CATIA V5 reference documentation for further information on SmartPick.

*Fig. 3-13.
SmartPick.*

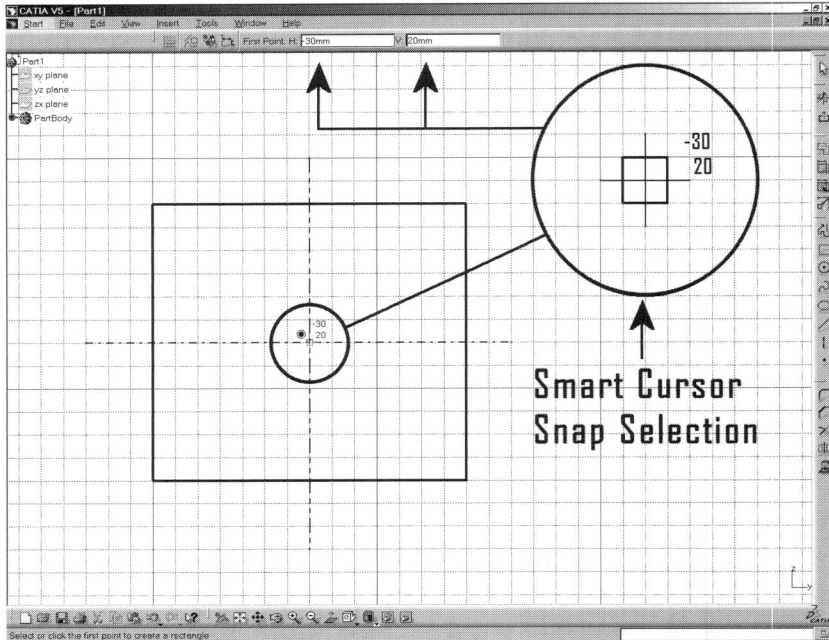

TIP: *Holding down the Shift key during the element creation process will temporarily deactivate the auto-detection feature.*

Profile Creation Exercises

In the following section you will create sketcher geometry using all of the predefined profile examples, and then using selected profiles from some of the other sections. Some of these sections (such as line, point, and circle) are very simple and easy to figure out, and require no elaboration. For all of the following examples, activate the Snap to Point, Geometrical Constraints, and Dimensional Constraints options (icons) on the Sketcher Tool toolbar. In all of these examples, we will be using the mouse to create the profiles.

NOTE: *Sketcher information may also be inputted into the Sketcher Tool toolbar to assist in the creation process for almost everything in the Profile toolbar. The only time this does not apply is when there is a need for preexisting geometry.*

Sketcher Profile Exercises

The sections that follow present brief exercises that explore use of various sketcher profiles. These include use of the Rectangle, Oriented Rectangle, Parallelogram, Elongated Hole, Cylindrical Elongated Hole, Keyhole Profile, Hexagon, and Profile icons.

Sketcher Profile Exercise 1: Rectangle

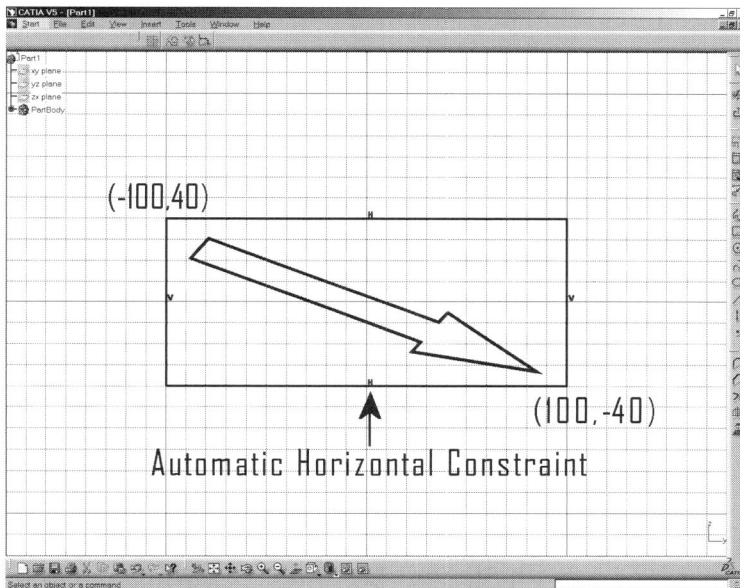

Fig. 3-14. Rectangle profiles.

To create a rectangle profile, perform the following steps.

1 In the Sketcher workbench, click on the Rectangle icon in the Predefined Profile section.

2 Select the coordinates -100,40 and 100,-40 and while holding down MB1 drag the cursor to 100,-40.

Note that the automatic creation of the horizontal and vertical geometric constraints, as shown in figure 3-14.

Sketcher Profile Exercise 2: Oriented Rectangle

To create an oriented rectangle profile, perform the following steps.

1 In the Sketcher workbench, click on the Oriented Rectangle icon in the Predefined Profile section.

2 Select the coordinates -60,40; 70,20; and 60.77,-40 to create the profile.

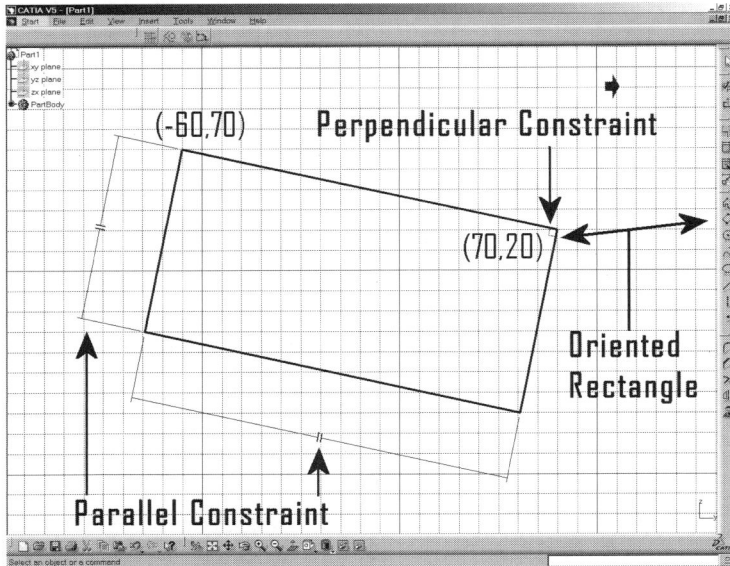

Note the automatic creation of parallelism and perpendicularity geometrical constraints. An oriented rectangle profile is shown in figure 3-15.

Fig. 3-15. Oriented rectangle profile.

Sketcher Profile Exercise 3: Parallelogram

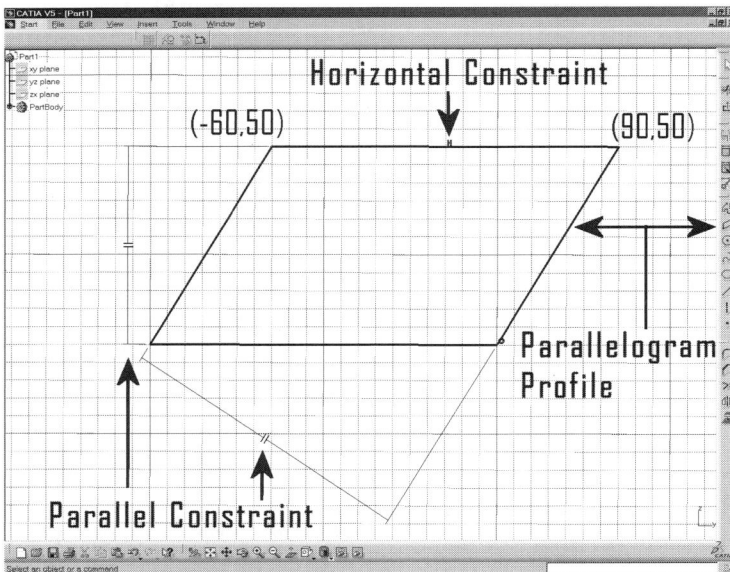

To create a parallelogram profile, perform the following steps.

1 In the Sketcher workbench, click on the Parallelogram icon in the Predefined Profile section.

2 Select the coordinates 60,-50; 90,50; and -60,50 to create the profile.

A parallelogram profile is shown in figure 3-16.

Fig. 3-16. Parallelogram profile.

Sketcher Profile Exercise 4: Elongated Hole

To create an elongated hold profile, perform the following steps.

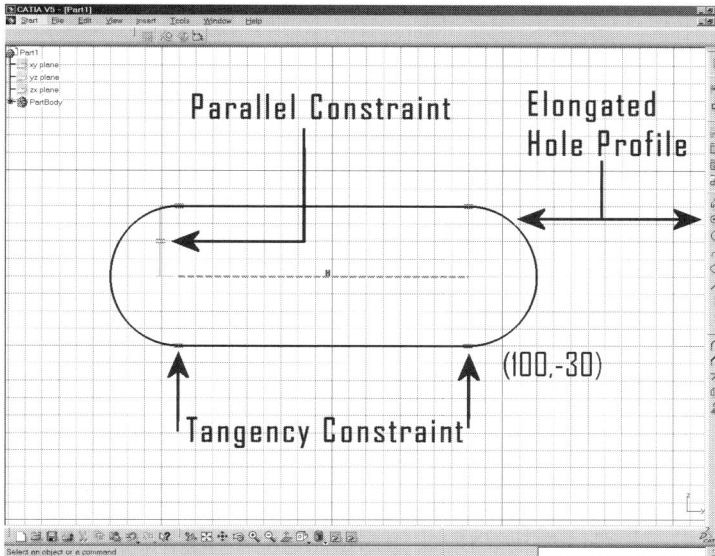

Fig. 3-17. Elongated hole profile.

1 In the Sketcher workbench, click on the Elongated Hole icon in the Predefined Profile section.

2 Select the coordinates—60,0 and -60,0 to establish the center-to-center distance, and then select 110,-30 to size and create the profile.

NOTE: *Remember to hold down MB1 while dragging the cursor.*

An elongated hole profile is shown in figure 3-17.

Sketcher Profile Exercise 5: Cylindrical Elongated Hole

To create a cylindrical elongated hole profile, perform the following steps.

1 In the Sketcher workbench, click on the Cylindrical Elongated Hole icon in the Predefined Profile section.

2 Select the coordinates 0,0; 50,100; and 110,20 to establish the center-to-center distance, and then select 120,40 to size and create the profile.

Note the coincidence, concentricity, and tangency geometrical constraints. A cylindrical elongated hole profile is shown in figure 3-18.

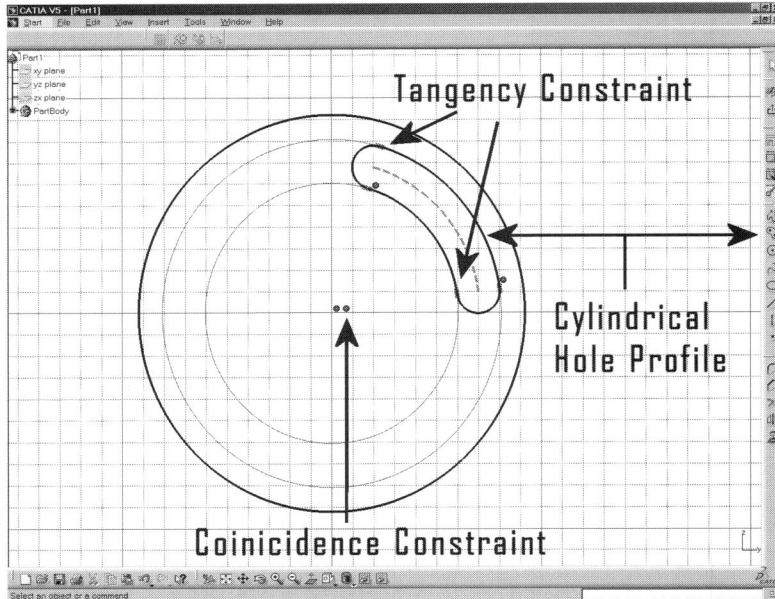

*Fig. 3-18.
Cylindrical
elongated hole
profile.*

Sketcher Profile Exercise 6: Keyhole Profile

To create a keyhole profile, perform the following steps.

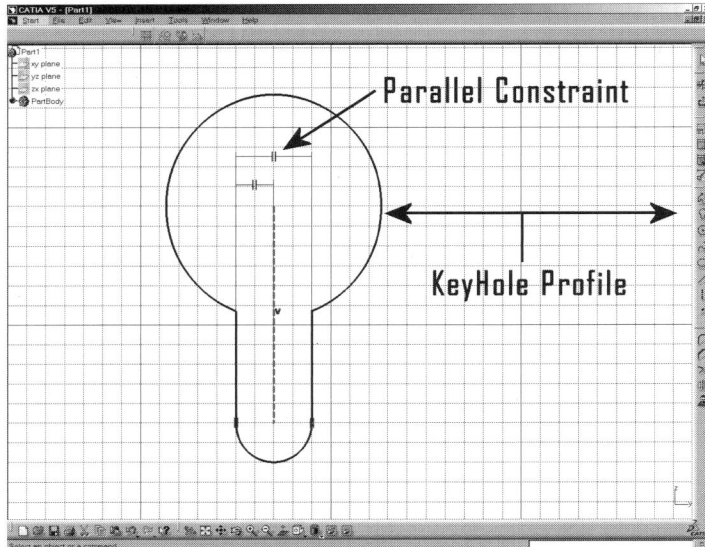

Fig. 3-19. Keyhole profile.

1   In the Sketcher work-bench, click on the Key-hole Profile icon in the Predefined Profile section.

2   Select the coordinates 0, 30 to define the center of the large radius, 0,-40 to define the center of the small radius, and then select 0,-60 to size the small radius and 0,70 to size the large radius.

3   Create the profile.

A keyhole profile is shown in figure 3-19.

Sketcher Profile Exercise 7: Hexagon

To create a hexagon profile, perform the following steps.

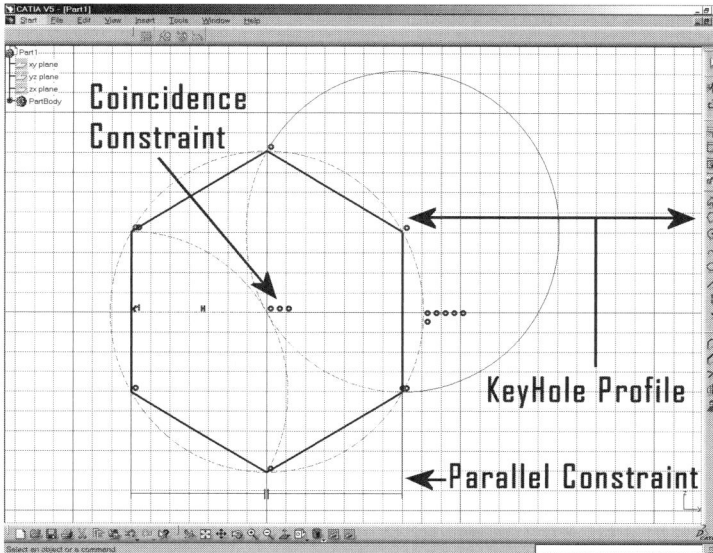

Fig. 3-20. Hexagon profile.

1 In the Sketcher work-bench, click on the Hexagon icon in the Predefined Profile section.

2 Select the coordinates 0,0 to define the center of the hexagon, and then select 0,60 to size the hexagon.

Note that before you select the size of the hexagon you can rotate it so that it is oriented the way you want. A hexagon profile is shown in figure 3-20.

Sketcher Profile Exercise 8: Using the Profile Icon

Now that we have created some profiles and have a feel for how functions work in this environment, let's create a sketch using the Profile icon. Perform the following steps.

1 In the Sketcher workbench, click on the Profile icon in the Predefined Profile section.

2 Select the coordinates 60,40 and place the cursor on and select the 60,-40 coordinate. Continuing to hold MB1 down, slide the cursor downward and let go to start the creation of an arc. Select coordinate 20,-80 to complete the arc.

NOTE: *After the arc is complete, the creation process will default back to Line.*

3 Create a horizontal line segment that ends at -20,-80.

4   Use -20,80 to start the creation of another arc, to be completed at -60,-40.

5   Create another line segment ending at -60,40. After that selection, click on the Three Point Arc icon in the Tools toolbar.

NOTE: *Your last selection is now the first point of this arc.*

6   Select 0,70, and then 60,40 to finish your profile.

This should provide you an idea as to how the Profile function works. Figure 3-21 shows the Profile icon.

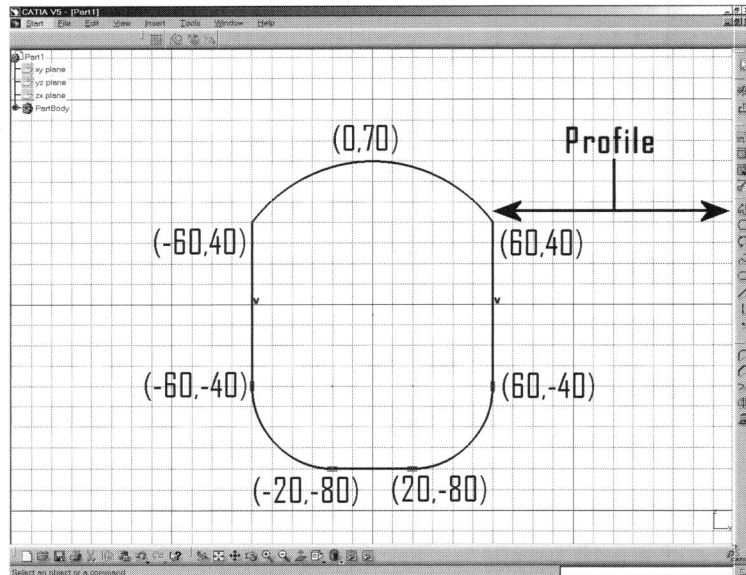

*Fig. 3-21.
Profile icon.*

Operations

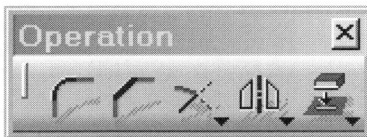

Fig. 3-22. Operations toolbar.

The sketcher Operations toolbar offers an array of functionality for performing editing and other operations on sketched profile elements. This toolbar, shown in figure 3-22, contains five main icons associated with editing elements.

The section that follows explores each operation function, depicted in figure 3-23.

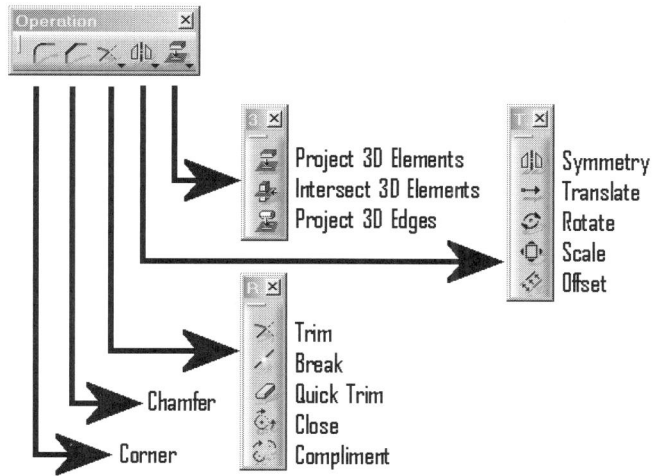

Fig. 3-23. Operation functions.

Project 3D Elements
Intersect 3D Elements
Project 3D Edges

Symmetry
Translate
Rotate
Scale
Offset

Chamfer

Corner

Trim
Break
Quick Trim
Close
Compliment

Sketcher Operation Exercises

Sketcher operations involve manipulating and editing sketched elements. These operations refine the sketch to further meet its design intent. The sections that follow explore various operations.

Sketcher Operation Exercise 1: Corner

The following steps take you through the process of performing a corner operation.

1 In the Sketcher workbench, recreate the rectangle from Sketcher Profile Exercise 1.

2 Double click on the Corner icon in the Operation toolbar, and then in the Tools toolbar click on the icon Trim All Elements.

3 Starting with the lower left-hand corner of the rectangle, select the two lines that form that corner. A radius appears.

4 Locate the Tools toolbar and enter a value of *25* as the size of the radius.

Note that both lines are trimmed to the corner.

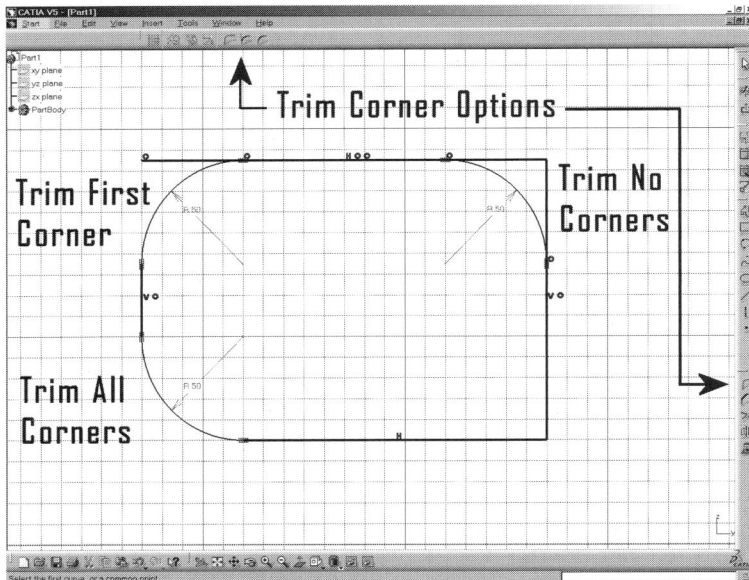

Fig. 3-24. Corner operation result.

5 Click on the Trim First Element icon in the Tools toolbar. Moving clockwise, go to the next corner and select the lines that form that corner.

6 Once again, enter a value of *25* as the size of the radius.

7 Perform this same routine with the No Trim icon activated.

When complete, the result should look like that shown in figure 3-24.

Sketcher Operation Exercise 2: Chamfer

The steps that follow take you through the process of performing a chamfer operation.

1 In the Sketcher workbench, recreate the rectangle created earlier.

2 Double click on the Chamfer icon in the Operation toolbar, and then in the Tools toolbar click on the icon Trim All Elements.

3 Starting with the lower left-hand corner of the rectangle, select the two lines that form that corner. A chamfer appears.

4 Go to the Tools toolbar and enter a value of *20* as the size of the chamfer.

Note that both of the selected lines are trimmed to the chamfer.

5 Now click on the Trim First Element icon in the Tools toolbar. Moving clockwise, go to the next corner and select the lines that form that corner.

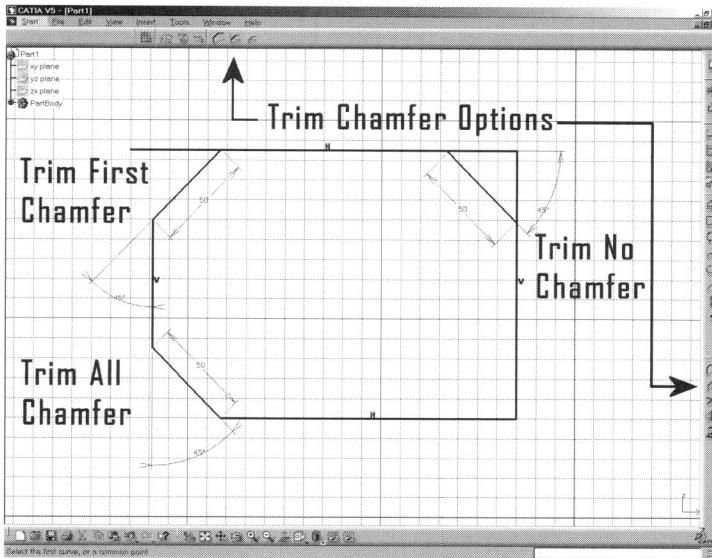

6 Once again, enter the value of *20* as the size of the chamfer.

7 Perform the same routine with the No Trim icon activated.

8 When finished with the No Trim example, see if you can get your last corner to have a chamfer like that shown in figure 3-25.

Fig. 3-25. Chamfer operation.

Sketcher Operation Exercise 3: Relimitations

The middle area of the Operation toolbar is the Relimitations area. The icons found here are Trim, Break, Quick Trim, Close, and Complement. The Trim icon works very much in the same way as the Corner and Chamfer icons. Options here include Trim All Elements and Trim First Element. In each of these cases, if multiple solutions exist, CATIA V5 will display a preview of the different solutions just by moving your cursor onto a different second element or by moving to a different location on the same element. When you see the result you want, select the second element.

Sketcher Operation Exercise 4: Break

Break is very easy to use. Simply select the element you want to break, and select a piece of geometry intersecting it, or select on the element itself where you would like it to break. This relimitations operation is shown in figure 3-26.

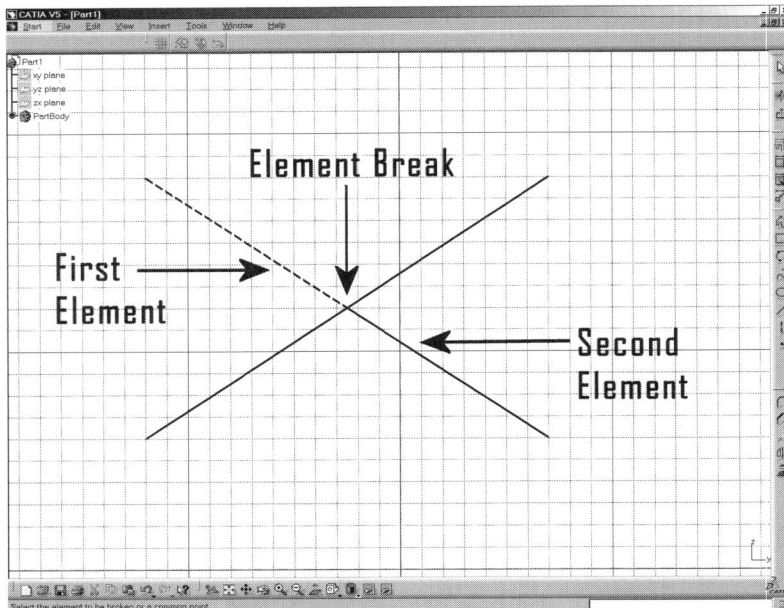

Fig. 3-26. Break operation.

Sketcher Operation Exercise 5: Quick Trim

Quick Trim can be a very useful tool if your sketch has many elements that intersect with each other and need to be trimmed. To use Quick Trim, perform the following steps.

1   In the Sketcher workbench, recreate the rectangle created earlier.

2   Now add a circle in the center that is larger than the height of the rectangle.

3   Double click on the Quick Trim icon and then click on the Break and Rubber In icon in the Sketch Tools toolbar.

4   Select the top line of the rectangle inside the circle and view the result.

5   Keeping the Break and Rubber In icon active, select the bottom line of the rectangle inside the circle and view the result.

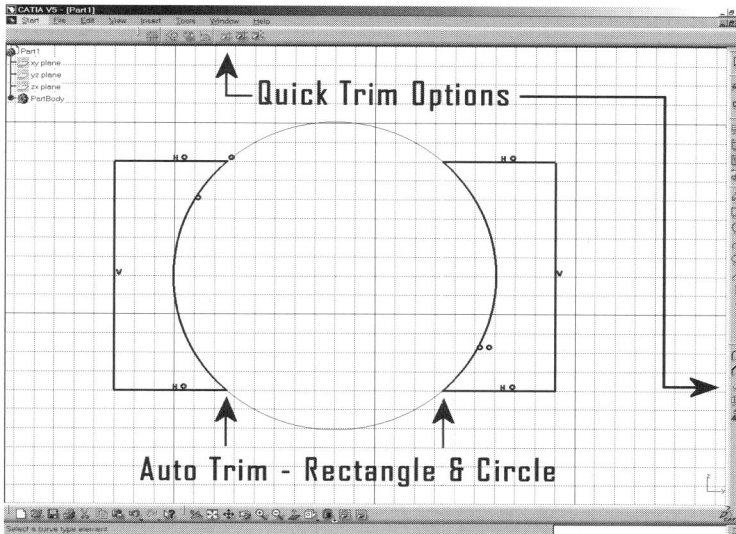

Fig. 3-27. Quick trim operation result.

Your result should look like that shown in figure 3-27.

NOTE: *The Break and Keep icon is similar to the Break and Rubber In and Break and Rubber Out icons in the Sketch Tools toolbar except that there is no erasing of any elements associated with Break and Keep.*

Sketcher Operation Exercise 6: Close and Complement

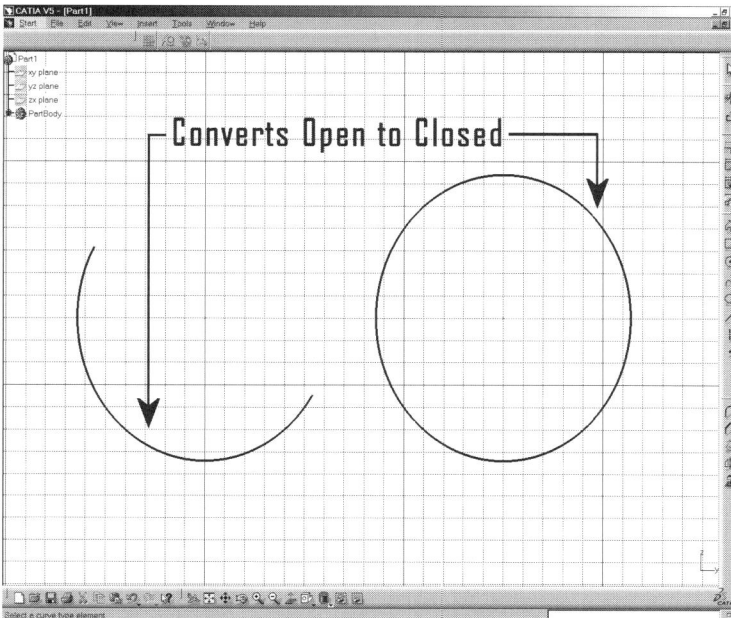

Close and Complement, shown in figure 3-28, are the two remaining icons of the Relimitations area. Close simply adds back the portion of the element that is missing. Complement inverts the remaining portion of the element.

Fig. 3-28. Close and Complement icons.

Constraints

As discussed in Chapter 1, CATIA V5 is a variational modeling system in that it offers the flexibility of constraining or not constraining geometric elements. This accommodates either a quick and simple unconstrained sketch or a more complex constrained sketch. The sketcher environment offers both geometrical and dimensional types of constraints.

Constraint Type	Symbol
Perpendicular	⌐
Parallel	⁄⁄
Coincident	◎
Concentric	◉
Vertical	V
Horizontal	H
Fixed	⚓
Radius Dimension	R50
Diameter Dimension	D100

Fig. 3-29. Geometric constraint symbols.

Geometrical

A geometrical constraint is a rule or condition that exists between geometric elements. Geometric constraints might be relationships such as parallelism and tangency. Constraints may be applied to an individual element or between multiple elements. Figure 3-29 shows key constraint symbols. These symbols indicate when 2D elements are constrained and the type of constraint imposed.

NOTE: *These symbols appear on the screen next to the appropriate geometric element and may be selected for modification.*

Dimensional

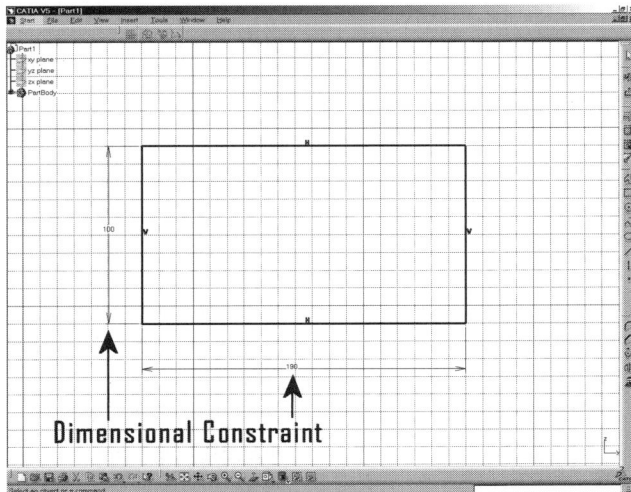

A dimensional constraint, an example of which is shown in figure 3-30, is a constraint in which a numerical value determines the exact measurement of the sketched element. This might be a radius on a circle or the length of a line.

NOTE: *Once a dimensional constraint is applied to a geometric element, it may be selected and modified to parametrically drive the geometry to a desired new location.*

Fig. 3-30. Dimensional constraint.

Sketcher Element Color	Description
Under Constrained	White
Selected	Red / Orange
Protected	Yellow
Over Constrained	Violet / Purple
Vertical	Green
Inconsistent	Red
No Change	Brown

Fig. 3-31. Constraint color code legend.

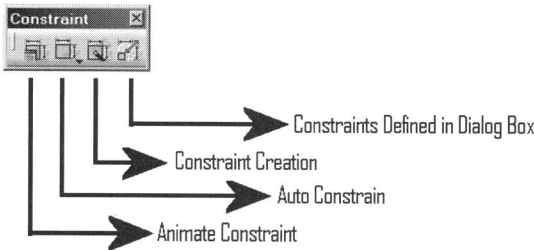

Fig. 3-32. Constraint toolbar.

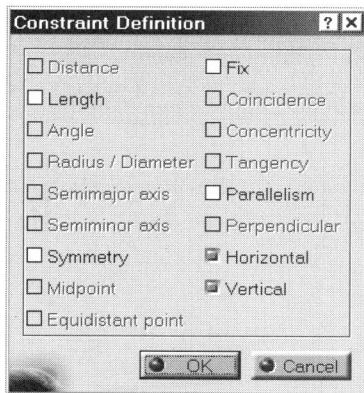

Fig. 3-33. Constraints Defined in Dialog Box icon.

Constraint Color Codes

The color of a sketched element indicates the status of that element's constraint condition or assumption. This is a great visualization aid in quickly determining the constraint relationships of sketched elements and dimensions. Figure 3-31 shows a legend that associates the color of the element with the constraint status.

Constraint Toolbar

Constraints may be automatically applied to a sketched profile prior to initiation by making sure that the Geometrical and/or Dimension icon is activated within the Sketcher Tools toolbar. Additional functionalities for constraining sketcher elements are located within the sketcher Constraint toolbar. This toolbar, shown in figure 3-32, contains four main icons associated with constraining 2D geometric elements. These icons are explored in the sections that follow.

Constraints Defined in Dialog Box

The Constraints Defined in Dialog Box icon allows you to impose certain conditions on elements in the sketcher environment. The icon is inactive until one or more elements are selected. Depending on what elements have been selected, only the possible conditions for those elements will be functional. There are 17 possible conditions (or constraints) available in this icon, shown in figure 3-33.

TIP: *Geometric constraints may also be imposed by first adding a dimensional constraint and converting by pressing MB3.*

Constraint Creation

The Constraint Creation icon, shown in figure 3-34, consists of two options: Constraint and Contact Constraint. The Constraint icon allows you to create dimensional constraints on or between elements in the sketcher environment. The Contact Constraint icon allows you to impose a contact constraint between elements in sketcher mode.

Fig. 3-34.
Constraint
Creation
icon.

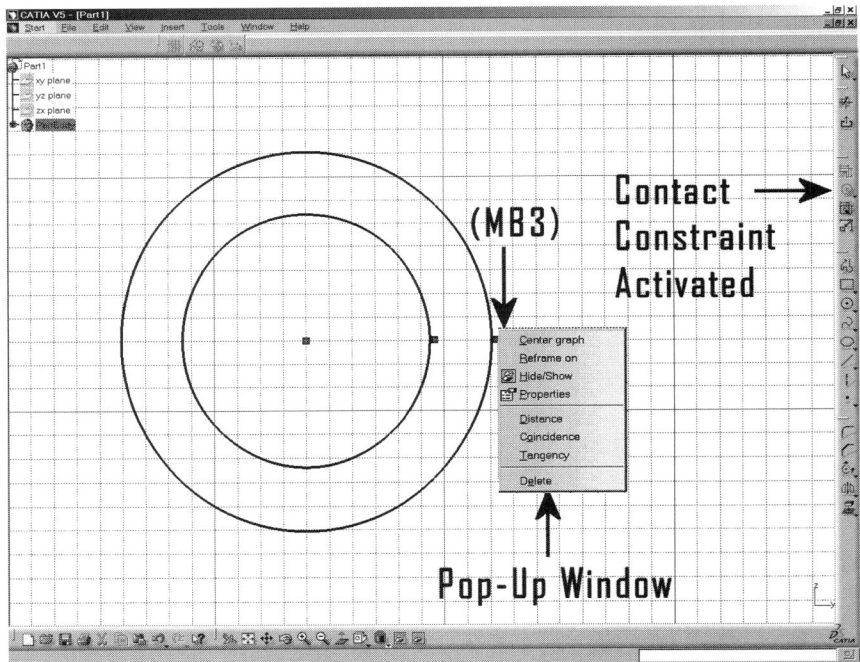

There are three possible contact constraints available: concentricity, coincidence, and tangency. These are dependent on what types of elements have been selected. These constraints can also be modified. To modify these, make sure the Contact Constraint icon is active, and then place the cursor on the constraint and hold down MB3 to view your choices for modification.

Auto Constrain

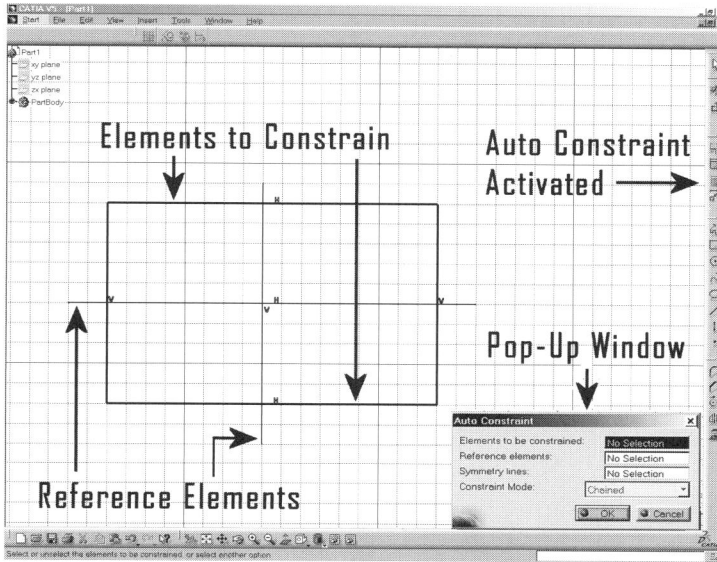

The Auto Constraint icon, shown in figure 3-35, allows CATIA V5 to automatically generate dimensional constraints on a range of geometric elements created in the sketcher.

Fig. 3-35. Auto Constraint icon.

Animate Constraint

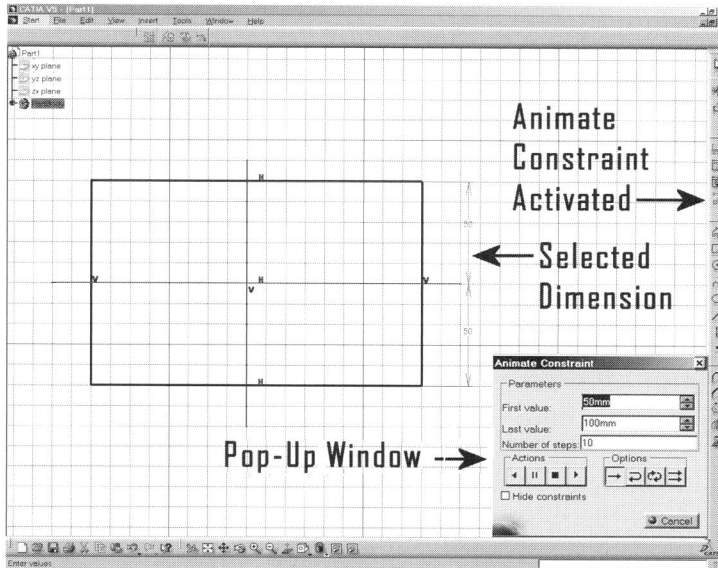

The Animate Constraint icon, shown in figure 3-36, allows the user to perform animations between dimensional constraints and their behavior with sketched profiles.

Fig. 3-36. Animate Constraint icon.

NOTE: *This is useful for flexing a 2D profile and evaluating its behavior under change.*

Sketcher Constraint Exercises

The following exercises are intended to demonstrate various methods of constraining and editing constraints within the sketcher environment.

Sketcher Constraint Exercise 1: Rectangle

To use the Rectangle constraint icon, perform the following steps.

1   Open a CATIA session and create a new sketched entity by selecting the Sketcher workbench. Make sure that both the Geometrical and Dimension icons in the Sketcher Tools toolbar are selected.

2   Click on the Rectangle (profile) icon and create a rectangular box, as shown in figure 3-37.

NOTE: *The sketcher elements should be white, indicating only horizontal and vertical constraint symbols.*

3   While holding down the first mouse button (MB1), select the left vertical side of the box and drag it around the screen.

Note that the box moves around the origin while maintaining the horizontal and vertical constraints, as indicated in figure 3-38.

Fig. 3-37.
Rectangular box.

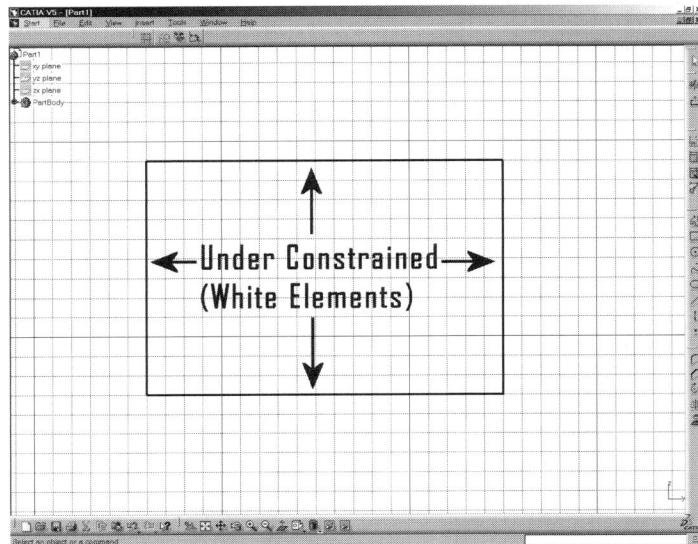

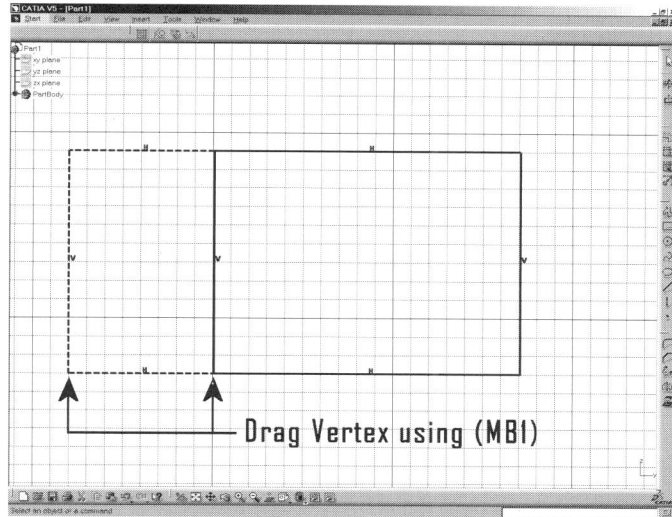

Fig. 3-38. Movement of rectangular box.

4 Select the top left-hand corner point of the box and attempt to drag it upward. Select and delete the top horizontal constraint symbol. (Note that the entire horizontal line still moves in relation to the point.)

5 Select the top left-hand corner point of the box and attempt to drag it upward.

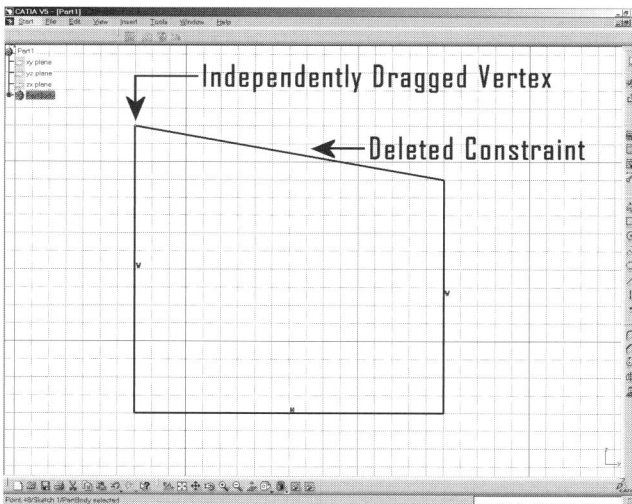

Note that the top corner now expands independently in relation to the rest of the box, as shown in figure 3-39.

6 Select the skewed top line and then select the constraint-defined dialog box. Check the Horizontal box to reapply the horizontal constraint on that element.

Fig. 3-39. Dragging a corner of the box.

Sketcher Constraint Exercise 2: Elongated Hole

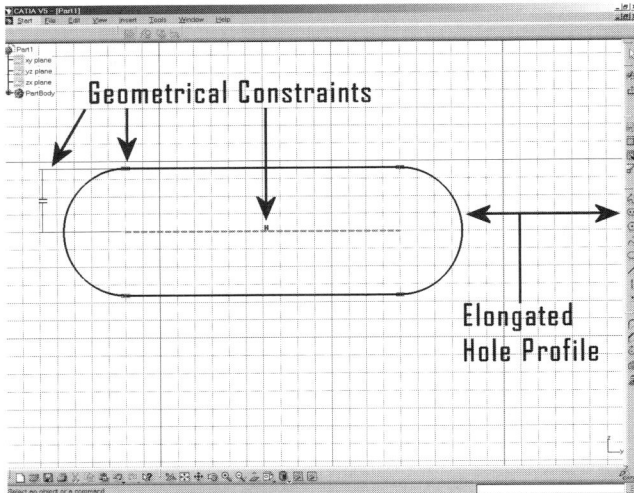

Fig. 3-40. Elongated Hole icon.

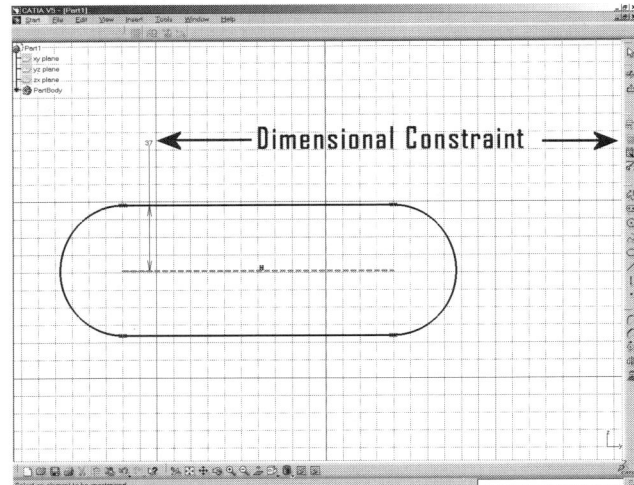

Fig. 3-41. Dimensional constraint.

To use the Elongated Hole constraint icon, perform the following steps.

1   Open a CATIA session and create a new sketched entity by selecting the Sketcher workbench. Make sure that both the Geometrical and Dimension icons in the Sketcher Tools toolbar are selected.

2   Click on the Elongated Hole icon, shown in figure 3-40, and sketch an elongated hole.

3   While holding down the Ctrl key, highlight and select the bottom line of the elongated hole and the sketcher origin. Move the cursor over and click on the Constraint icon. Place the dimension by clicking the first mouse button (MB1).

Note that the bottom line is now green, indicating that a dimensional constraint is present, as indicated in figure 3-41.

4   Finish the exercise by further constraining the sketch.

Sketcher Constraint Exercise 3: Corner Dome

To create the following geometry, perform the following steps.

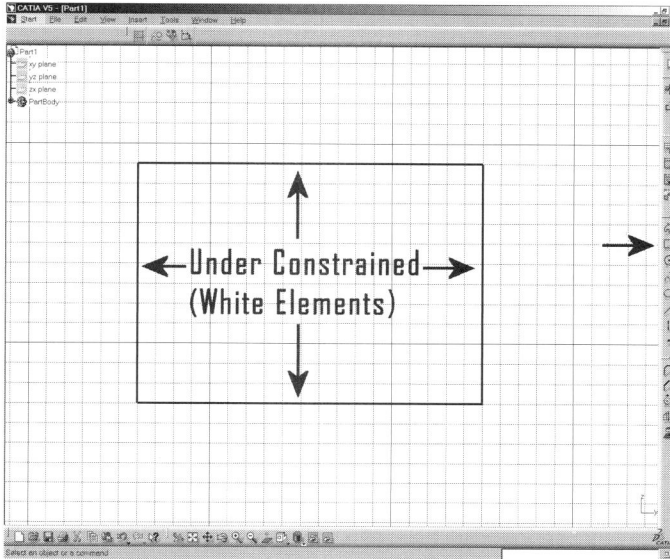

Fig. 3-42. Rectangle icon.

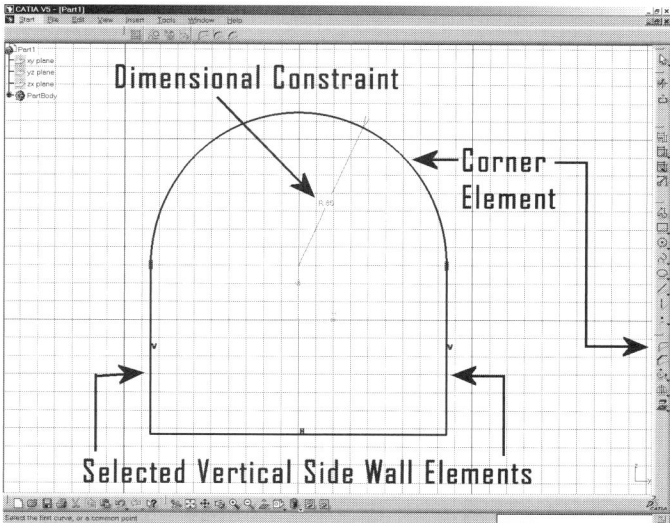

Fig. 3-43. Round created on rectangle.

1 Open a CATIA session and create a new sketched entity by selecting the Sketcher workbench. Make sure that both the Geometrical and Dimension icons in the Sketcher Tools toolbar are selected.

2 Click on the Rectangle (profile) icon, shown in figure 3-42, and create a rectangular box.

3 Select and delete the top horizontal line of the rectangle. Select the corner icon and pick the remaining two side vertical lines. Dynamically drag the created round (dome) and place it using MB1.

Note that a dimensional constraint is automatically placed on the round, as shown in figure 3-43.

4 While holding down the Ctrl key, highlight and select the center of the round and the sketcher origin.

5 Activate the Constraint Defined in Dialog Box icon and select the Coincidence option from the box.

Note that the sketcher automatically snaps the two points together and that the sketched elements are moved to the origin, as shown in figure 3-44. In addition, note that the round and the two side walls are *not* green, indicating that they are constrained. See figure 3-44.

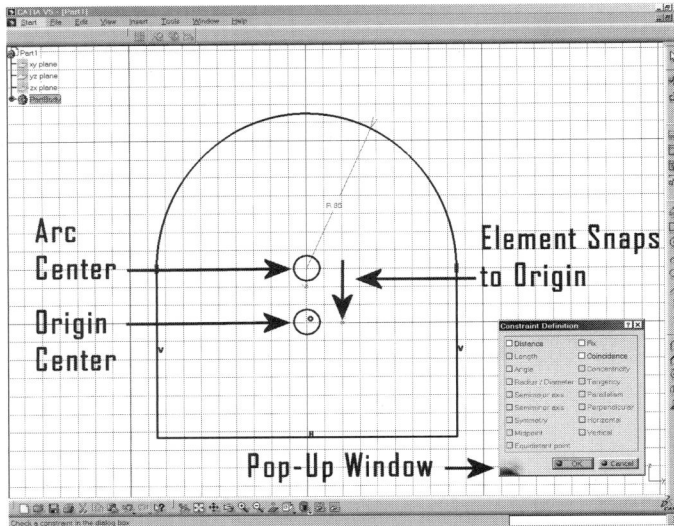

Fig. 3-44. Origin snap.

6 Select one of the green constrained elements and attempt to drag it. Select the bottom white unconstrained horizontal line and drag it within the relationships and boundaries of the adjacent elements.

Note the difference between constrained and unconstrained elements.

Sketcher Tips

The section that follows explore points to keep in mind when working with the sketcher.

Insert New Body Prior to Initial Sketch

Sketches in CATIA V5 reside in either a body or open body. Sketches should not be created in the default part body, which is located within the part model. It is recommended that a new body be inserted into the model prior to creating any new geometry. This is because the default part body is predefined and used for the following reasons.

- It is a permanent body that is created when the document is initialized.

- This body is the root of the specification tree. It cannot be reordered.

- It should remain an empty body that has all of the other bodies assembled to it when the design is complete.

- It defines the result of all operations and assemblies.

- Mass properties and analysis are located only in this body.

- All drawing feature links are located only in this body.

Keep Sketches Simple

Keep sketches as simple as possible. However, keeping sketches simple does not mean that the design has to be simple. Building a model via multiple simple sketches and features offers more model flexibility when the model becomes very large and typically needs some level of modification further along in the process. A rule of thumb here is to place one important piece at a time. Adding two simple features, rather than one complicated feature, is often more effective in terms of design flexibility and change management.

Build Flexibility into a Sketch

It is important to build flexibility into a sketch to promote desired reactions. Dimensional constraints are the means by which you accomplish this flexibility. The key to effective dimensioning is understanding which is the most desirable way of driving the design, and then to control behavior via dimensional constraints. For example, if a table manufacturer were designing a new table, a key dimensional constraint would be from the floor to the top of the table, not from the ceiling to the top of the table.

Flexing the Sketch with Animate Feature

Animating the dimensional constraints of a sketched profile is good practice for evaluating the constraint behavior between and among dimensional limits. This allows you to evaluate and make adjustments to the sketch prior to leaving the environment.

Avoid Rounds, Fillets, and Draft in Sketch

Rounds, fillets, and draft should be avoided within a sketch. These elements should be added as separate features to allow for downstream model flexibility. This also cuts down on unwanted parent/child relationships when future geometric reorganization is necessary.

Summary

The CATIA V5 sketcher is a powerful tool that offers the flexibility of quickly generating 2D elements. The ability to constrain or not constrain is a key in increasing the productivity of new users. A rich set of tools is offered to assist a user in capturing the desired design intent. This chapter has introduced you to the CATIA V5 sketcher workbench. You should have at this point a basic understanding of how to create and edit 2D sketched profiles. You should also be familiar with the following topics.

- The Sketcher environment
- Sketcher tools
- Profiles
- Geometric dimension constraints
- Sketcher operations

Review Questions

1   What is the sketcher's primary function in developing geometry within CATIA V5?

2   How is the sketcher identified within the graphical user interface (GUI)?

3   Name five geometric element creation types located in the Geometry Profile toolbar.

4   How are these geometric elements created within the Sketcher workbench?

5   How does the color display of the elements relay important information about the state of a sketched profile?

6   Name the four primary functions of the Sketcher Tools toolbar.

7   List the two primary types of constraints that may be applied to elements within the Sketcher workbench.

8   How can a dimensional constraint be turned into a geometrical constraint?

9   Must all elements within the Sketcher workbench be constrained?

10   What are some ways to check the robustness of a sketched profile?

CHAPTER 4

THE PART DESIGN WORKBENCH

Introduction

THE OBJECTIVE OF THIS CHAPTER is to introduce new users of CATIA V5 to the Part Design workbench. Understanding the basics of the Part Design workbench is the key to building robust models.

The Part Design workbench is the starting point for creating parametric solid models in CATIA V5. This application combines the power of feature-based design with the flexibility of a Boolean approach. The Part Design workbench allows for the creation of associative feature-based models. Solid and surface features are created through the use of reference elements, sketches, sketcher-based features, dress-up features, and transform operations.

This hybrid approach to modeling offers great flexibility in giving the user the ability to constrain or not constrain geometry. This flexibility also offers the ability to offer post-design and local 3D parameterization to assist in capturing part design intent.

Objectives

The following are the key points of this chapter.

- Reference elements
- Geometry-based features

- Sketcher-based features
- Dress-up features
- Surface-based features
- Transform features
- Boolean operations

Reference Elements

Fig. 4-1. Reference Element toolbar.

Reference elements are used to assist in the construction of solid features within CATIA V5. Points, planes, and lines are all reference elements located on the Reference Element toolbar, shown in figure 4-1.

These elements are used to help set up the creation and construction of solid or surface element features by further capturing the design intent. Reference elements reside in open bodies located within the configuration tree.

TIP: *It is generally considered not good practice to create a sketched feature that resides under the open bodies portion of the configuration tree.*

Point Elements

Reference points are located within the Reference Element toolbar, shown in figure 4-2. Points are used as an anchor position for the creation of other elements, such as lines, planes, and rounds. CATIA V5 offers several methods of creating a point, as shown in figure 4-3.

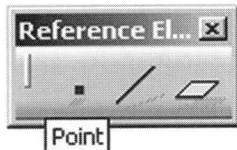

Fig. 4-2. Point Elements icon.

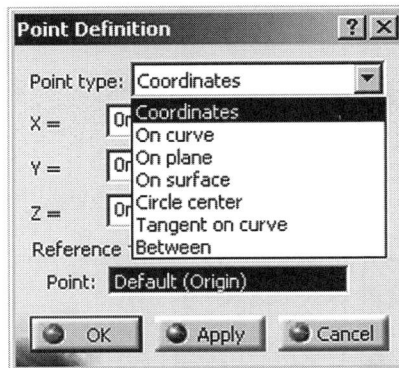

Fig. 4-3. Reference point creation types.

Coordinates

This creates a point in the X-Y-Z coordinate located from the current system origin axis. Optionally, a point may also be located from an existing point location. An example of a coordinate point is shown in figure 4-4.

Fig. 4-4. Coordinate point.

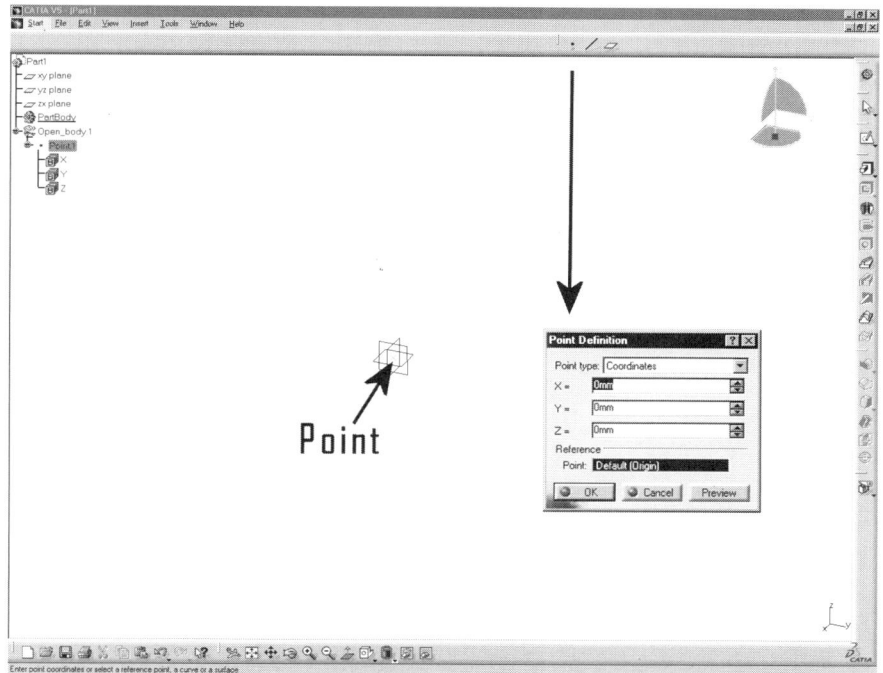

On Curve

This creates a point on an existing curve at the point of selection using the mouse button. This point may be either located by the exact distance on the curve or a ratio of length on the curve. Two additional options (Nearest Extremity and Middle Point) are also available to quickly move the point to common positions. An example of a point on a curve is shown in figure 4-5.

A reference point may also be selected to assist in the creation of a point. If this point is not on a curve, CATIA V5 will automatically project the point onto the curve. If no point is selected, the curve's extremity is used as reference.

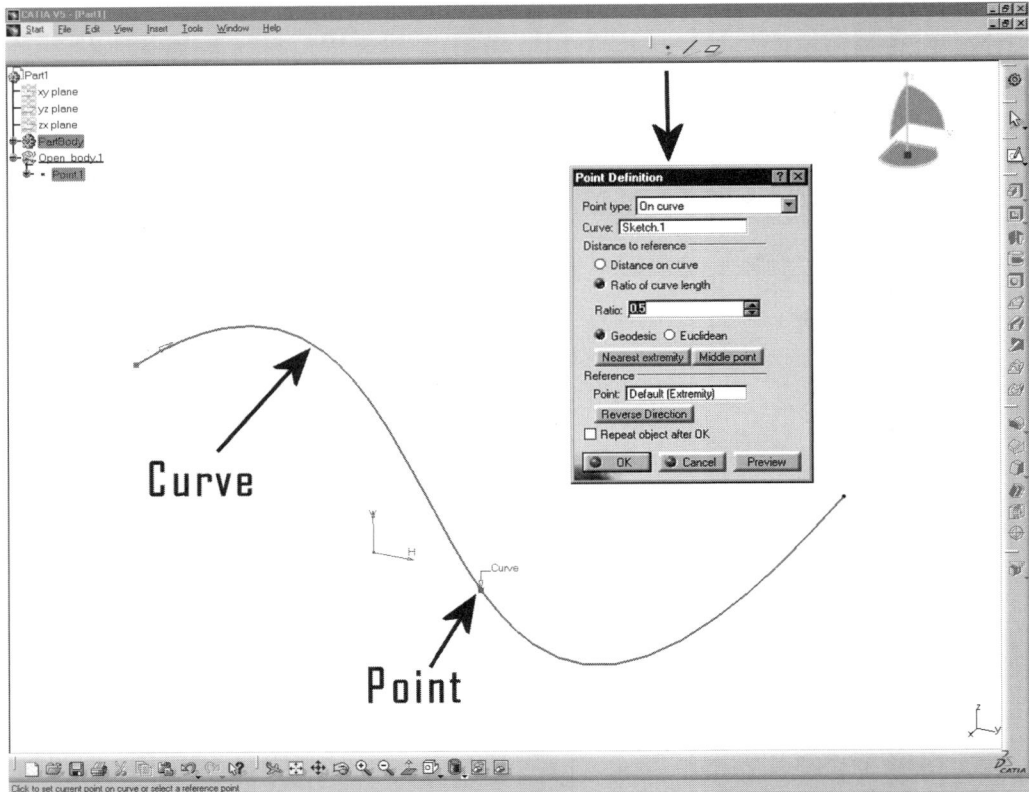

Fig. 4-5. Point on a curve.

TIP: *Selecting the Repeat Object option at the bottom of the Form menu allows for the creation of multiple equidistant points on the curve.*

On Plane

The On Plane option creates, as shown in figure 4-6, a point on a plane with reference to an H and V vector position relative to the system origin axis. A point may also be optionally selected to define the starting point of the two vectors. When the reference plane is selected, you have the ability to drag the point around until the final mouse button selection (MB1).

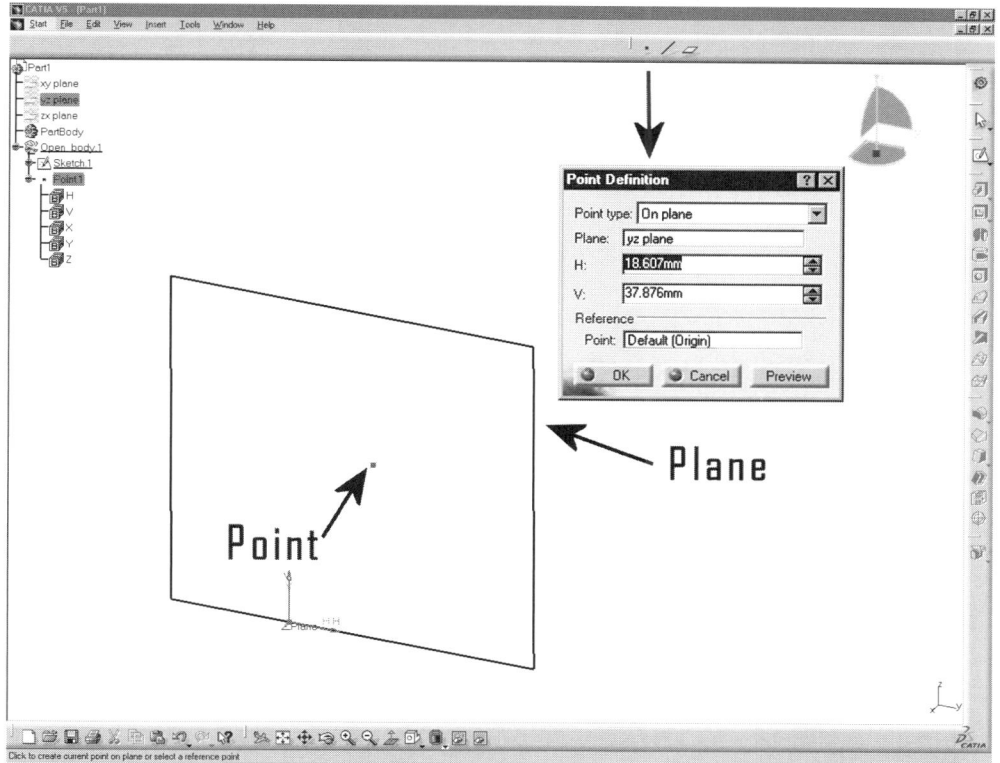

Fig. 4-6. On Plane option.

On Surface

The On Surface option creates, as shown in figure 4-7, a point on an existing face or surface relative to the selection of the mouse button (MB1). By default, CATIA V5 selects the surface's middle point as a reference position. A direction may also be selected to orient the point. A reference point may also be selected as the starting point of reference instead of the default system origin axis.

Circle Center

The Circle Center option creates, as shown in figure 4-8, a point at the center of an existing circle. This may be a curve, surface, or solid face.

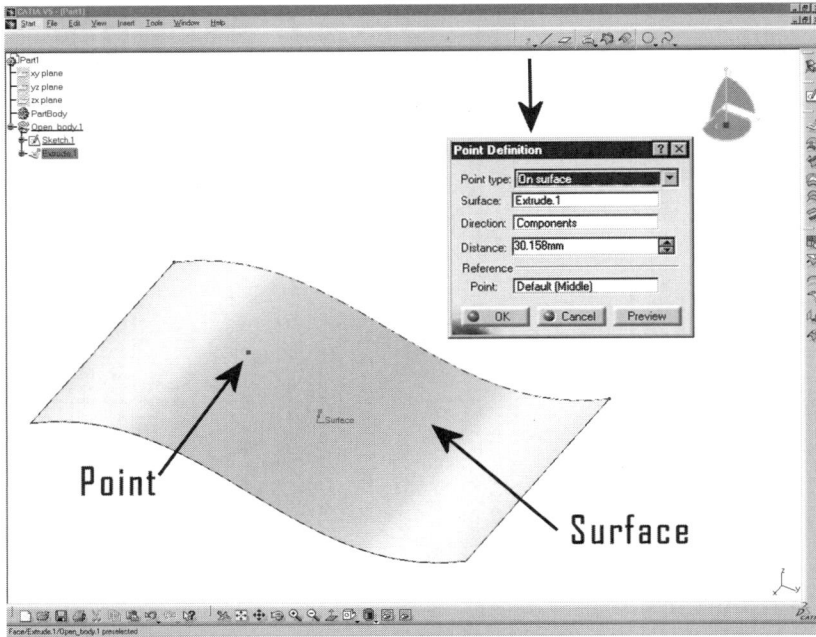

Fig. 4-7. On Surface option.

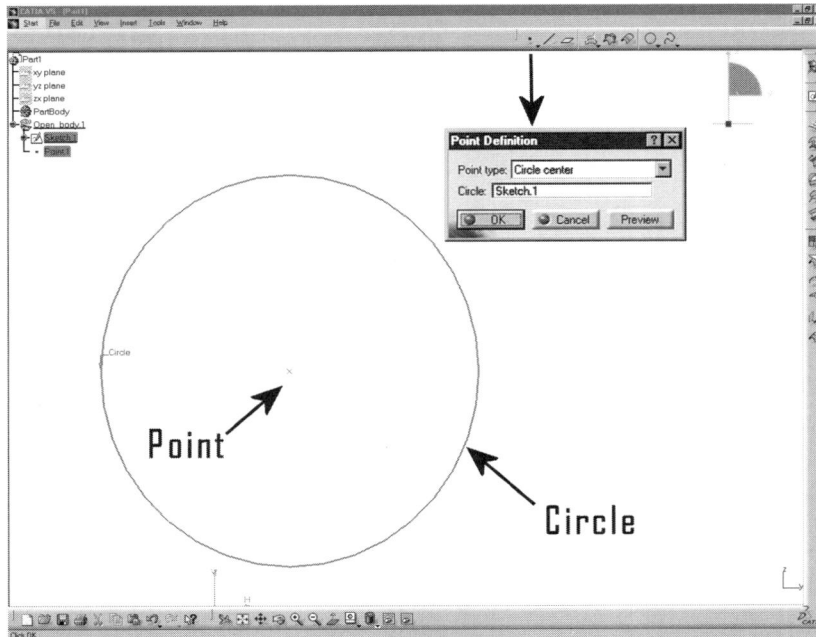

Fig. 4-8. Circle Center option.

Tangent on Curve

The Tangent on Curve option creates, as shown in figure 4-9, a point on a planar curve with respect to a direction line.

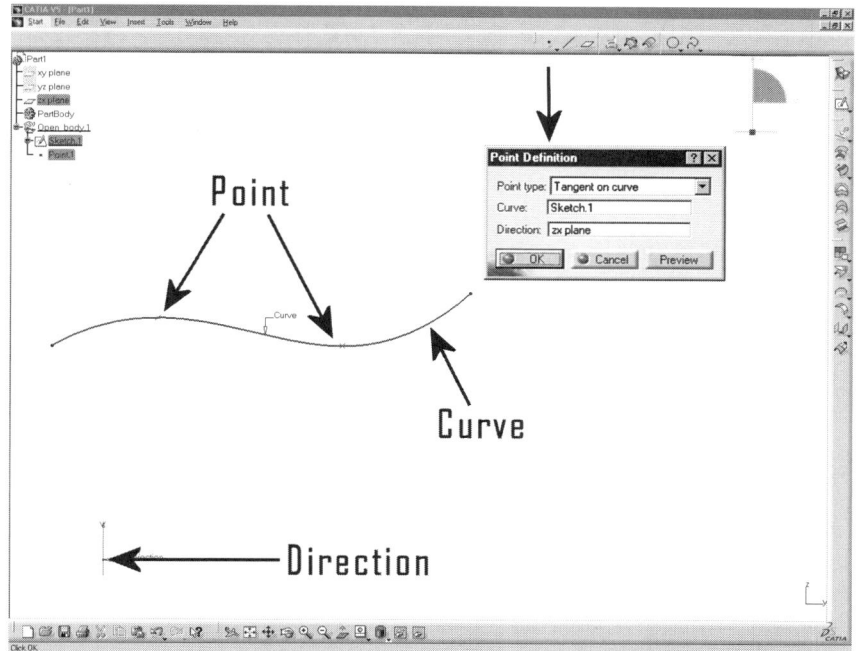

Fig. 4-9. Tangent on Curve option.

Between

The Between option creates, as shown in figure 4-10, a point between two existing points at a determined ratio. Two additional options (Reverse Direction and Middle Point) are available for quickly moving the point to common positions.

Plane Elements

Reference planes are important in the creation of solid features via the Plane Elements icon, shown in figure 4-11. They may be used as reference in the creation of elements but are primarily used as the starting place for sketched profiles. Once a plane is defined, it is represented as a red square, which may be dragged to its final position using the graphic manipulator. CATIA V5 offers several methods of creating a plane, as indicated in figure 4-12.

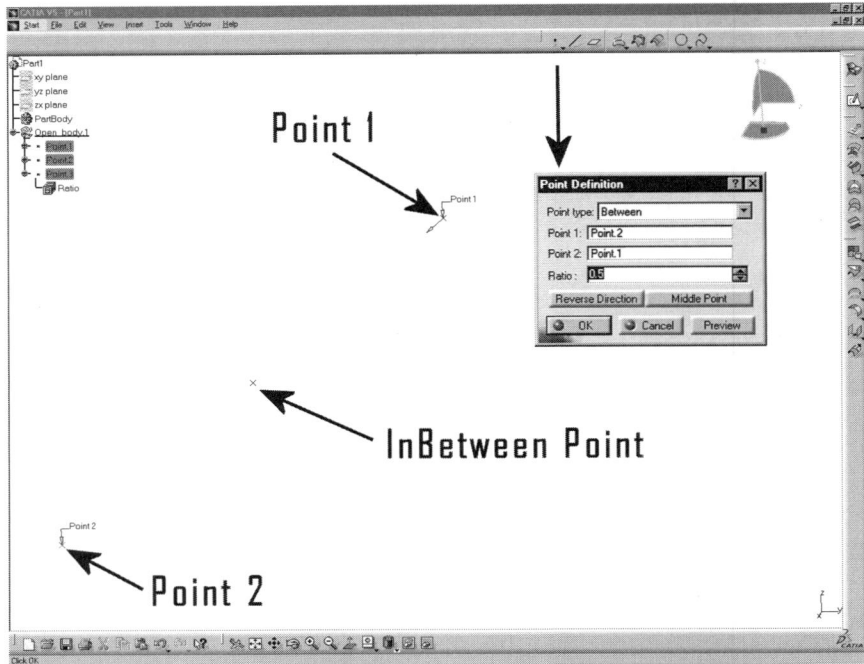

Fig. 4-10. Between option.

Fig. 4-11. Plane Elements icon.

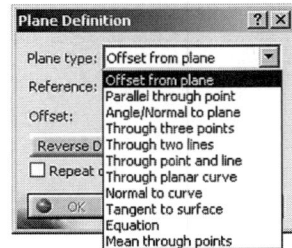

Fig. 4-12. Reference plane creation methods.

NOTE: *The locations of plane dimensions may be updated in real time using the cursor. Geometry attached to these planes will also update in real time.*

Offset

The Offset option creates a plane offset from an existing plane for some given distance. The direction of the offset may be toggled back and forth by clicking on the Reverse Direction button or by selecting the arrow in the Display window. An example of an offset plane is shown in figure 4-13.

TIP: *Selecting the Repeat Object option at the bottom of the Form menu allows for the creation of multiple equidistant planes.*

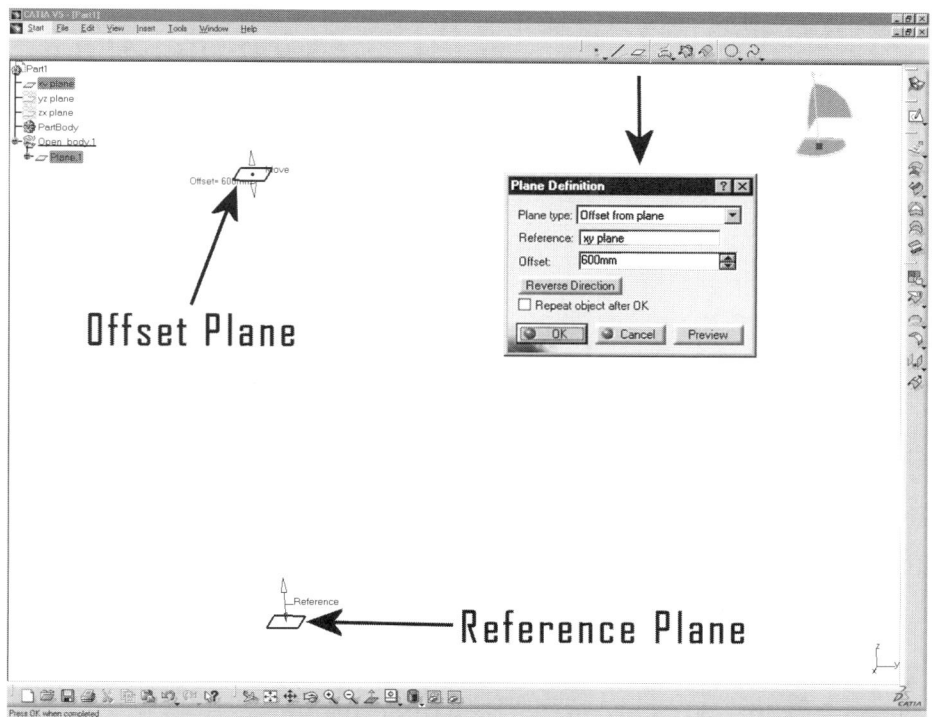

Fig. 4-13. Offset plane.

Parallel through Point

The Parallel through Point option creates a plane through a given point. The plane is parallel to an existing reference plane, as shown in figure 4-14.

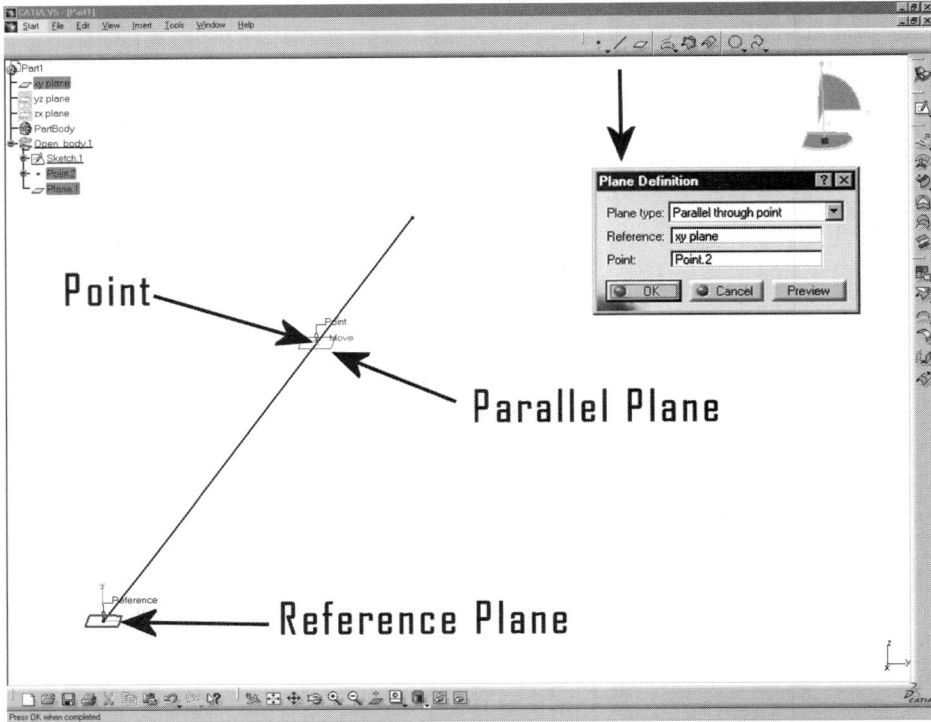

Fig. 4-14. Plane created using Parallel through Point.

Angle or Normal to Plane

The Angle or Normal to Plane option creates a plane through a rotation axis. The plane is normal or on an angle to an existing reference plane, as shown in figure 4-15.

Through Three Points

The Through three Points option creates, as shown in figure 4-16, a plane through three existing points. Once the plane is created, the plane symbol may be dynamically dragged to any location prior to the final selection.

Fig. 4-15.
Plane created
using Angle
or Normal to
Plane.

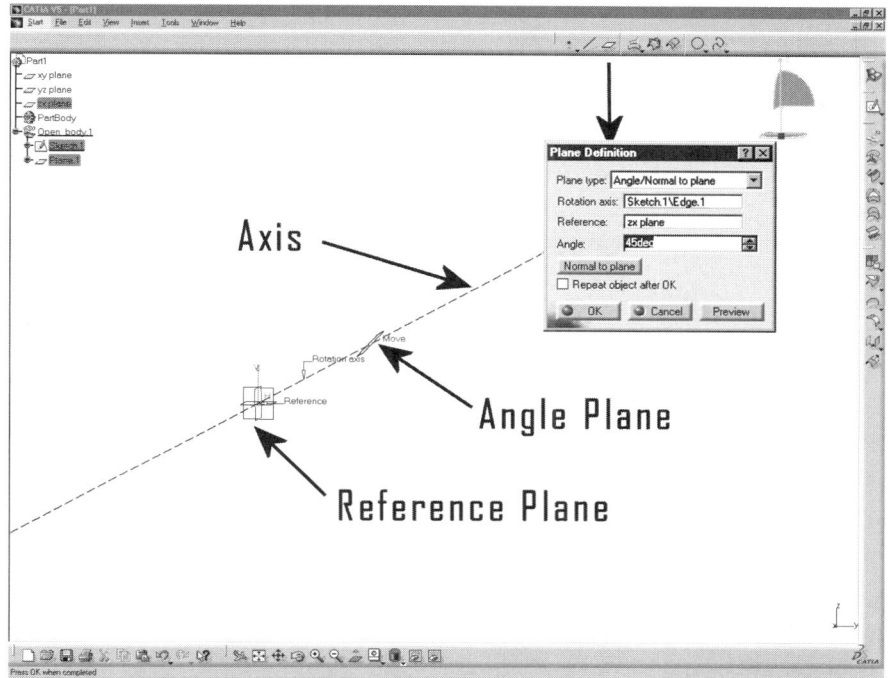

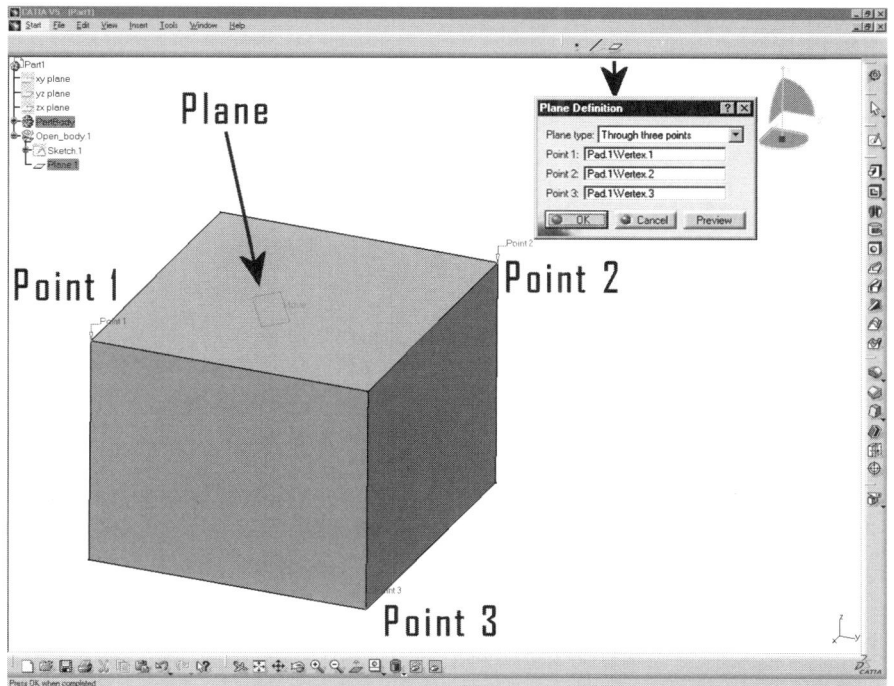

Fig. 4-16.
Through Three
Points option.

Through Two Lines

The Through Two Lines option creates, as shown in figure 4-17, a plane through two line segments. When the two selected lines are not coplanar, CATIA automatically moves the vector of the second line to the first line's location.

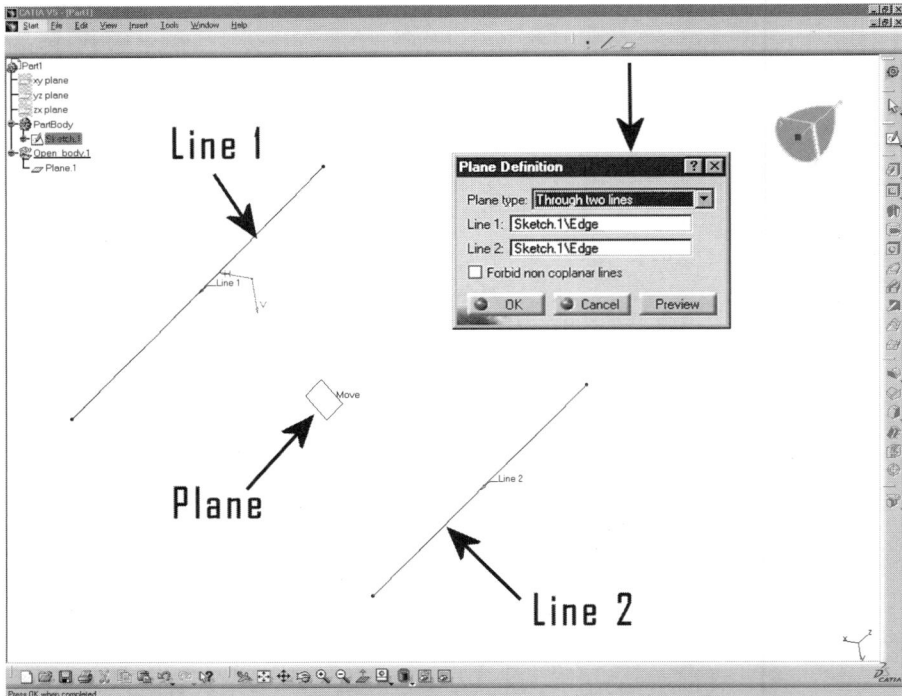

Fig. 4-17. Through two Lines option.

Through Point and Line

The Through Point and Line option creates, as shown in figure 4-18, creates a plane that passes through an existing line and point. Once the plane is created, the plane symbol may be dynamically dragged to any location prior to the final selection.

Through Planar Curve

The Through Planar Curve option creates, as shown in figure 4-19, a plane through an existing planar curve element.

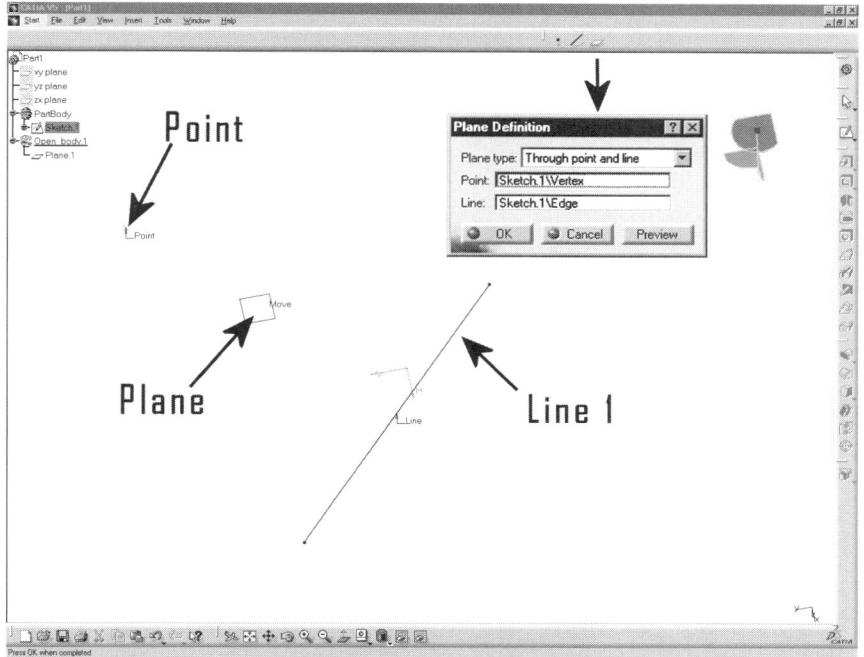

Fig. 4-18.
Through Point
and Line
option.

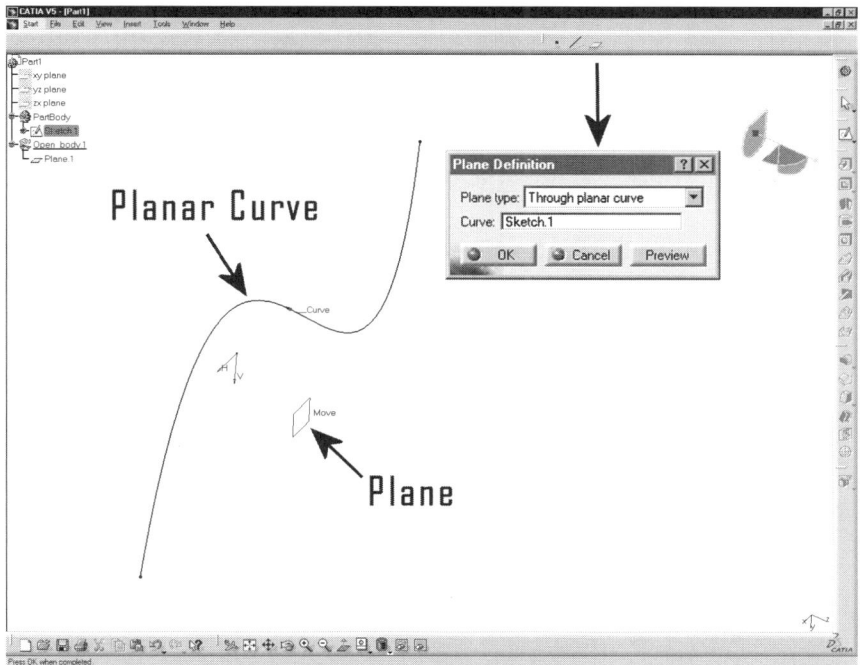

Fig. 4-19.
Through Planar
Curve option.

Normal to Curve

The Normal to Curve option creates, as shown in figure 4-20, a plane at a reference point. The plane is normal to the selected curve. CATIA V5 automatically selects the middle of the curve as its default location.

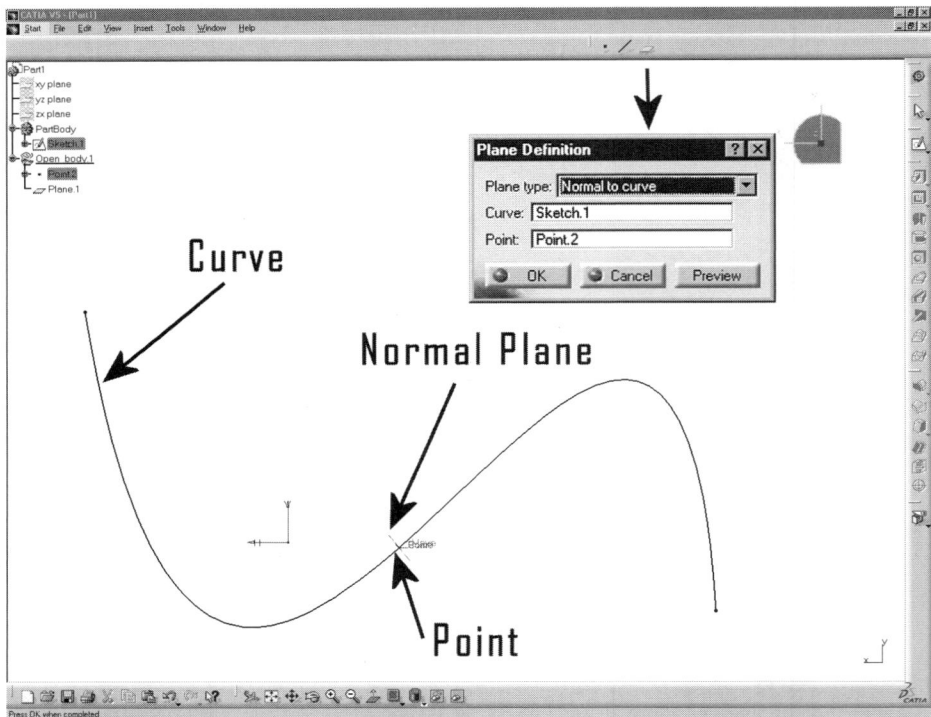

Fig. 4-20. Normal to Curve option.

Tangent to Surface

The Tangent to Surface option creates, as shown in figure 4-21, a plane tangent to a surface at a selected point.

Mean through Points

The Mean through Points option, shown in figure 4-22, creates a plane through the average mean of three selected reference elements.

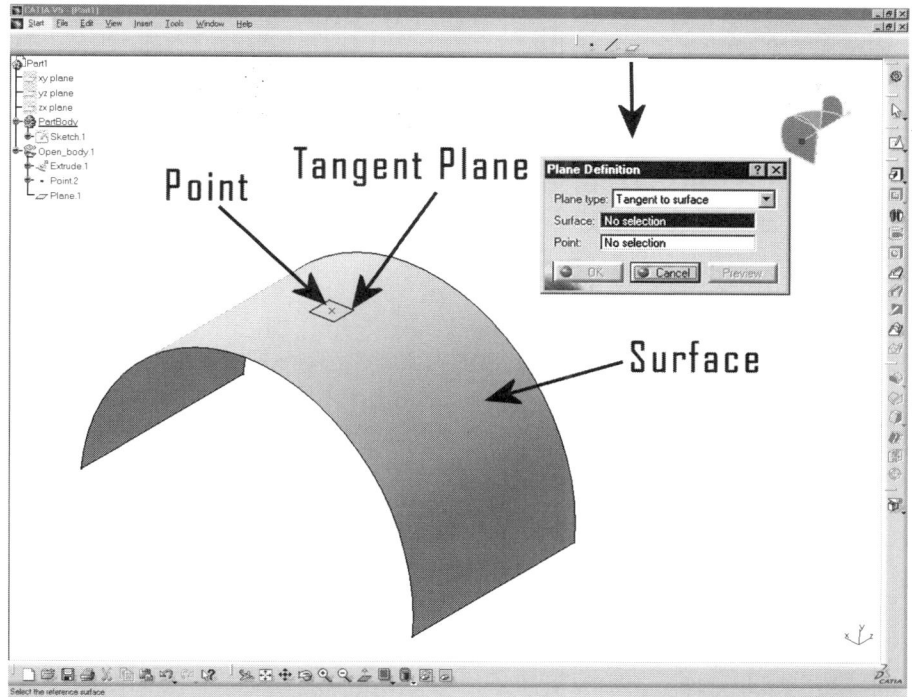

Fig. 4-21.
Tangent to
Surface
option.

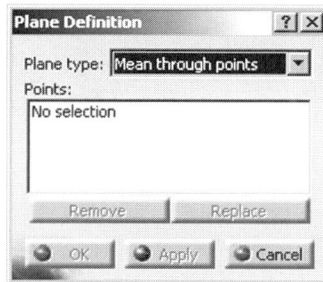

Fig. 4-22. Mean through Points
option.

Equation

The Equation option, shown in figure 4-23, creates a plane using an equation to drive the location. This is accomplished by entering values for the A, B, C, and D input areas. These entries define a plane via the equation $Ax + By + Cz = D$. Two additional options (Normal to Compass and Parallel to Screen) are also available for quickly moving the plane to common positions.

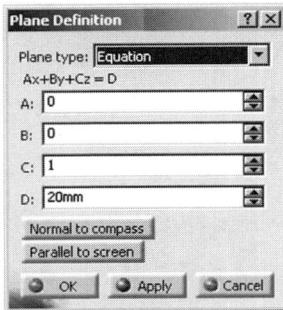

Fig. 4-23. Equation option.

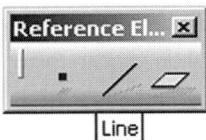

Fig. 4-24. Line Element icon.

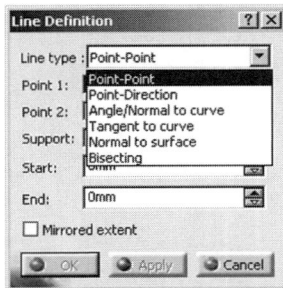

Fig. 4-25. Line creation methods.

Line Element

Via the Line Element icon, shown in figure 4-24, lines are used as construction elements for the creation of both solid and surface geometry features. CATIA V5 offers several methods of creating a line, as indicated in figure 4-25.

Point-Point

The Point-Point option creates, as shown in figure 4-26, a line between two existing points. The Start and End point condition fields are initially set at 0, and may be modified in the dialog box or dynamically dragged in the Display window using the mouse cursor. The line length may never be less than the distance of the original reference points. A support plane may be selected, which will project the line segment normal to the plane, while maintaining reference to the initial point locations.

Point-Direction

The Point-Direction option creates, as shown in figure 4-27, a line at a point normal to the reference element selected. The Start and End point condition fields are initially set at 0, and may be modified in the dialog box or dynamically dragged in the Display window using the mouse cursor. The line may also be mirrored about the reference element via the Mirror Extent box.

Angle or Normal to Curve

The Angle or Normal to Curve option creates, as shown in figure 4-28, a line, on a point, at an angle to another line with respect to a reference plane. The Start and End point condition fields are initially set at 0, and may be modified in the dialog box or dynamically dragged in the Display window using the mouse cursor. Two additional options (Normal to Curve and Reverse Direction) are also available for quickly moving the line to common positions. The Geometry on Support option will create a geodesic line on a support surface.

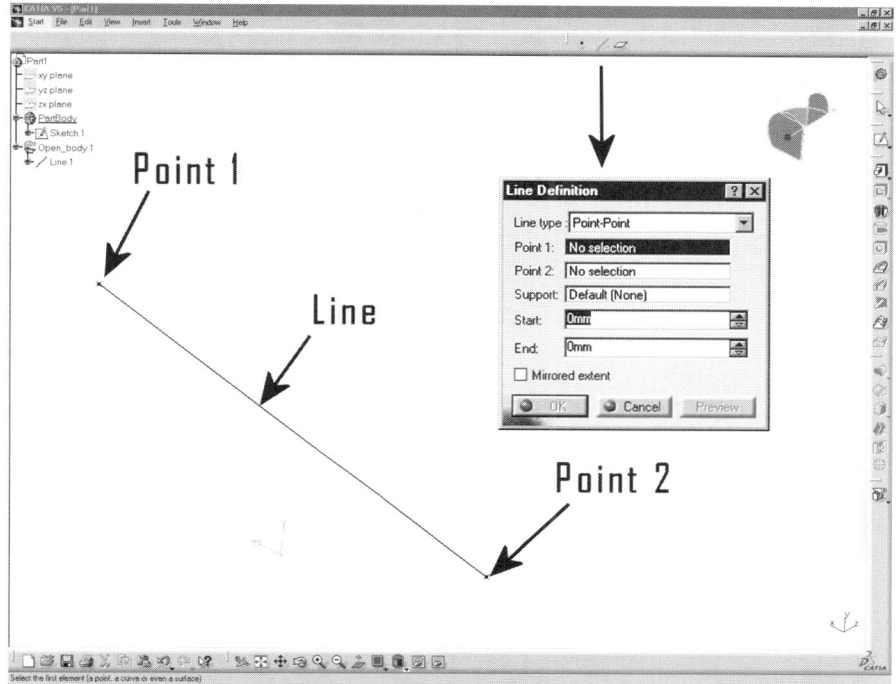

Fig. 4-26. Point-Point option.

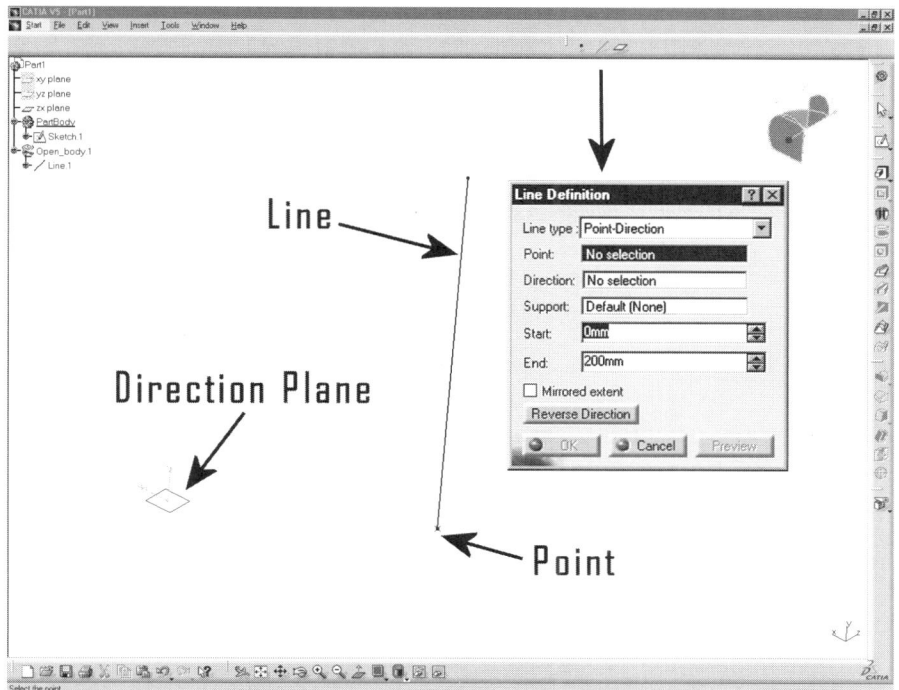

Fig. 4-27. Point-Direction option.

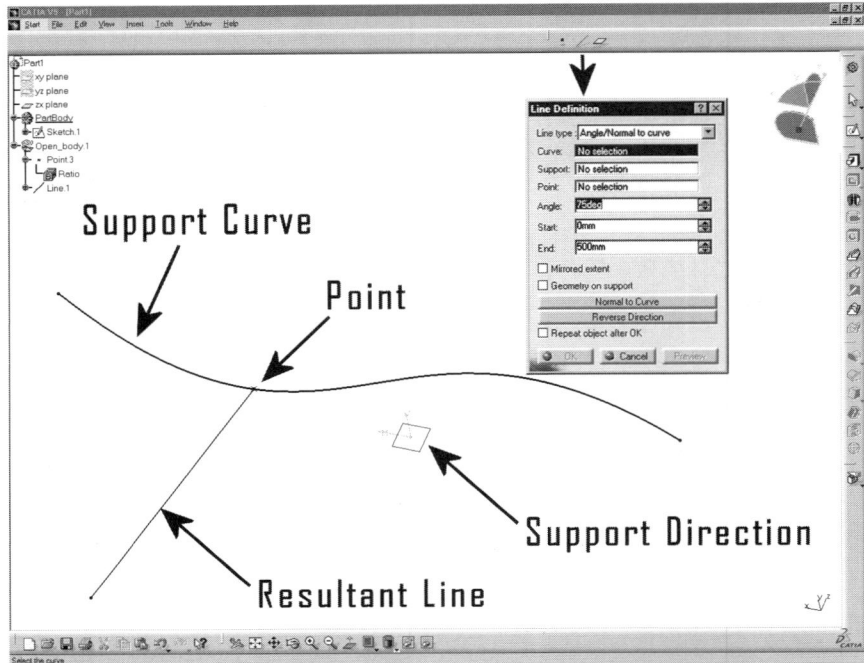

Fig. 4-28. Angle or Normal to Curve option.

Tangent to Curve

The Tangent to Curve option creates, as shown in figure 4-29, a line tangent to a curve element relative to another curve or point. A mono-tangent type is imposed if the second element selected is a point. A bi-tangent type is imposed if the second element selected is another curve. A support plane is required when selecting two curves. The Start and End point condition fields are initially set at 0, and may be modified in the dialog box or dynamically dragged in the Display window using the mouse cursor.

Normal to Surface

The Normal to Surface option creates, as shown in figure 4-30, a line normal to a surface at a selected point. The Start and End point condition fields are initially set at 0, and may be modified in the dialog box or dynamically dragged in the Display window using the mouse cursor.

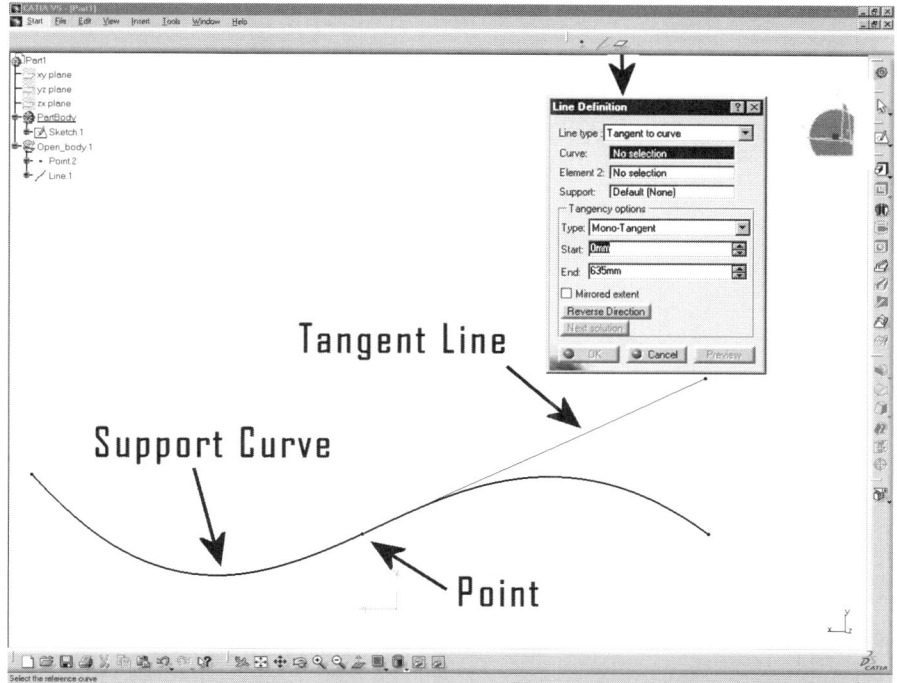

Fig. 4-29. Tangent to Curve option.

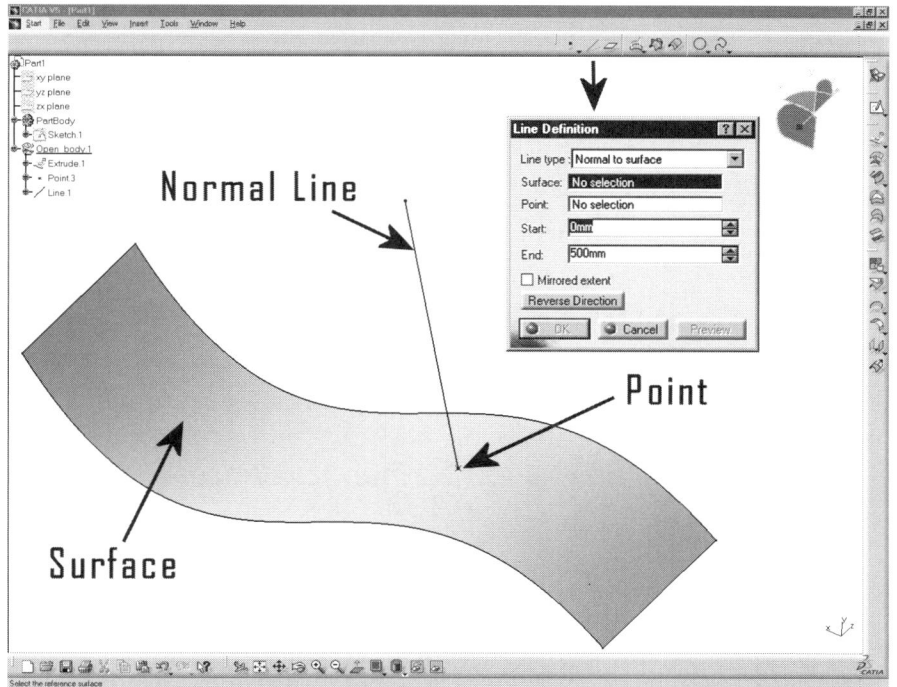

Fig. 4-30. Normal to Surface option.

Bisecting

The Bisection option creates, as shown in figure 4-31, a line that bisects two selected lines. The bisecting line represents a supporting surface that splits the angle between the two selected lines. The Start and End point condition fields are initially set at 0, and may be modified in the dialog box or dynamically dragged in the Display window using the mouse cursor.

*Fig. 4-31.
Bisecting option.*

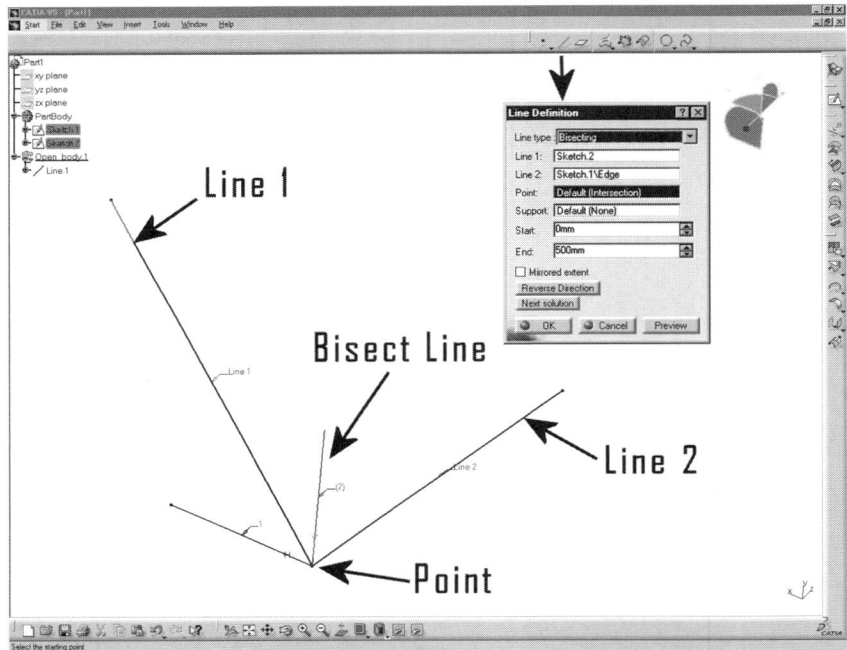

Reference Element Exercises

The sections that follow present tutorials intended to provide practice in working with reference elements.

Reference Element Exercise 1: The Offset Plane

This tutorial creates an offset plane from an existing plane. Perform the following steps.

1   Start a new CATIA session and create a new part design, shown in figure 4-32.

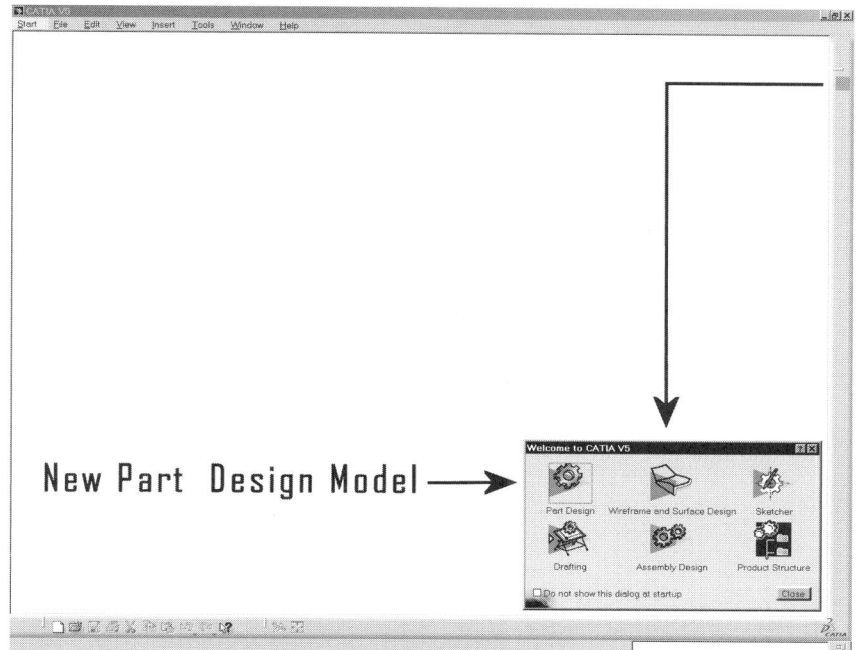

*Fig. 4-32.
New part
design to
be created.*

*Fig. 4-33.
Plane icon.*

2 Click on the Plane icon (shown in figure 4-33) on the Reference Element toolbar. The Plane Definition dialog box will appear, with the default plane type set at Offset from Plane.

3 Select the YZ plane using MB1.

Note that CATIA V5 initially creates a green plane, indicating that the plane may be reviewed prior to selecting the final offset dimension.

4 Toggle the Reverse Direction button a few times to switch the plane back and forth.

5 Increase or decrease the offset plane position (shown in figure 4-34) by selecting the Up or Down arrow in the offset area of the dialog box or by dynamically dragging, while holding down MB1, the offset green dimension located in the window display area.

6 Click on the OK button to complete the creation of the plane.

Fig. 4-34.
Offset plane
position.

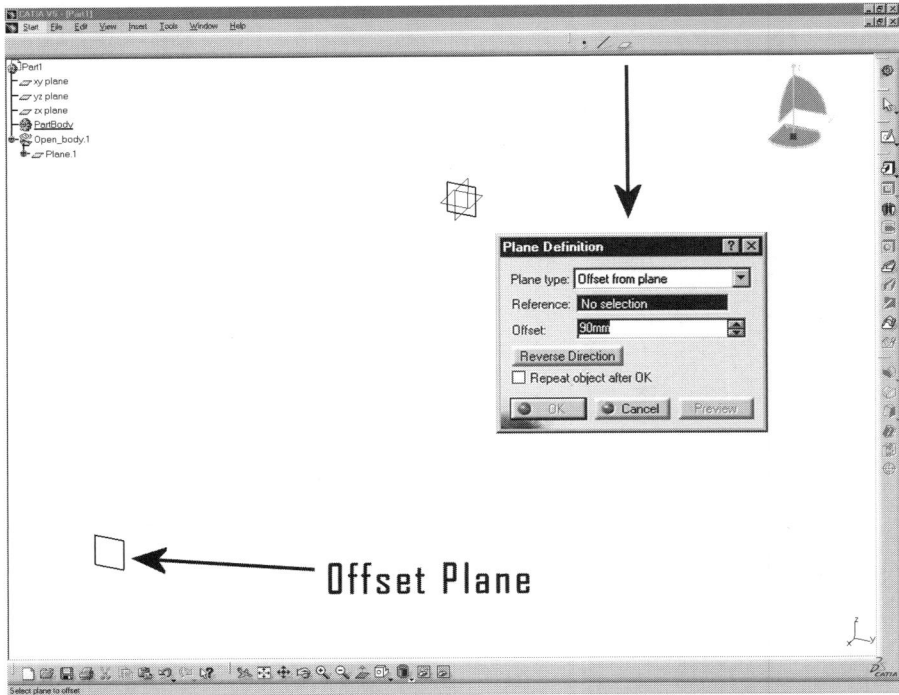

Reference Element Exercise 2: The Coordinate Point

This tutorial creates a point offset from the default system origin axis. Perform the following steps.

1   Start a new CATIA session and create a new part design, shown in figure 4-35.

2   Click on the Point icon (shown in figure 4-36) on the Reference Element toolbar. The Point Definition dialog box will appear, with the default point type set at Coordinates.

3   In the X (directional) input box, enter a value of *100* (mm).

4   Click on the OK button to complete the creation of the point.

The newly created coordinate point is shown in figure 4-37.

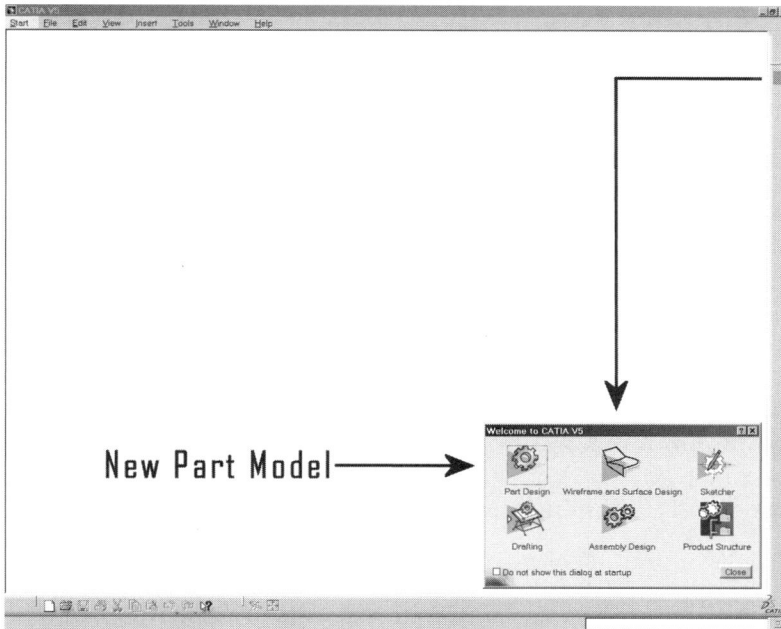

Fig. 4-35. New part design to be created.

Fig. 4-36. Point icon.

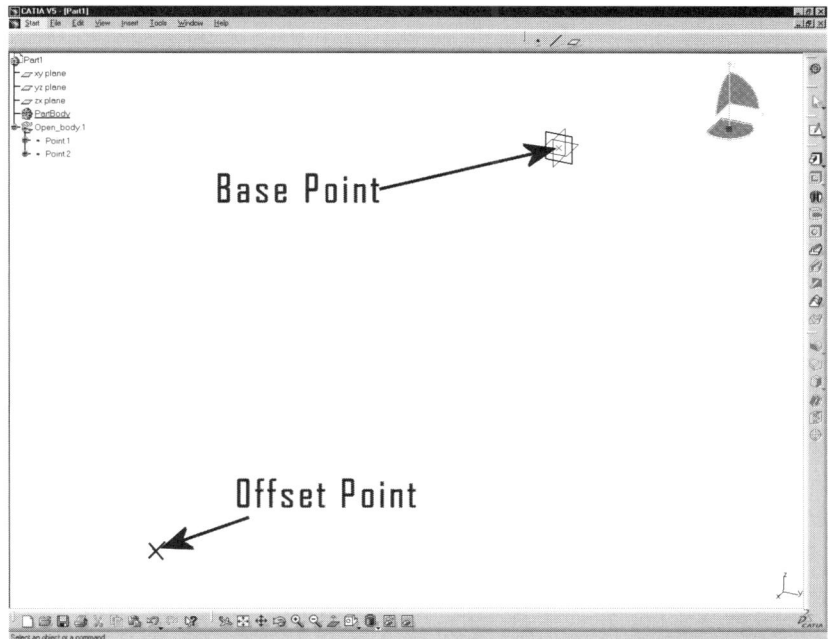

Fig. 4-37. Coordinate point.

Reference Element Exercise 3: The Point-Point Line

This tutorial creates a line between two points. Perform the following steps.

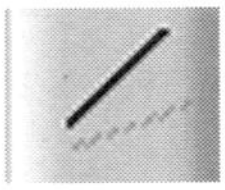

*Fig. 4-38.
Line icon.*

1   Use the existing part model from the previous tutorial. Click on the Line icon (shown in figure 4-38) on the Reference Element toolbar. The Line Definition dialog box will appear, with the default line type set at Point-Point.

2   Select the default axis origin and the 100-mm offset point.

3   Increase the start or end point condition setting by selecting the Up or Down arrow in the Start or End areas of the dialog box or by dynamically dragging, while holding down MB1, the green offset dimensions in the Display window.

4   Click on the OK button to complete the creation of the line.

The newly created Point-Point line is shown in figure 4-39.

Geometry-based Features

Geometry-based features are 3D solid features primarily created from 2D element sketches or from dress-up features. These features are the building blocks of the model and assist in capturing design intent. CATIA V5 offers a variety of geometry-based features, which are explored in the sections that follow.

Sketch-based Features

Sketch-based features are the building blocks of the design model. They are the fabric of the entities that make up the part. Some sketch-based features create material within the design model, whereas others may remove material. Sketch-based features are created from sketched profile sections.

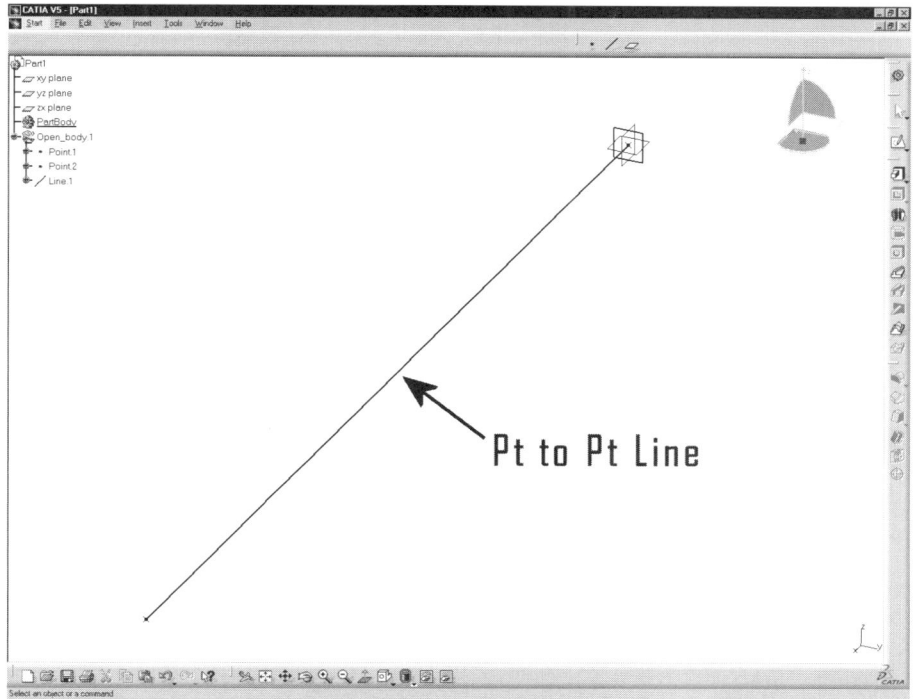

Fig. 4-39. Point-Point line.

Sketch-based features include pads, pockets, shafts, grooves, holes (standard and threaded), ribs, stiffeners, and lofts. These features, discussed in the sections that follow, are created via the Sketch-Based Feature toolbar, shown in figure 4-40.

Fig. 4-40. Sketch-Based Features toolbar.

Pad

A pad, created via the Pad icon (shown in figure 4-41), is a solid feature that generates material by extruding a 2D sketched profile. It is typically one of the first solid features added to a model when creating a new design. Each feature group covers a range of types. For example, pads include all extrusion-type features (with various limiting options), all controlled via the Pad dialog box, shown in figure 4-42.

Fig. 4-41. Pad icon.

*Fig. 4-42.
Pad dialog
box.*

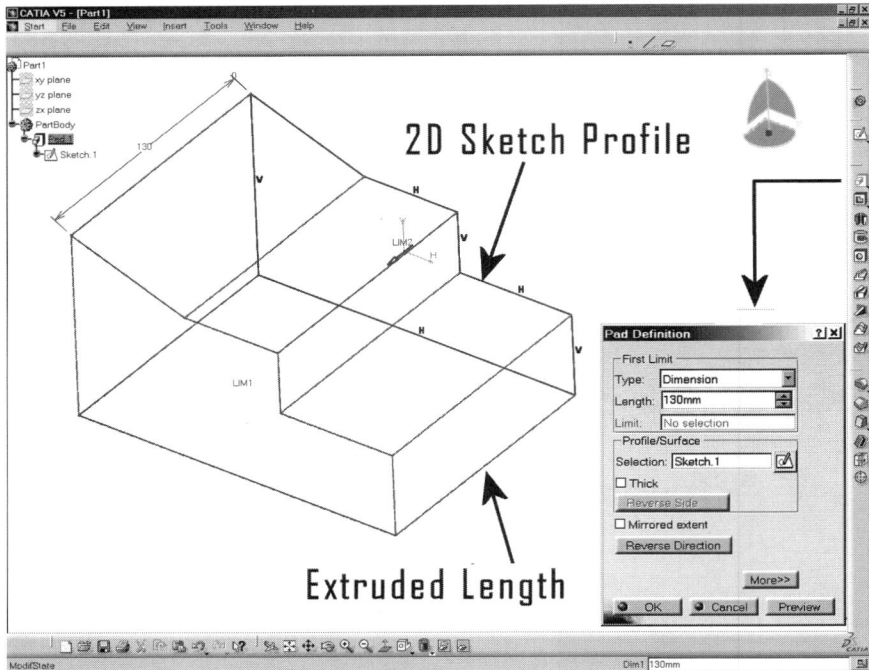

Pad-limiting Options

In the creation of a pad feature, it is important that you understand the various ways in which you can limit the length of the extruded feature. This is very important toward building in the proper associativity in the design model as it relates to potential downstream changes. V5 currently offers the following five limiting types in regard to controlling the extruded distance.

- *Dimension:* Extrudes a pad to an exact dimension. An example of the Dimension pad limit is shown in figure 4-43.

- *Up to Next:* Extrudes a pad to the next detected surface or face. An example of the Up to Next pad limit is shown in figure 4-44.

- *Up to Last:* Extrudes a pad to the last detected surface or face. An example of the Up to Last pad limit is shown in figure 4-45.

- *Up to Plane:* Extrudes a pad to a selected plane. An example of the Up to Plane pad limit is shown in figure 4-46.

- *Up to Surface:* Extrudes a pad to a selected surface. An example of the Up to Surface pad limit is shown in figure 4-47.

Fig. 4-43.
Dimension
limit.

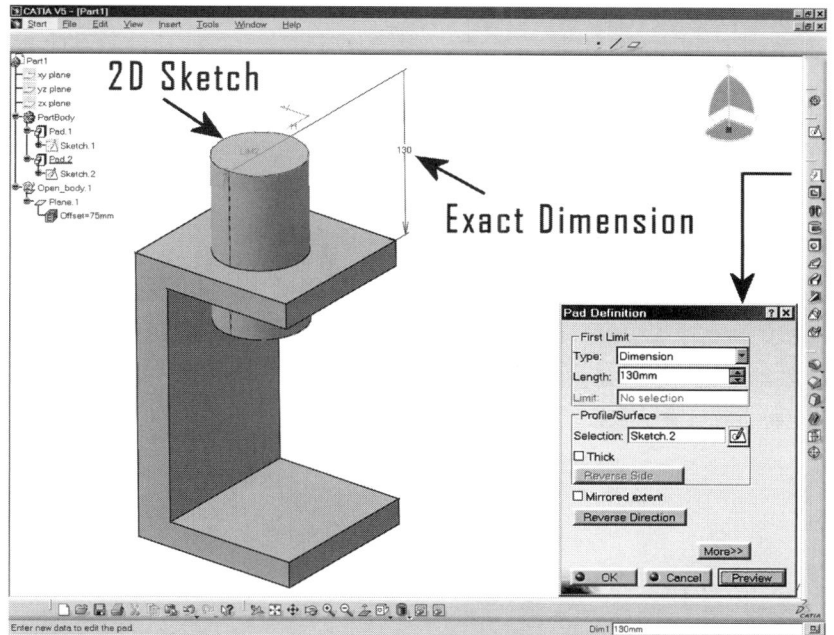

Fig. 4-44.
Up to
Next limit.

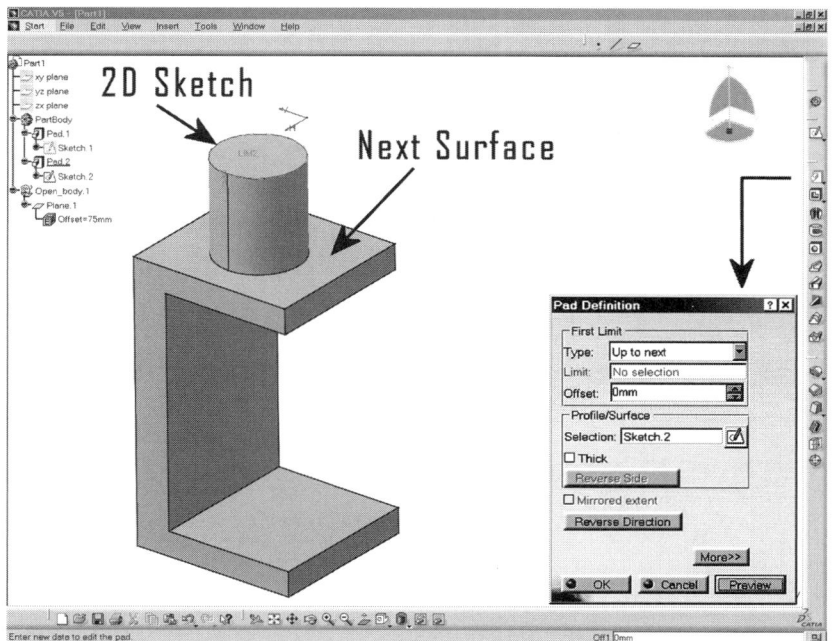

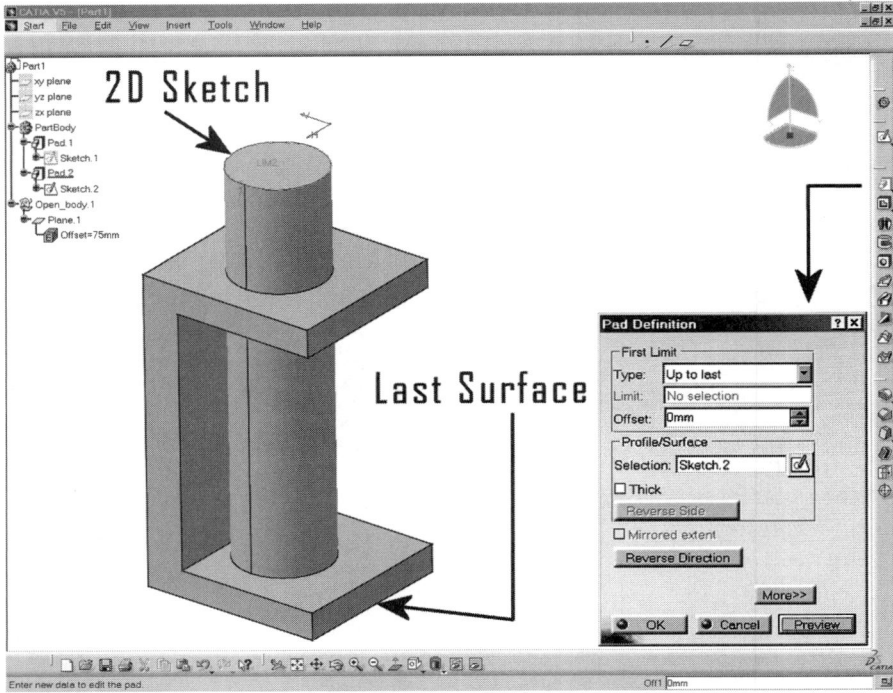

*Fig. 4-45.
Up to Last
limit.*

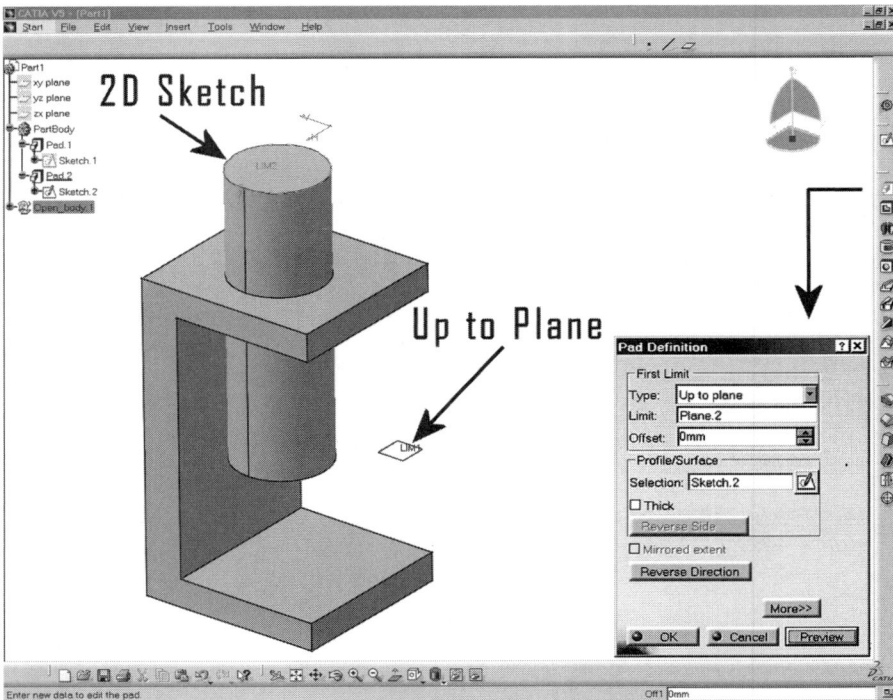

*Fig. 4-46.
Up to Plane
limit.*

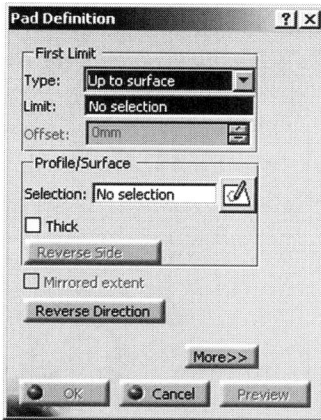

Thickness

The Thickness option applies a modifiable thickness value to the selected sketched profile.

Mirrored Extent

The Mirrored Extent option mirrors the pad feature around the selected sketched profile plane. Pad limiting options may be independently applied in both directions from the sketch profile plane.

Fig. 4-47. Up to Surface limit.

Reverse Direction

The Reverse Direction option toggles the pad feature about the selected sketched profile plane. The Pad dialog box creates an extruded pad using a reference line as the extrude direction, as opposed to one that is normal to the profile. At least one sketch profile must exist in the model to create a pad feature.

NOTE 1: *CATIA V5 will allow the creation of an open profile, provided existing geometry can trim the pad feature.*

NOTE 2: *Pads can be created from multiple profiles, as long as they do not intersect.*

Drafted Filleted Pad

Fig. 4-48. Drafted Filleted Pad icon.

This derivative of the Pad command, accessed via the Drafted Filleted Pad icon (shown in figure 4-48), allows for the addition of a draft-and-fillet feature within the pad creation function. These features are all created simultaneously, rather than independently. This type of pad feature is limited in options compared to the standard pad feature. All controlling inputs are located within the Drafted Filleted Pad dialog box, shown in figure 4-49.

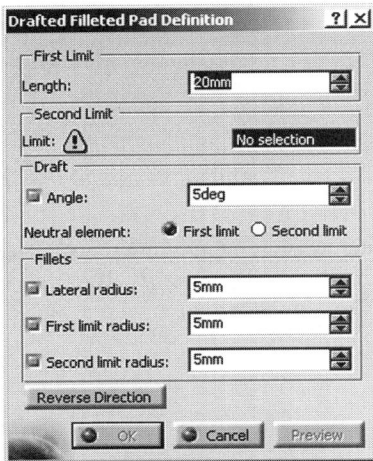

Fig. 4-49. Drafted Filleted Pad dialog box.

Fig. 4-50. Multi-Pad icon.

Multi-Pad

This derivative of the Pad command, accessed via the Multi-Pad icon (shown in figure 4-50), allows for the creation of multiple pads at one time. All controlling inputs are located within the Multi-Pad dialog box, shown in figure 4-51.

Pocket

A pocket, created via the Pocket icon (shown in figure 4-52), is a solid cut feature that removes material by extruding a 2D sketched profile. A pocket feature is identical in every respect to a pad feature, except that it removes material instead of adding material. The Drafted Filleted Pocket and Multi-Pocket options are also identical in function, except that they also remove material instead of creating material.

NOTE: *CATIA V5 will create material if a pocket feature is the first feature in a new body. This is identified by a different icon located within the configuration tree.*

Fig. 4-51. Multi-Pad dialog box.

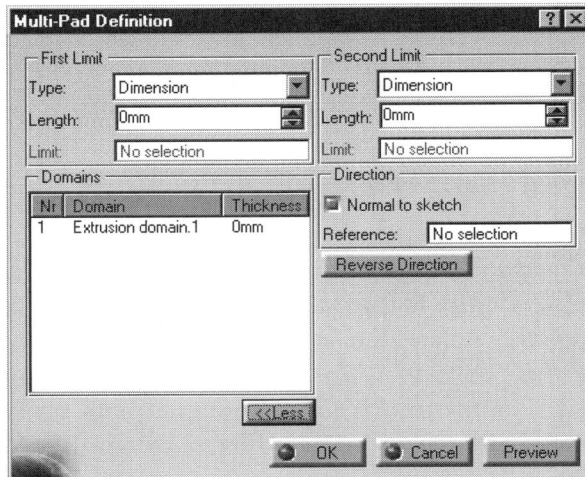

Fig. 4-52. Pocket icon.

Shaft

A shaft, created via the Shaft icon (shown in figure 4-53), is a solid model feature that generates material from the revolution of a 2D sketched profile. Within the dialog box, the angle limit options First and Second are available for independently controlling the clockwise and counterclockwise directions of the revolution from the sketched profile plane. An example of a shaft is shown in figure 4-54.

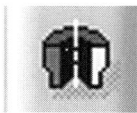

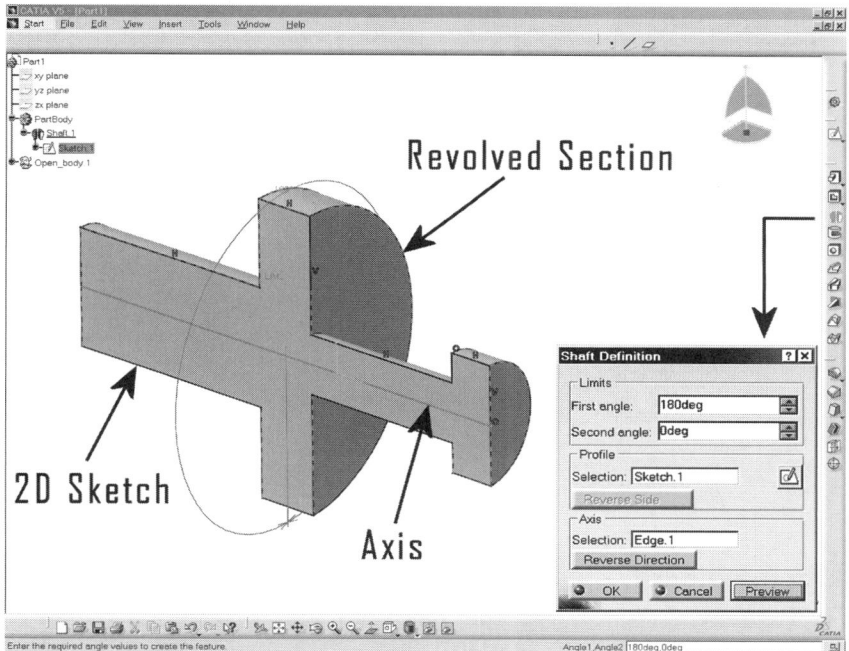

Fig. 4-53.
Shaft icon.

Fig. 4-54. Shaft.

Groove

Fig. 4-55.
Groove icon.

The groove feature, created via the Groove icon (shown in figure 4-55), is identical to the shaft feature in every respect, except that it is a solid cut feature that removes material by the revolution of a 2D sketched profile. Within the dialog box, the angle limit options First and Second are available for independently controlling the clockwise and counterclockwise directions of the revolution from the sketched profile plane. An example of a groove feature is shown in figure 4-56.

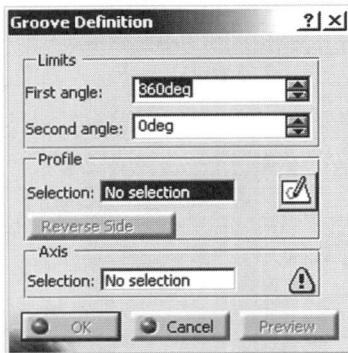

Fig. 4-56. Groove feature.

Fig. 4-57.
Hole icon.

Hole

A hole, created via the Hole icon (shown in figure 4-57), is a solid cut feature that removes circular material from an existing feature with some defined depth option. Its depth can be defined by dimension or with respect to existing 3D geometry. The dialog box supports a wide range of geometric configurations, extrusions types, and thread attributes. Holes can be roughly located or precisely positioned on existing geometry.

Holes do not require a preexisting sketch. CATIA V5 automatically creates the sketch profile on the fly during the initial creation process. This sketch may later be modified by editing the feature.

Hole types include simple, tapered, counterbored, counterdrilled, and countersunk, all from a single dialog box. Extension types include the options Blind, Up to Next, Last, Plane, and Surface. You can also add thread attributes via the Hole Definition dialog box, shown in figure 4-58.

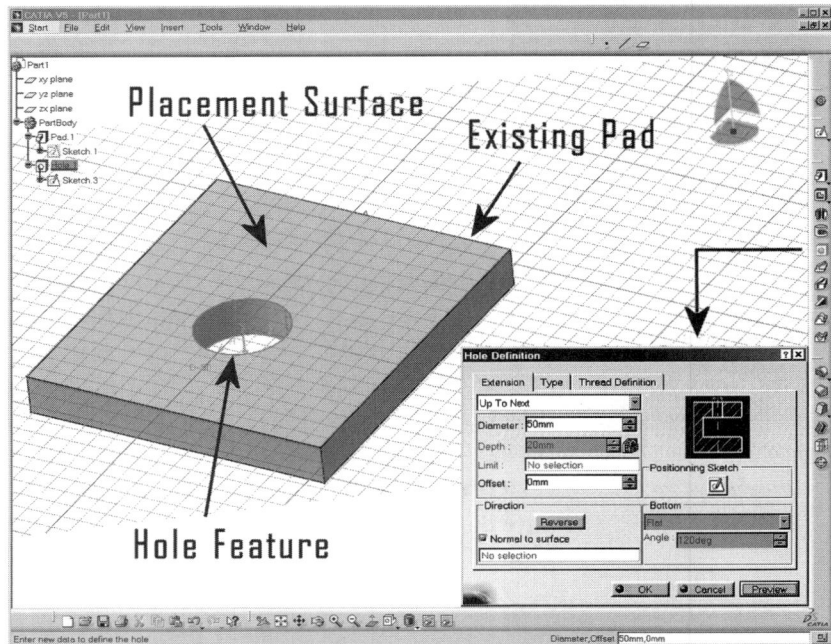

Fig. 4-58. Hole Definition dialog box.

Figure 4-59 shows hole extension types. Figure 4-60 shows hole geometry types. Figure 4-61 shows thread definition types.

NOTE: *All holes are true to a plane or surface. Thus, if you want a hole that is not perpendicular to a surface, you must create and position a reference axis ahead of time.*

TIP: *A pocket should not be used in place of a hole, because holes contain characteristics not present in a pocket.*

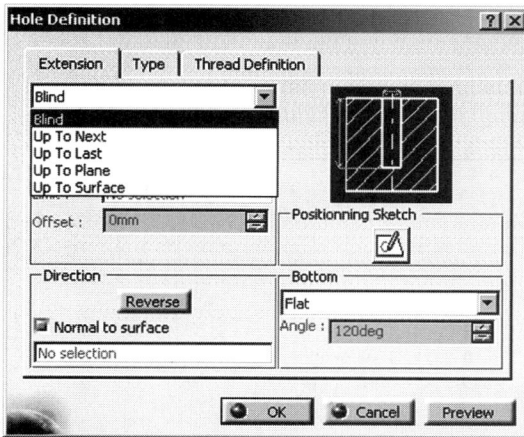

Fig. 4-59. Hole extension types.

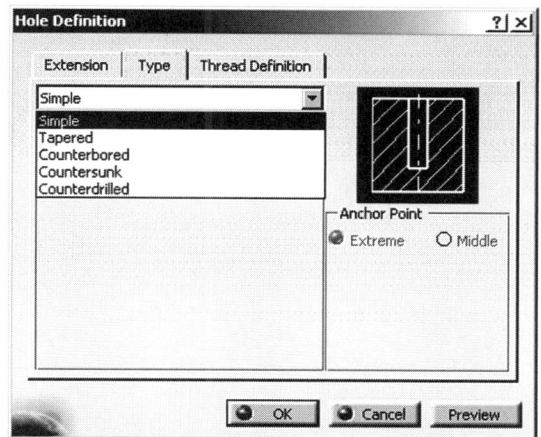

Fig. 4-60. Hole geometry types.

Fig. 4-61. Thread definition types.

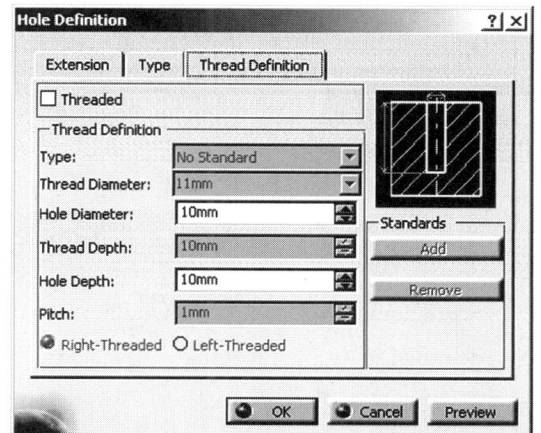

Ribs and Slots

Fig. 4-62.
Ribs and
Slots icon.

The Rib command, accessed via the Ribs and Slots icon (shown in figure 4-62), creates material through the application of a sweep function. The Slot function removes material through the application of a sweep function. Rib and slot features provide options for center curves (sweeps), as well as profile control. A center trajectory curve and sketched planar profile are required in the creation of a rib or slot. Several sketched profiles may be used in the creation of a rib feature. Reference elements and pulling directions are optional constraints that can be added to these features. An example of a Rib and Slot definition is shown in figure 4-63.

NOTE: *All 3D trajectory curves must be continuous in tangency, while 2D planar trajectory curves may be discontinuous in tangency.*

Fig. 4-63.
Rib and Slot
definition.

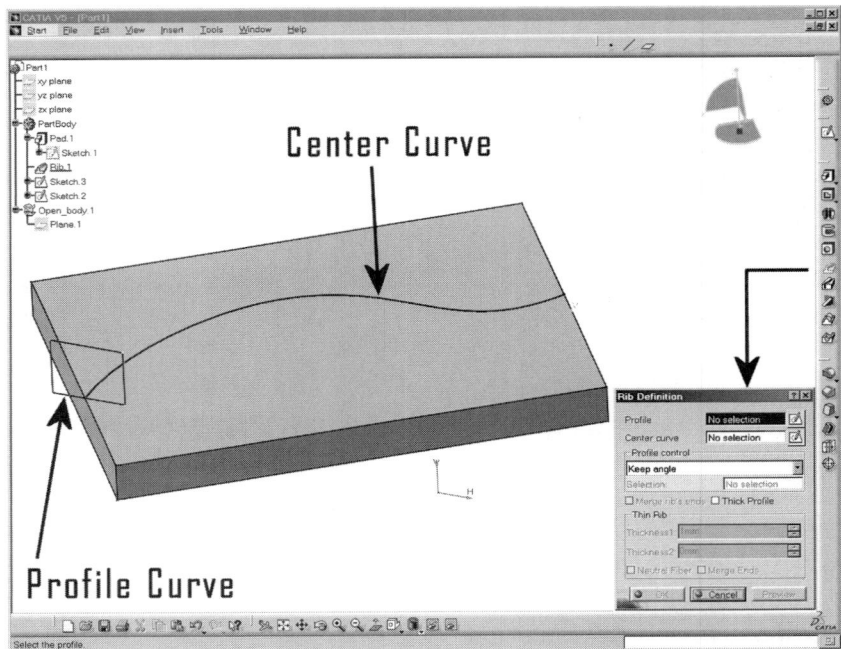

Stiffener

Fig. 4-64.
Stiffener icon.

A stiffener, created via the Stiffener icon (shown in figure 4-64), is a solid feature that creates material in the form of a gusset or rib. The stiffener feature can be created the From Side or From

Top option, depending on the desired geometry. Either option extrudes the sketched section to a desired distance.

From Side

A From Side feature, shown in figure 4-65, can only be created from an open sketched profile. This feature is created normal to the sketcher profile plane and is extruded a given distance.

From Top

A From Top feature, shown in figure 4-66, can be created from either an open or closed sketched profile. Both the extrusion and the thickness feature are created normal to the sketcher profile plane.

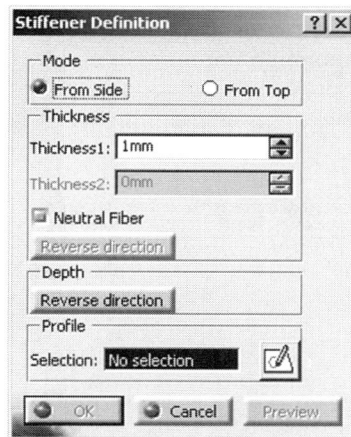

Fig. 4-65. Stiffener From Side feature.

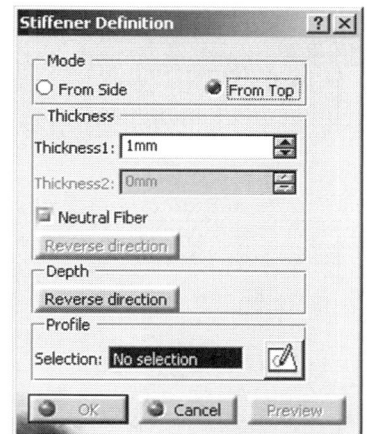

Fig. 4-66. Stiffener From Top feature.

Loft and Remove Loft

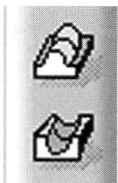

Fig. 4-67. Loft icon.

The Loft function, accessed via the Loft icon (shown in figure 4-67), creates material, whereas the Remove Loft function removes material. A loft is generated by sweeping one or more planar sections along a user-defined trajectory or spine. The loft functions present a good selection of options, including the use of guides curves, spine curves, tangency and curvature constraints, and limiting options. Examples of loft features are shown in figure 4-68.

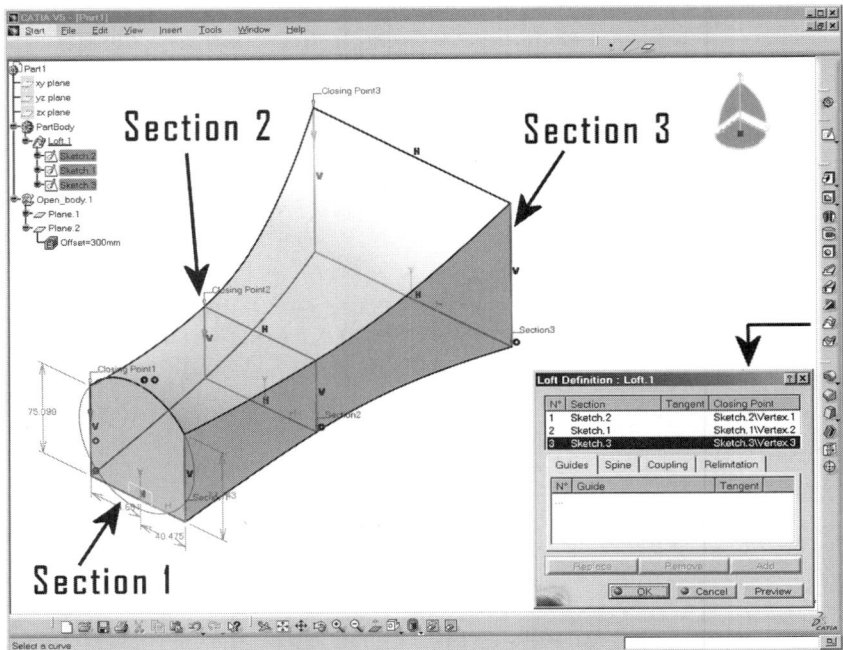

Fig. 4-68. Loft features.

Sketch-based Feature Exercises

The following tutorials explore sketch-based features.

Sketch-based Feature Exercise 1: Shaft

This exercise creates a shaft feature part. Perform the following steps.

1   Start a new CATIA session and create a new part design, shown in figure 4-69.

2   Click on the Sketcher icon, shown in figure 4-70, on the Sketcher toolbar, and then select the YZ plane using MB1. Note that CATIA V5 now automatically enters into the sketcher environment.

3   Sketch the profile shown in figure 4-71, and then exit the sketcher environment.

4   Click on the Shaft icon, shown in figure 4-72, and then select the previous sketched profile.

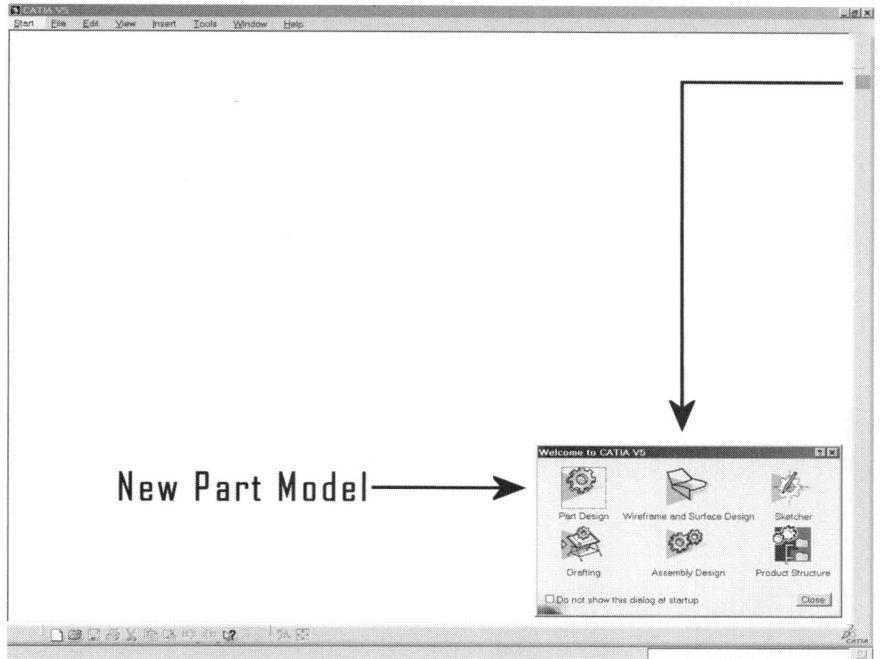

Fig. 4-69. Part design to be created.

Fig. 4-70. Sketcher icon.

New Part Model

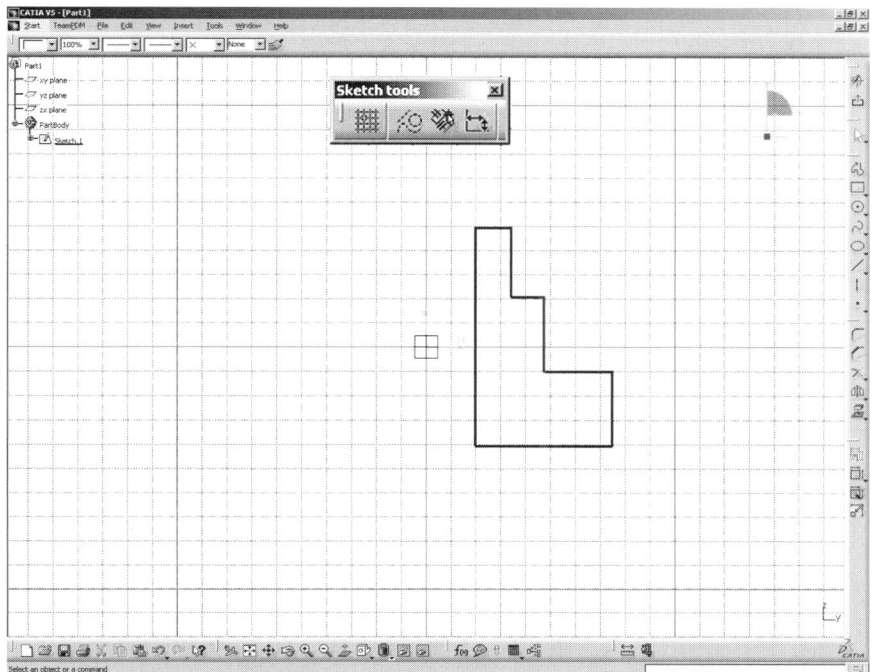

Fig. 4-71. Shaft profile.

You may select the sketch profile either from the Display window or from the configuration tree.

Fig. 4-72.
Shaft icon.

5   Click on the OK button to complete the creation of the shaft feature.

The shaft part design should look like that shown in figure 4-73.

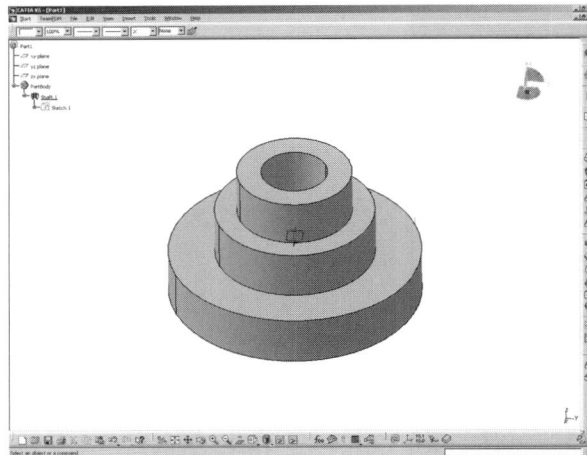

Fig. 4-73. Shaft part.

Sketch-based Feature Exercise 2: Rib

This exercise creates a rib feature part. Perform the following steps.

1   Start a new CATIA session and create the part design shown in figure 4-74.

2   Click on the Sketcher icon, shown in figure 4-75, on the Sketcher toolbar, and then select the YZ plane using MB1.

3   Sketch the spine profile shown in figure 4-76, and then exit the sketcher environment.

4   Click on the Plane icon, shown in figure 4-77, and set the type to Normal to Curve.

5   Select the near end point and curve of the previously sketched spine profile, to create a normal plane at the end of the spine profile.

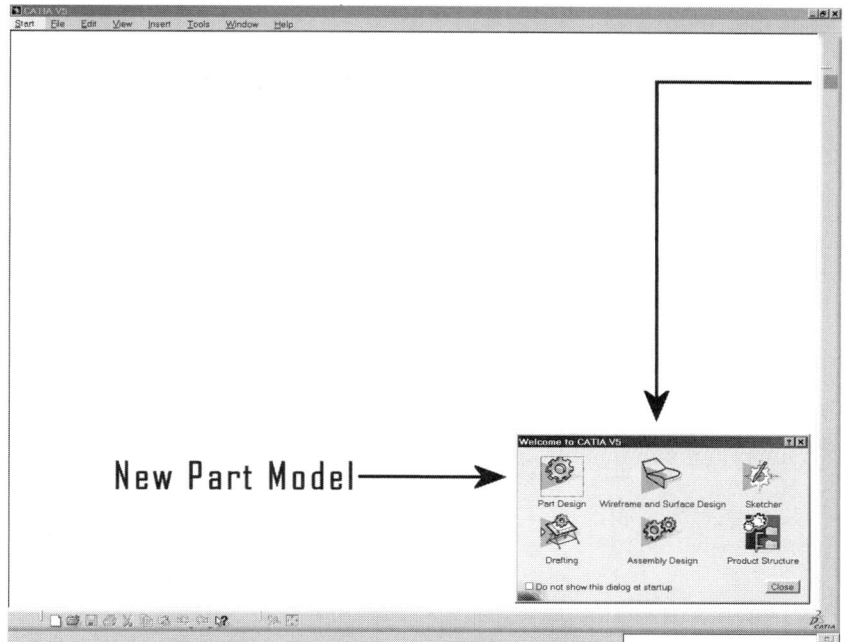

(Right):
Fig. 4-74.
Part design to
be created.

New Part Model

Fig. 4-75.
Sketcher
icon.

(Right):

Fig. 4-76.
Spine profile.

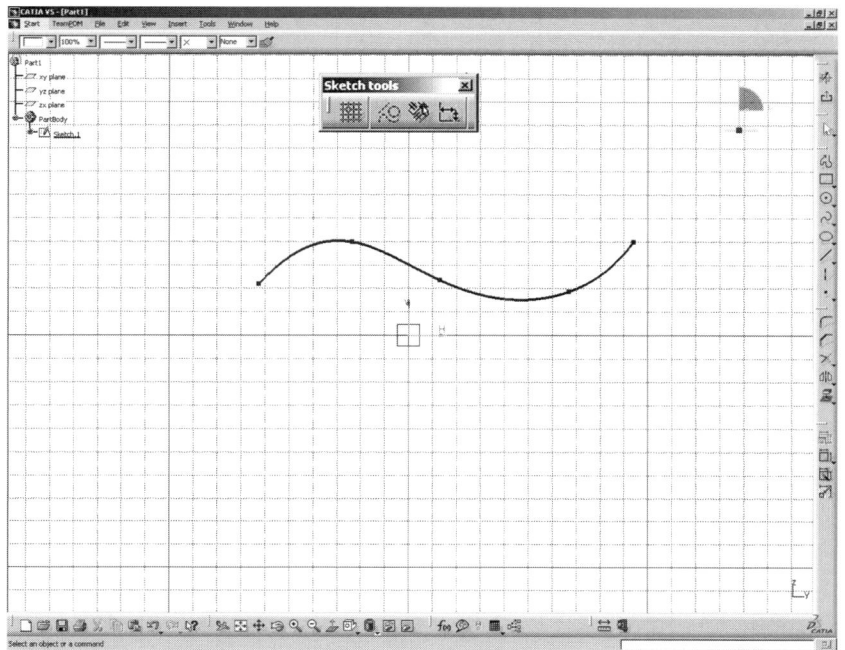

Fig. 4-77.
Plane icon.

6 Sketch the shape profile shown in figure 4-78, and then exit the sketcher environment.

7   Click on the Rib icon, shown in figure 4-79, on the Sketched-Based Feature toolbar. Select the spine and shape profiles, per the requirements of the dialog box.

8   Click on the OK button to complete the creation of the shaft feature.

The part should look like that shown in figure 4-80.

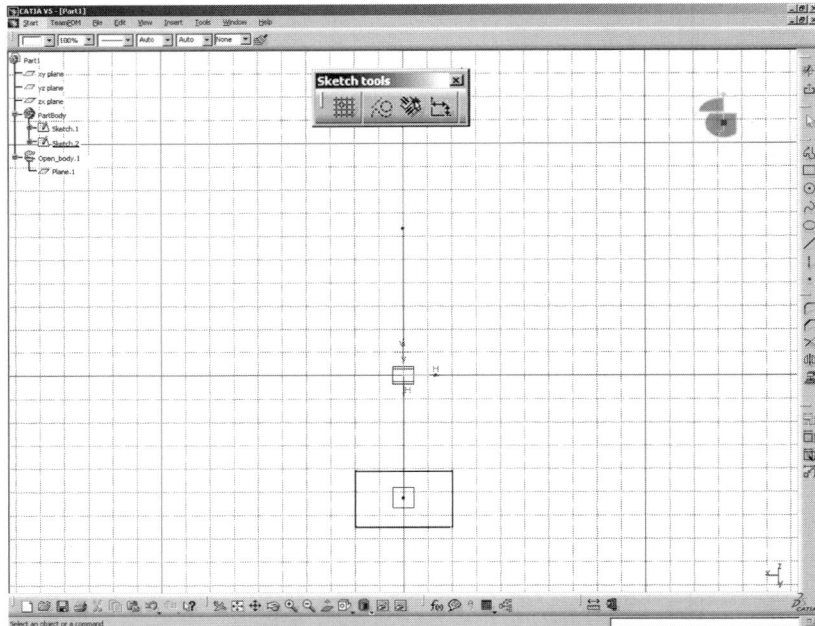

Fig. 4-78.
Sketched
shape profile.

Fig. 4-79.
Rib icon.

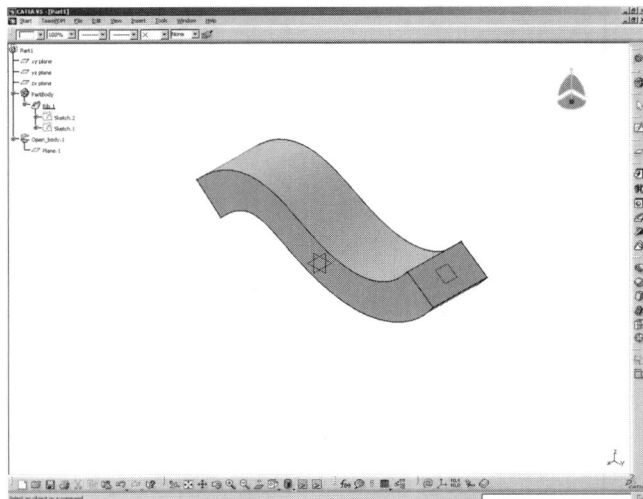

Fig. 4-80. Rib feature.

Sketch-based Feature Exercise 3: Bearing Block

This exercise creates a bearing block part via the Pad, Pocket, and Hole commands. Perform the following steps.

1 Start a new CATIA session and create a new part design, shown in figure 4-81.

2 Click on the Sketcher icon, shown in figure 4-82, on the Sketcher toolbar, and then select the YZ plane using MB1.

3 Sketch the profile shown in figure 4-83, and then exit the sketcher environment.

4 Click on the Pad icon, shown in figure 4-84, and select the previous sketched profile. Extrude the pad to a depth of 50 mm.

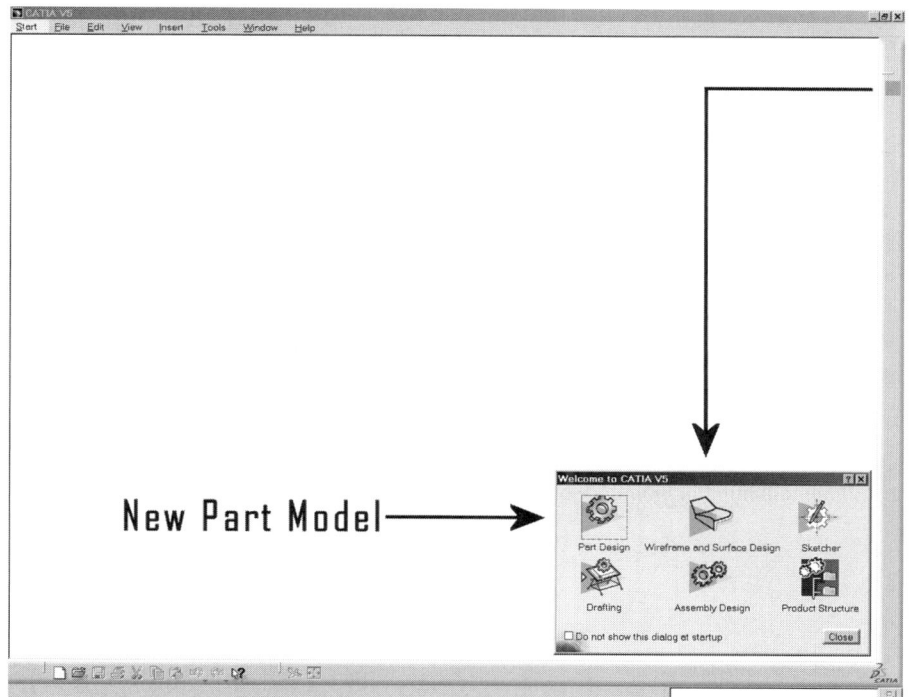

Fig. 4-81. Part design to be created.

Fig. 4-82. Sketcher icon.

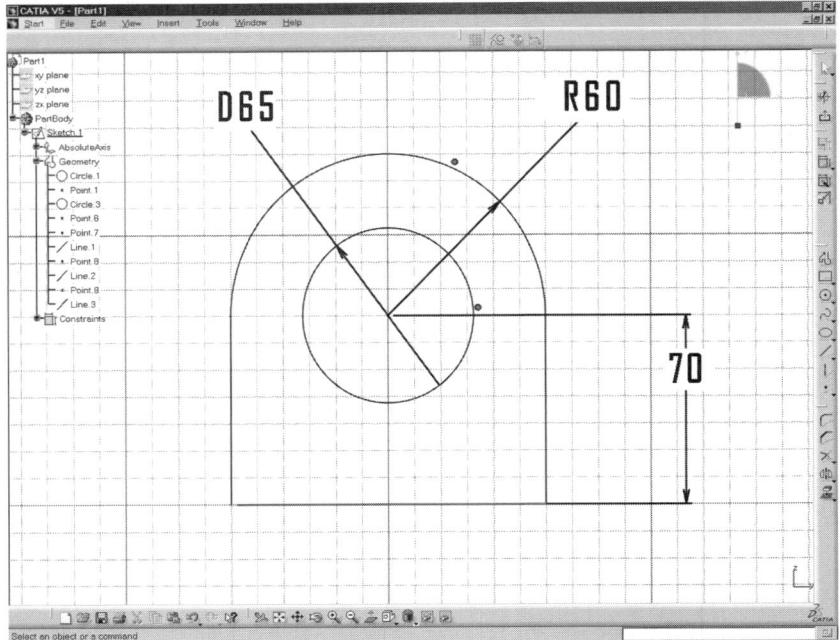

Fig. 4-83. Bearing block pad profile.

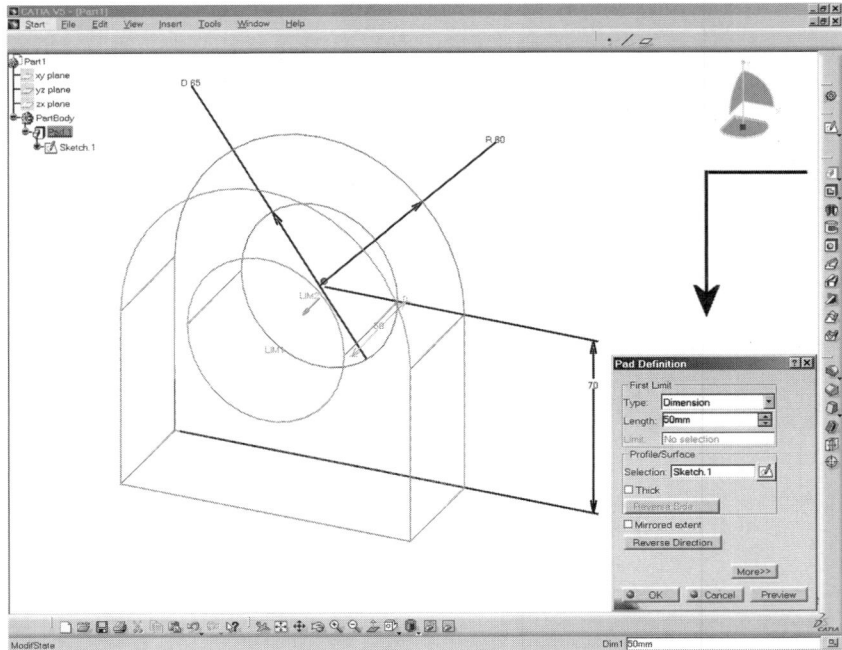

Fig. 4-84. Pad icon.

5 Click on the OK button to complete the creation of the pad feature.

The part should look like that shown in figure 4-85.

6 Click on the Sketcher icon, shown in figure 4-86, and then select the bottom face of the pad feature.

7 Sketch the profile shown in figure 4-87, and then exit the sketcher environment.

8 Click on the Pad icon, shown in figure 4-88, and select the previous sketched profile. Extrude the pad a length of 15 mm.

9 Click on the OK button to complete the creation of the pad feature.

The part should look like that shown in figure 4-89.

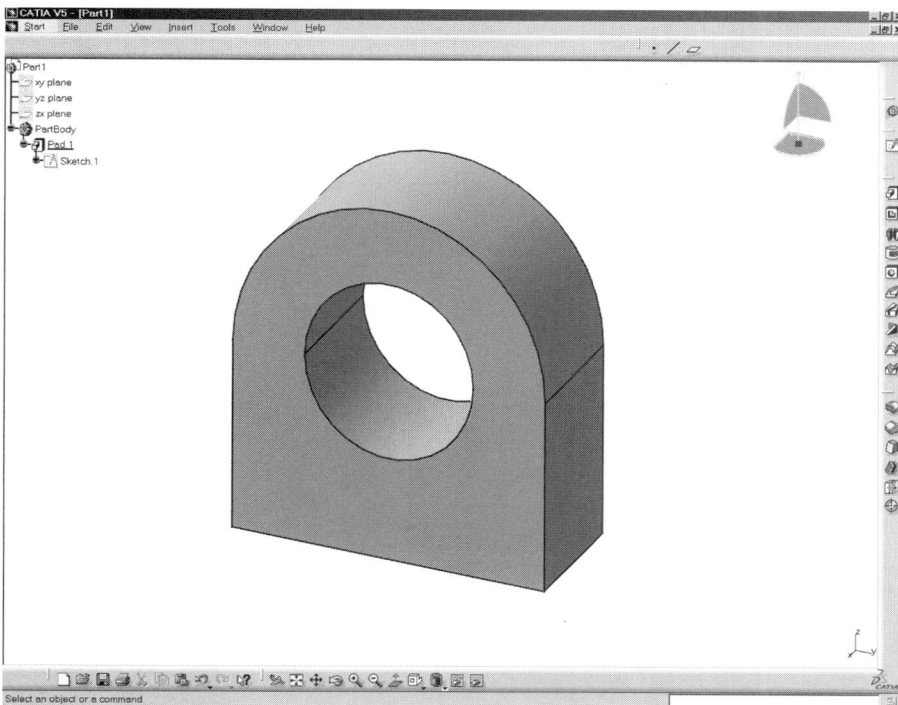

Fig. 4-85. Bearing block pad feature.

Fig. 4-86. Sketcher icon.

Fig. 4-87.
Bearing block
pocket profile.

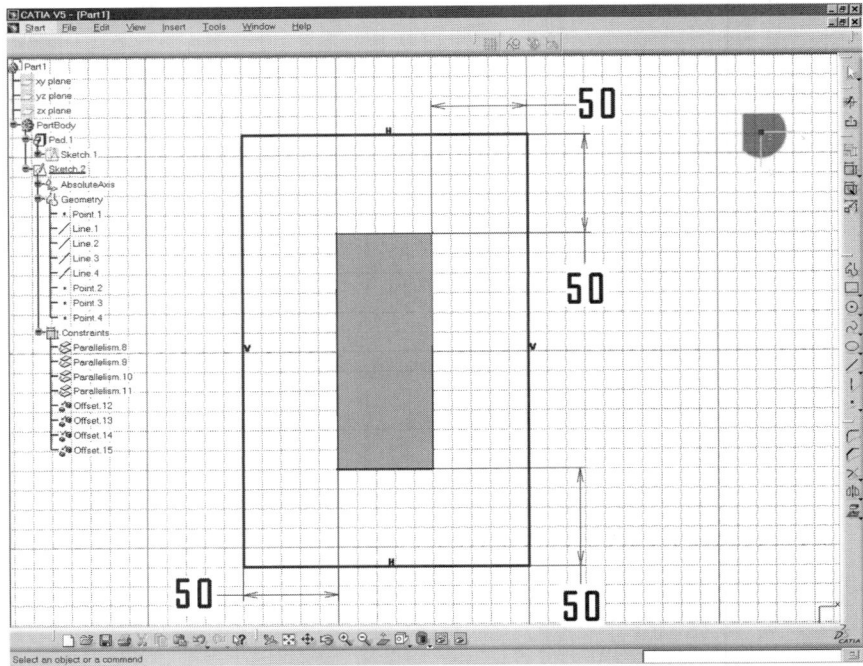

Fig. 4-88.
Pad icon.

Fig. 4-89.
Bearing block
pocket feature.

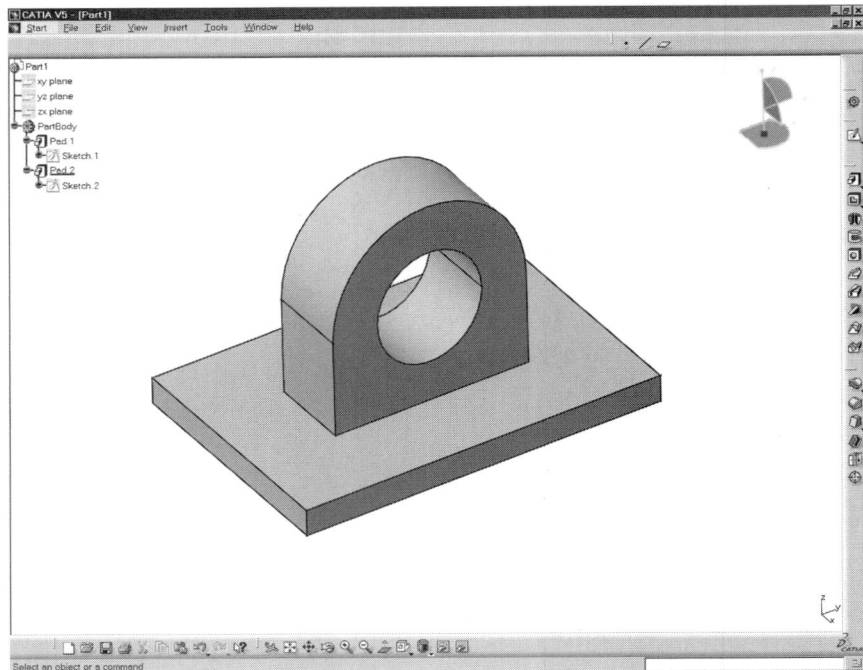

Fig. 4-90.
Hole icon.

10 Click on the Hole icon, shown in figure 4-90, and then select the upper right-hand corner of the bearing block face. Enter a value of *15* (mm) for the hole diameter and select the *Up to next* option. Click on the Sketch icon and enter the Sketcher workbench. Constrain the center of the hole to be 25 mm from each edge and then exit the Sketcher workbench.

11 Click on the OK button to complete the creation of the hole feature.

The part should look like that shown in figure 4-91.

12 Save the part by clicking on the Save icon, shown in figure 4-92, located on the main menu toolbar.

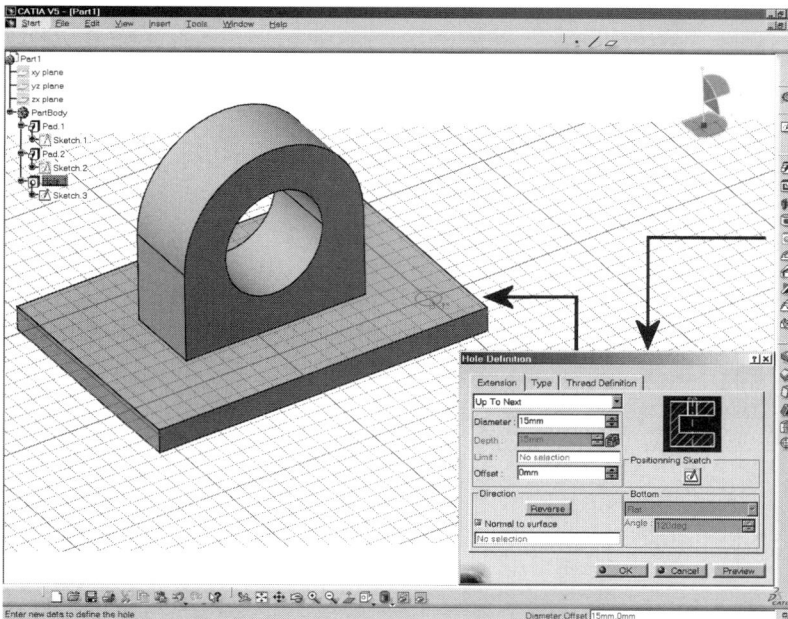

Fig. 4-91. Bearing
block finished part.

Fig. 4-92.
Save icon.

Dress-up Features

Dress-up features are features created by applying commands to existing model geometry such as points, edges, and faces. Dress-up features are quickly applied and do not require a sketched profile. These "pick and place" features (such as fillets, chamfers, draft, and shell) are quickly added to a model to further mature the design.

Fig. 4-93. Dress-up Feature toolbar.

Fig. 4-94. Fillet types.

Fig. 4-95.
Edge Fillet
command.

CATIA V5 offers a large number of possibilities in applying dress-up features to a design model. These features, discussed in the sections that follow, are applied via the Dress-up Feature toolbar, shown in figure 4-93.

Fillets

Fillet features are round elements used to break sharp edges within the design model. Fillets are typically tangent to the two faces of a sharp edge joint. An outside round is called a "round," and an inside round is often called a "fillet." CATIA V5 offers a variety of options for creating fillets and rounds. Figure 4-94 shows examples of fillet types, explored in the sections that follow.

Edge Fillet

The Edge Fillet command, shown in figure 4-95, is the most basic and widely used filleting option within CATIA V5. Edge fillets, an example of which is shown in figure 4-96, are created by selecting a sharp edge and applying a controlling radius dimension. This blends a constant smooth radial transition surface between two adjacent surfaces. The Edge Fillet dialog box offers commands that further define edge fillets. These are discussed in the sections that follow.

Propagation and Trim Ribbons. CATIA V5 offers two propagation options for an edge fillet: Tangency and Minimal. The Tangency option automatically propagates a radius-blended surface along all edges tangent to the original selected edge. The Minimal option creates a radius-blended surface along the length of the selected edge only. The Trim Ribbons options, shown in figure 4-97, are used to blend radius surfaces that overlap. These options are used when selecting multiple edges that are closer in proximity than the sum of the two radius dimensions.

Edges to Keep. The Edges to Keep option is used when the desired radius needs to terminate at a selected edge. CATIA V5 will automatically detect the edge and calculate the required radius, as shown in figure 4-98.

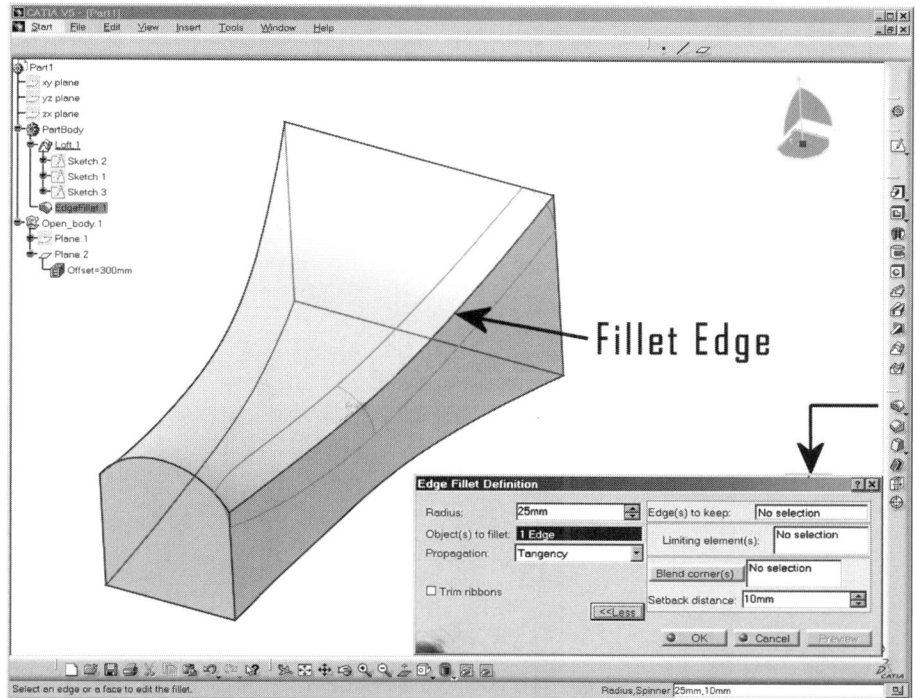

Fig. 4-96.
Edge fillet.

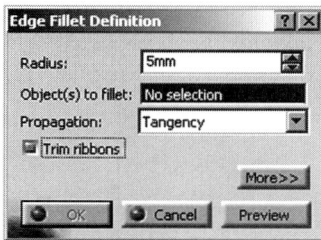

Fig. 4-97. Trim Ribbons
options.

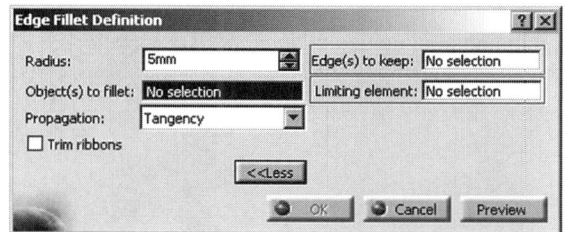

Fig. 4-98. Edge termination.

Limiting Element. The Limiting Element option is used when the desired radius needs to terminate at a limiting element such as a face, edge, or point. CATIA V5 will create the radius feature and propagate until it reaches the selected limiting element.

TIP 1: *Multiple edges may be filleted by holding down the Ctrl key during the selection process.*

TIP 2: *Selecting a face will create fillets along all edges of that face.*

Variable Fillet

Variable radius fillets are created, via the Variable Fillet icon (shown in figure 4-99), by defining multiple radius dimensions at the start point, end point, and any intermediate point along the selected edge. CATIA V5 automatically blends a transition surface along the selected edge using the dimensions established at these points. A cubic or linear blending variation option is available for controlling the tangent blending lines between the points. An example of a variable radius fillet is shown in figure 4-100.

Fig. 4-99.
Variable
Fillet icon.

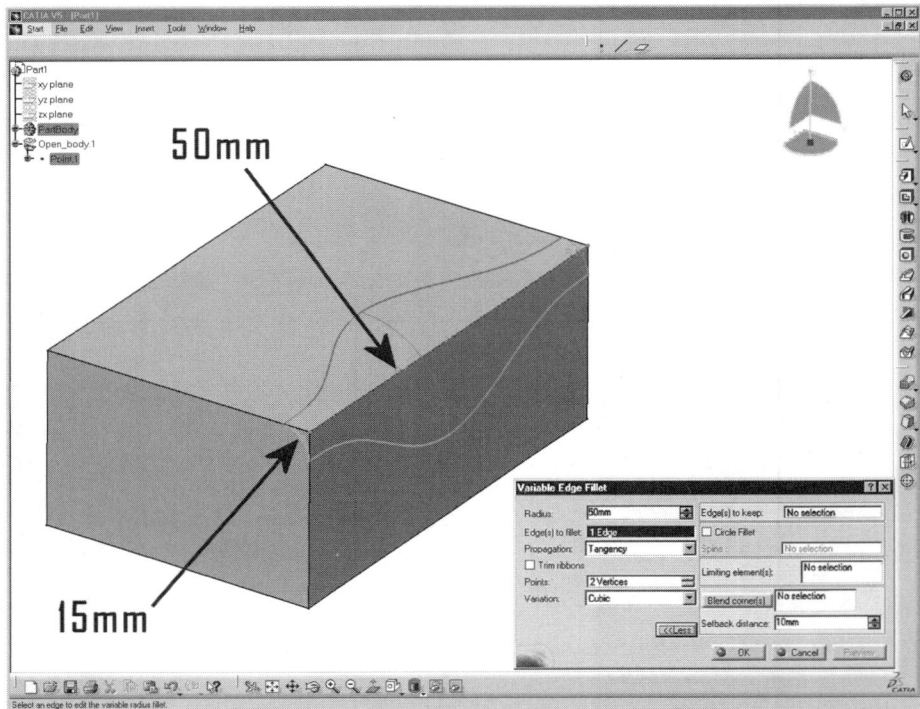

Fig. 4-100. Variable radius fillet.

TIP: *If a preexisting point is not present, you can select a plane for the purpose of creating an intermediate point along a selected edge.*

Face-Face Fillet

A face-face fillet, created via the Face-Face Fillet icon (shown in figure 4-101), blends a smooth rounded transition surface between two selected faces. Face-face fillets are typically used to create transition blends when there are no sharp edges present or when multiple edges exist between two faces. An example of a face-face fillet is shown in figure 4-102.

Fig. 4-101.
Face-Face
Fillet icon.

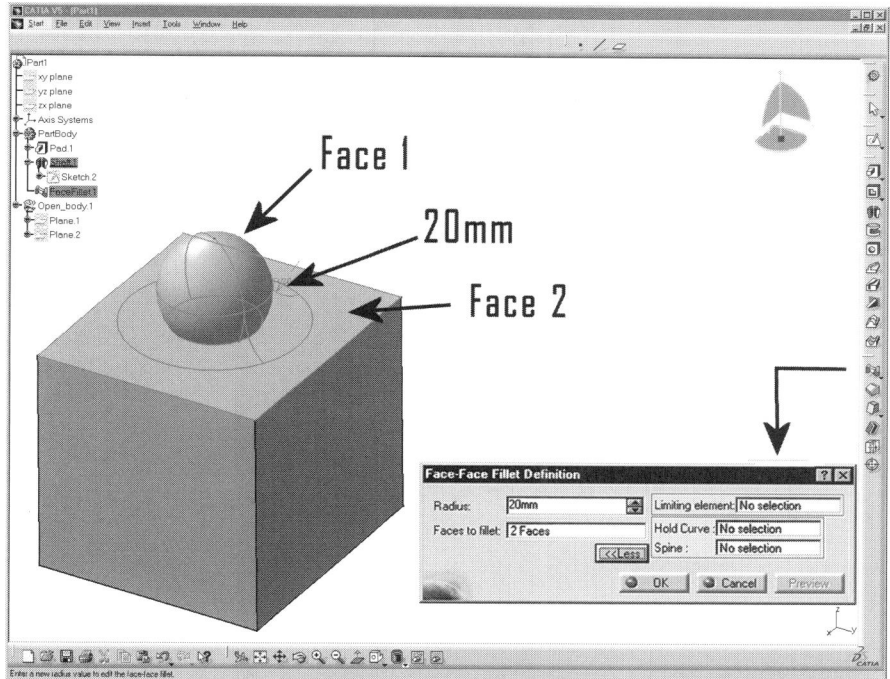

Fig. 4-102. Face-face fillet.

Tri-Tangent Fillet

Fig. 4-103.
Tri-Tangent
Fillet icon.

A tri-tangent fillet, created via the Tri-Tangent Fillet icon (shown in figure 4-103), creates a blended radius fillet among three selected faces. This option removes the middle selected face and replaces it with a smooth blended transition surface. CATIA V5 automatically calculates the radius fillet required to blend the appropriate transition. An example of a tri-tangent fillet is shown in figure 4-104.

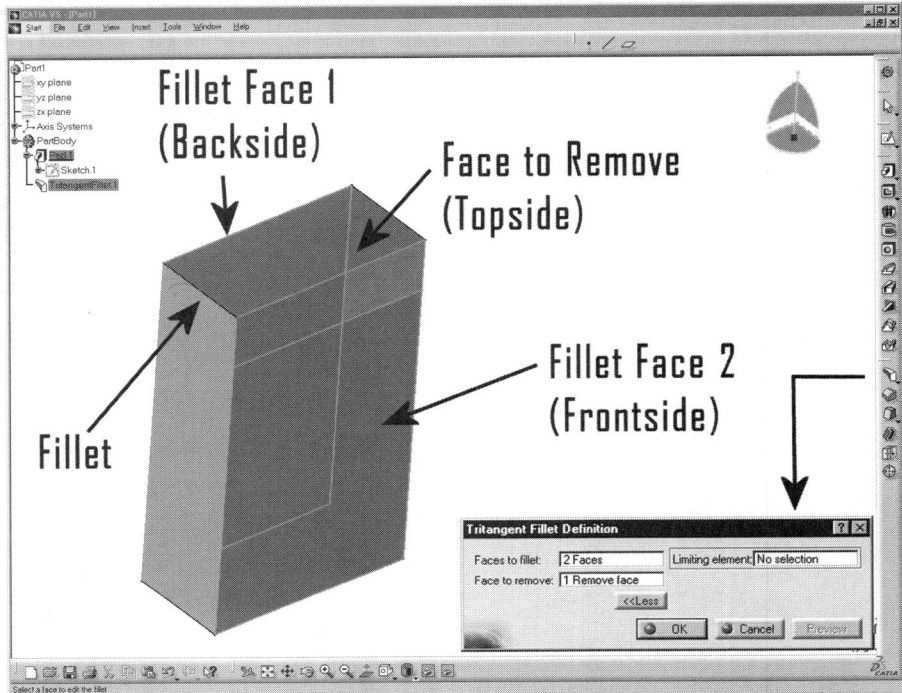

Fig. 4-104.
Tri-tangent
fillet.

Draft

Fig. 4-105.
Draft icon.

Drafts, created via the draft icon (shown in figure 4-105), are angled faces in a solid model that define the draft angle on a molded part for easier release from a mold. Material either gets removed or added, depending on the direction of the draft feature. Draft operations are performed from the Draft dialog box, shown in figure 4-106.

CATIA V5 offers a variety of basic and advance features for creating draft elements. Figure 4-107 shows the available features, which are discussed in the sections that follow.

Draft Angle

The Draft Angle option represents the angular dimension that controls the amount of draft applied to a face or multiple faces. This may be a constant or variable draft angle. Negative values may be entered for draft.

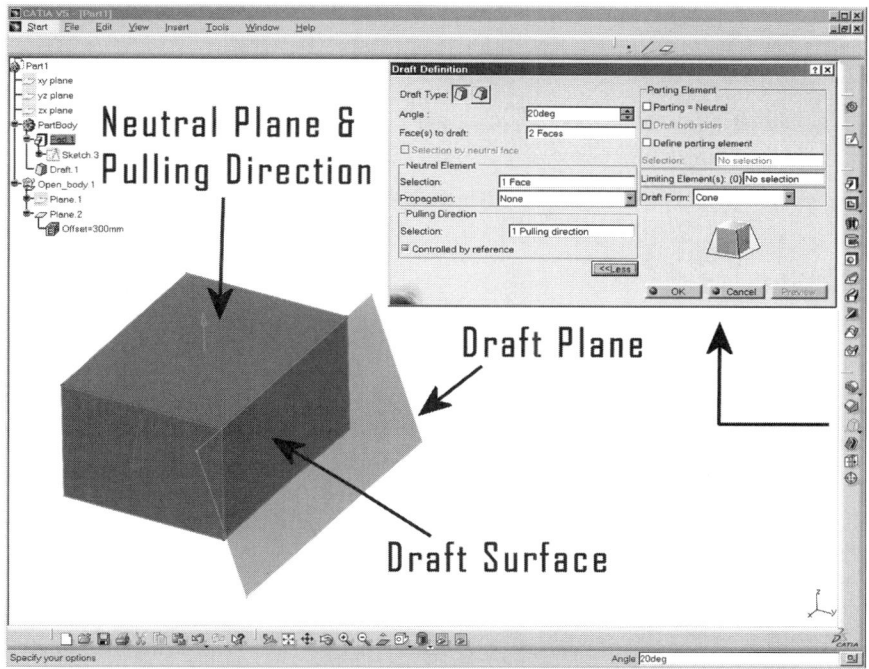

Fig. 4-106. Draft dialog box.

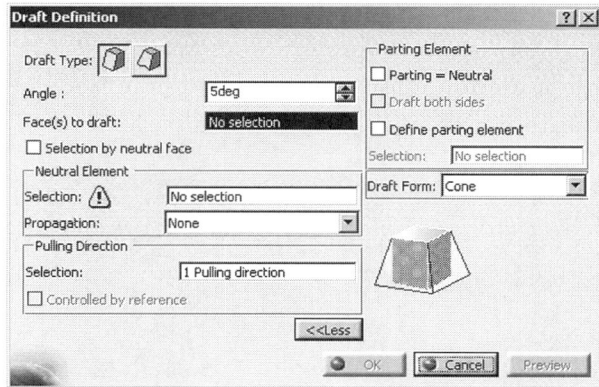

Fig. 4-107. Features available for creating draft.

Face(s) to Draft

The Face(s) to Draft option represents the faces to be drafted. Multiple faces may be drafted by holding down the Ctrl key during the selection process.

Pulling Direction

The Pulling Direction option represents the reference direction in which the draft angle is defined.

Neutral Element

Draft propagation typically begins at the neutral element. The Neutral Element option specifies the neutral element, which may be a face, curve, or plane.

NOTE: *The placement of draft features within the configuration tree is a key in building a robust model, especially when creating edge fillets from the resultant of a drafted face.*

Draft with Reflect Line

This command, accessed via the Draft with Reflect Line icon (shown in figure 4-108), creates a draft feature from the tangent or reflect intersection position located on a fillet radius. The face adjacent to the fillet is drafted and blended into the existing fillet feature. An example of this type of draft feature is shown in figure 4-109.

Fig. 4-108. Draft with Reflect Line icon.

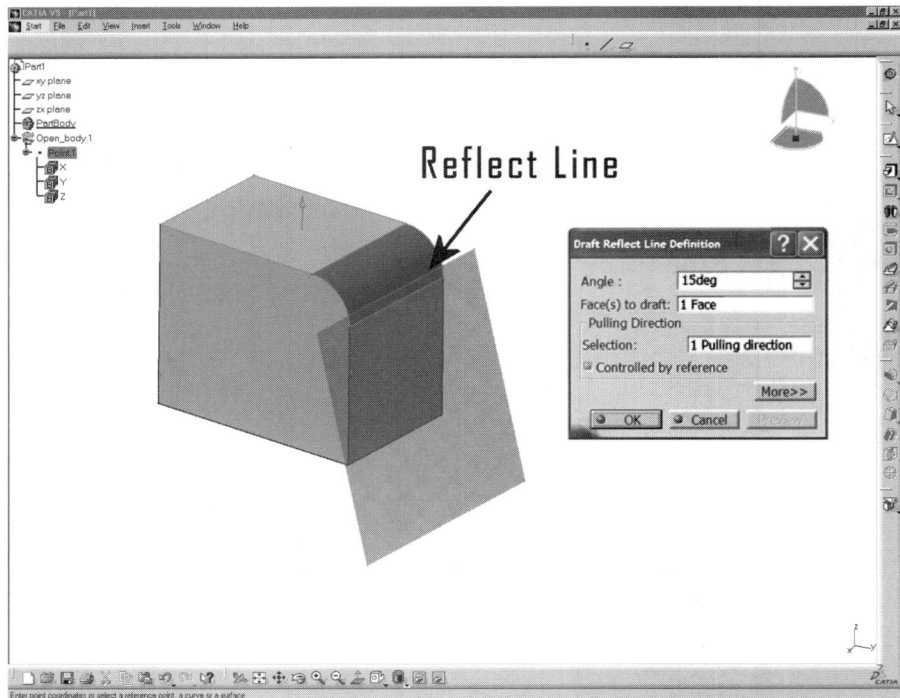

Fig. 4-109. Draft feature created using Draft with Reflect Line.

Chamfer

This command, accessed via the Chamfer icon (shown in figure 4-110), creates a beveled feature that adds or removes material in the form of a flat section at an intersecting edge joint. A chamfer may be propagated along one or multiple edges. Chamfer options are shown in figure 4-111.

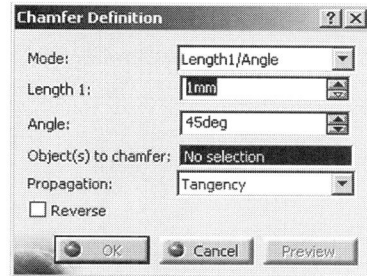

Fig. 4-110. Chamfer icon.

Fig. 4-111. Chamfer options.

Shelling

Fig. 4-112. Shelling icon.

The Shelling command, accessed via the Shelling icon (shown in figure 4-112), turns a solid model part into a thin-walled part by offsetting the outside walls to a desired thickness. The shelling feature may be applied to the inside, outside, or both sides of the part. Complete faces may be removed in the process to create an open cavity part. Wall thickness may be may be specified for individual faces, other than the default inside or outside thickness. An example of shelling is shown in figure 4-113.

NOTE 1: *CATIA V5 will create a shell with a thickness greater than the smallest fillet radius present on the part. These fillets will be consumed by the shell feature.*

NOTE 2: *The order of features within the configuration tree is very important when utilizing the shell feature. This is to ensure that the right feature content is captured within the shell feature.*

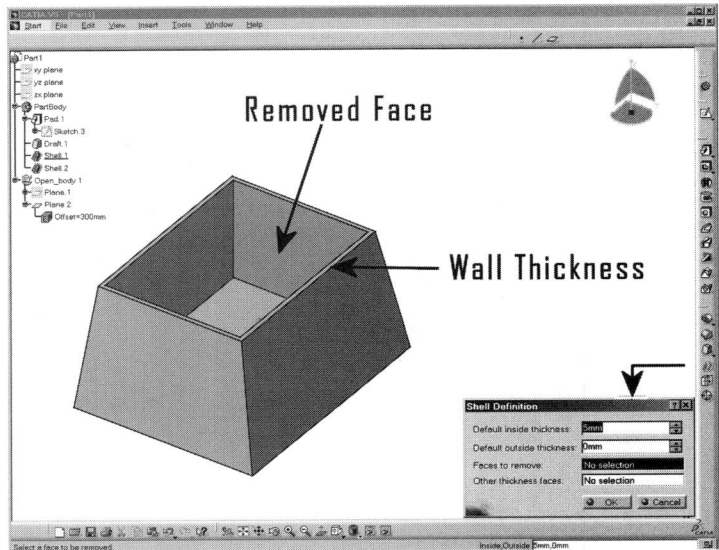

Fig. 4-113. Shelling.

Thickness

The thickness command, accessed via the Thickness icon (shown in figure 4-114), creates material by extruding the selected solid face normal to itself to a desired distance. Another thickness may be applied to independently drive multiple faces. Thickness command options are shown in figure 4-115.

Fig. 4-114. Thickness icon.

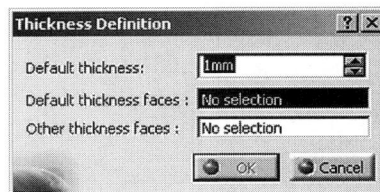

Fig. 4-115. Thickness command application.

Thread

The Thread command, access via the Thread icon (shown in figure 4-116), creates thread or tap features to existing holes or cylindrical faces. CATIA V5 offers many controlling factors for determining the outcome of the feature. An example of use of the Thread command is shown in figure 4-117.

Fig. 4-116.
Thread icon.

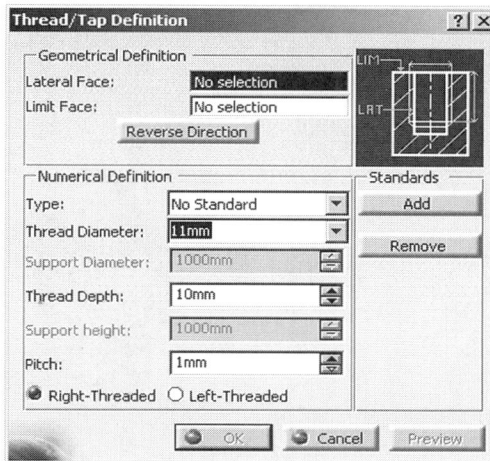

Fig. 4-117. Thread command
application.

Fig. 4-118.
Open icon.

Fig. 4-119.
Draft icon.

Dress-up Feature Exercise: Bearing Block Draft and Fillet

This exercise adds draft and round features to the bearing block part from the previous exercise. Perform the following steps.

1 Retrieve the bearing block part (via the Open icon, shown in figure 4-118) from the previous exercise.

2 Click on the Draft icon (shown in figure 4-119) from the Dress-Up toolbar. Within the Draft Creation dialog box, select the Faces to Draft area. While holding down the Ctrl key, select the four outside faces of the bearing block part using MB1. These are the faces to be drafted. Figure 4-120 shows the faces to be drafted.

3 Within the Draft Creation dialog box, select the Neutral Element area. Select the top face of the bearing block using MB1. This is the intended neutral element, as shown in figure 4-121.

4 Within the Draft Creation dialog box, select the Pulling Direction area. Select the top face of the bearing block using MB1. This is the intended pulling direction, as shown in figure 4-122.

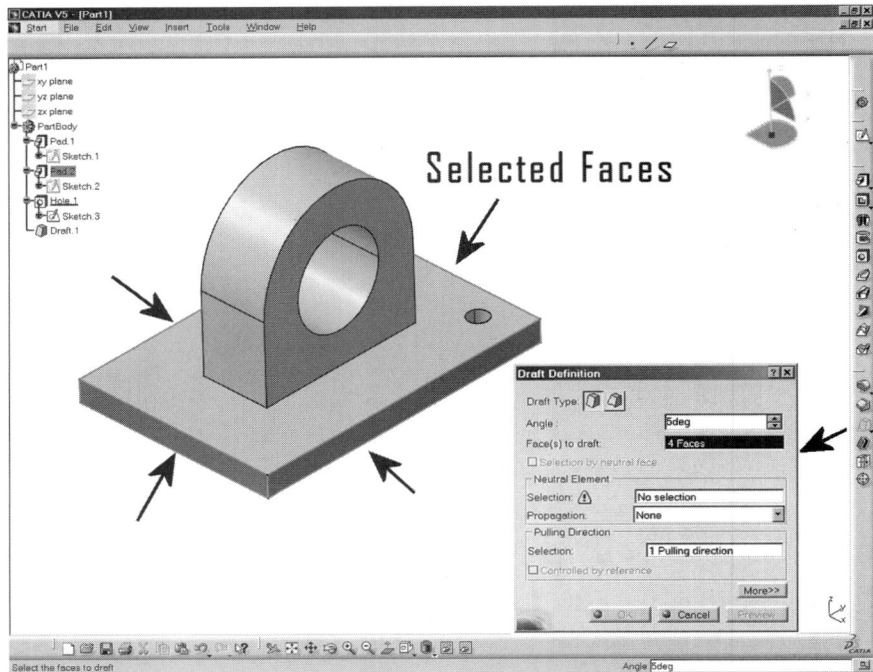

Fig. 4-120. Faces to be drafted.

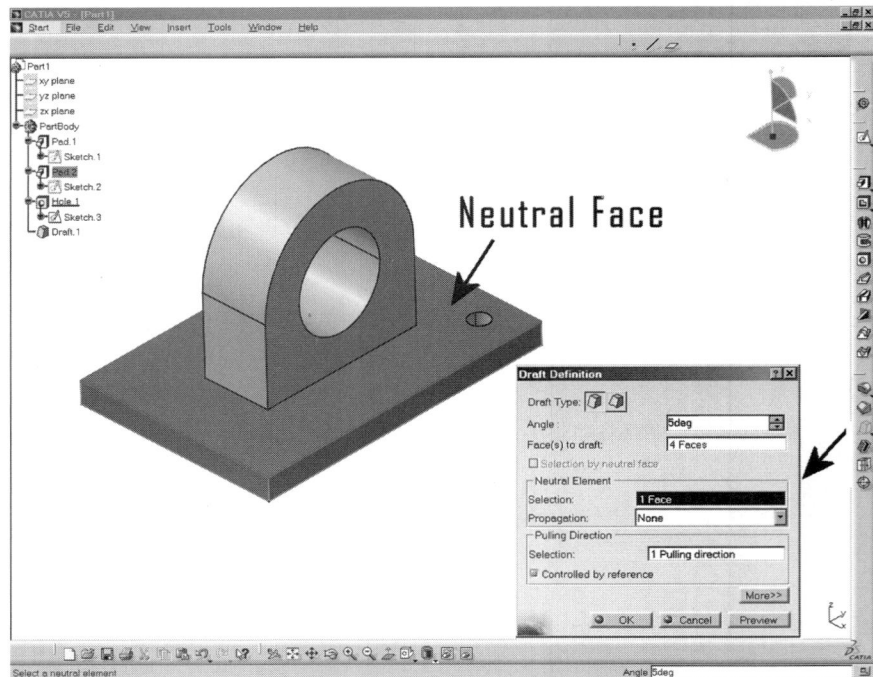

Fig. 4-121. Neutral element.

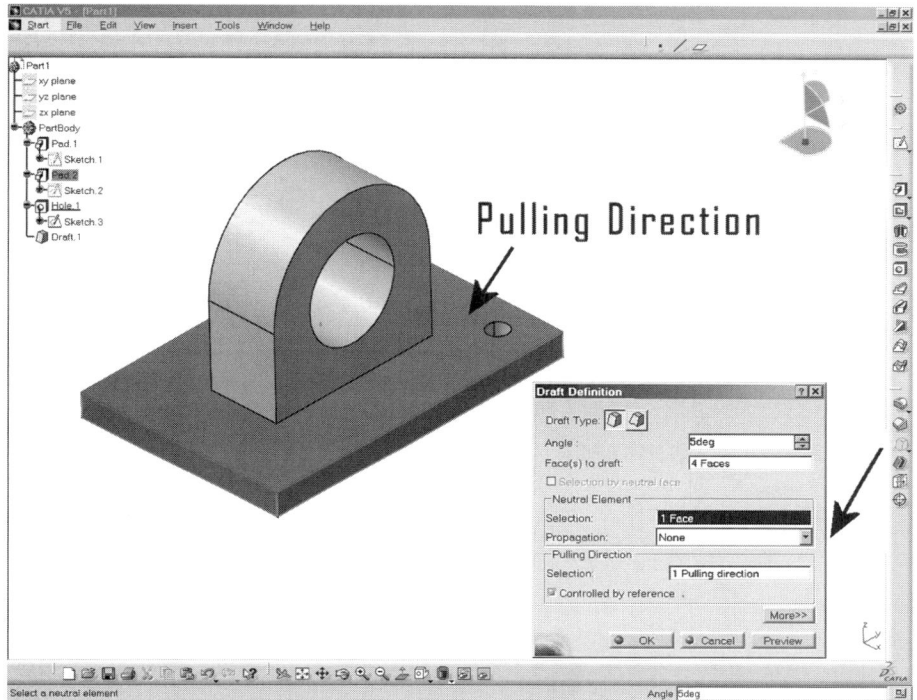

*Fig. 4-122.
Pulling
direction.*

5 Enter a draft angle of *5* (degrees).

6 Click on the OK button to complete the creation of the draft feature. The part should look like that shown in figure 4-123.

7 Click on the Edge Fillet icon (shown in figure 4-124), and while holding down the Ctrl key select the four outside edges of the bearing block part using MB1. Figure 4-125 shows the edges to be filleted.

8 Enter a radius value of *20* (mm).

9 Click on the OK button to complete the creation of the fillet feature.

The part should look like that shown in figure 4-126.

10 Save the part by clicking on the Save icon (shown in figure 4-127), located in the main menu toolbar.

Fig. 4-123.
Drafted bearing
block part.

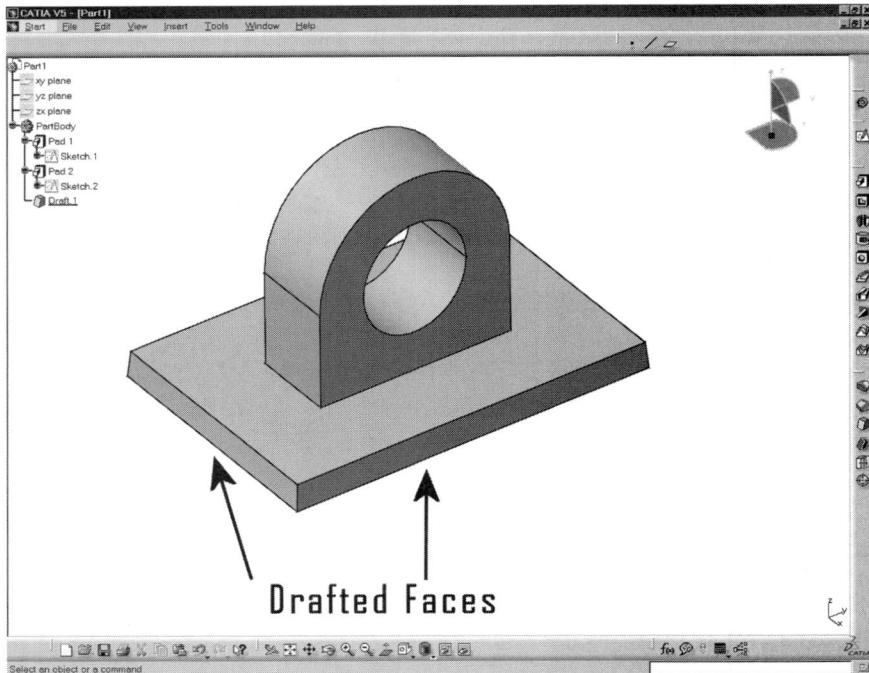

Drafted Faces

Fig. 4-124.
Edge Fillet
icon.

Fig. 4-125.
Edges to be
filleted.

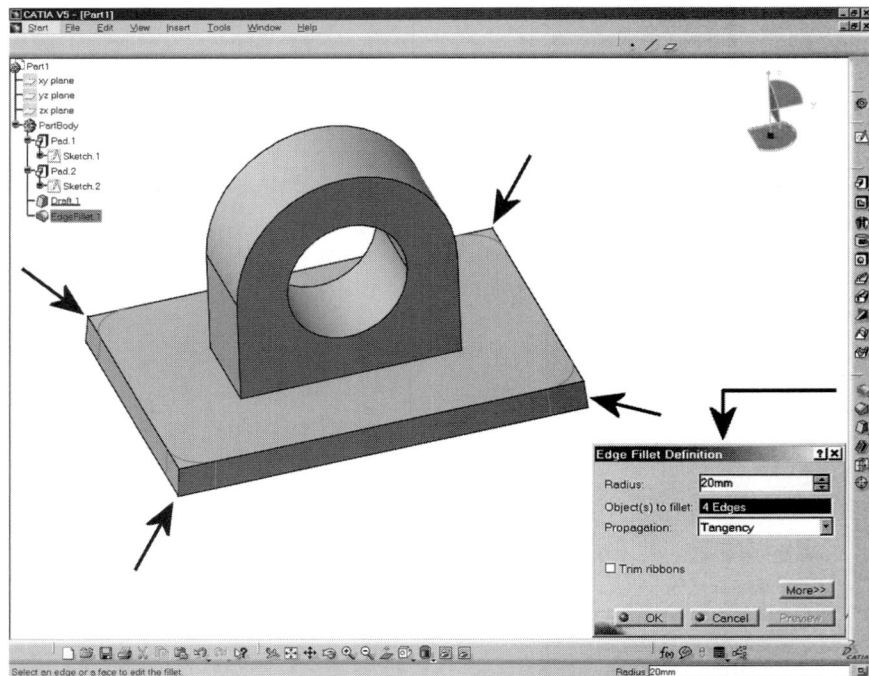

Fig. 4-126.
Bearing block part
with round.

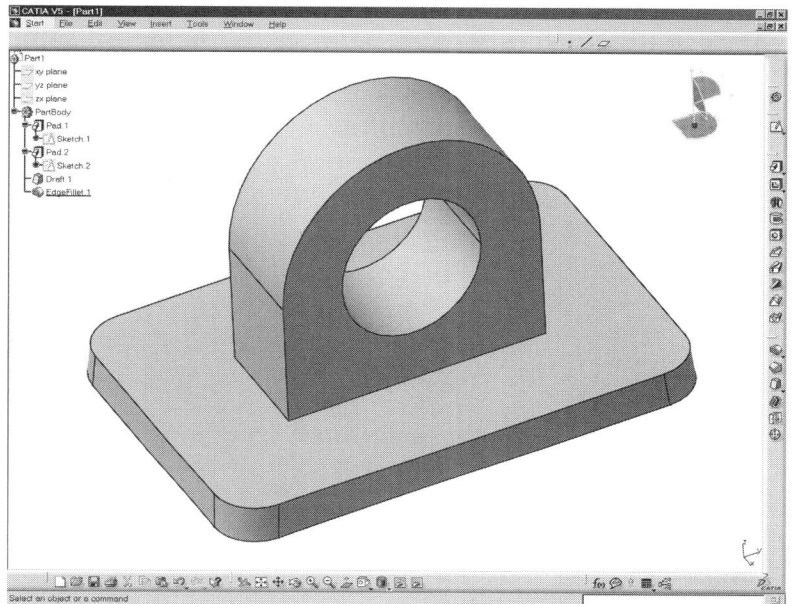

Fig. 4-127.
Save icon.

Surface-based Features

CATIA's surface-based features include splitting a solid with a surface, adding thickness to a surface, closing an open set of surfaces, and trimming a solid with a surface (which CATIA calls sewing). These features are beyond the scope of this book. See CATIA's online user documentation for more details on these commands.

Transform Features

Fig. 4-128. Transform
Feature toolbar.

Transform features are used to manipulate existing features within the design model. The three primary modes of operation involve moving, copying, and scaling existing features. CATIA V5 offers numerous ways of creating transform function features. These operations, discussed in the sections that follow, are performed from the Transform Feature toolbar, shown in figure 4-128.

Moving and Copying

Moving and copying transform functions incorporate the commands discussed in the sections that follow.

Fig. 4-129.
Translation icon.

Fig. 4-130.
Rotation icon.

Fig. 4-131.
Symmetry icon.

Fig. 4-132.
Mirror icon.

Fig. 4-133. Pattern
toolbar.

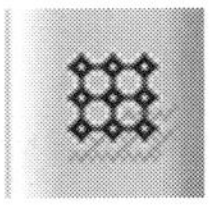

Fig. 4-134.
Rectangular pattern
icon.

Translation

This command, accessed via the Translation icon (shown in figure 4-129), translates the entire part body along a selected direction.

Rotation

This command, accessed via the Rotation icon (shown in figure 4-130), rotates the entire part body around a selected edge or axis.

Symmetry

This command, accessed via the Symmetry icon (shown in figure 4-131), mirrors the entire part body about a selected face or plane.

Mirror

This command, accessed via the Mirror icon (shown in figure 4-132), creates a duplicate mirror image of the entire part body about a selected face or plane.

Patterns

Pattern features allow for the creation of multiple identical features within the design model. They also allow for the simultaneous placement of the selected multiple feature, based on a rectangular, circular, or user-defined criterion. These functions, explored in the sections that follow, are performed from the Pattern toolbar, shown in figure 4-133.

Rectangular

Rectangular patterns, created via the Rectangular icon (shown in figure 4-134), are created in a linear fashion and are controlled through multiple choices within the Rectangular Pattern dialog box, shown in figure 4-135.

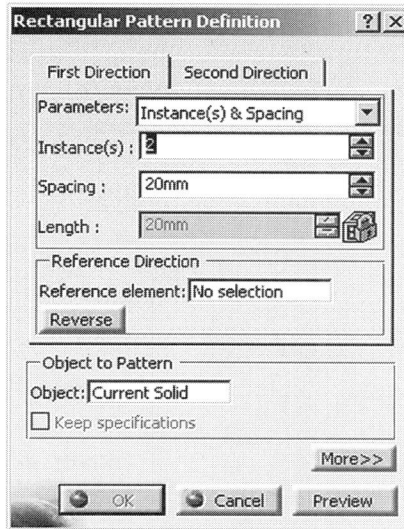

Fig. 4-135. Rectangular Pattern dialog box.

Circular

Circular patterns, accessed via the Circular icon (shown in figure 4-136), are created in a radial fashion and are controlled through multiple choices within the Circular Pattern dialog box, shown in figure 4-137.

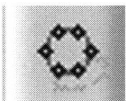

Fig. 4-136. Circular icon.

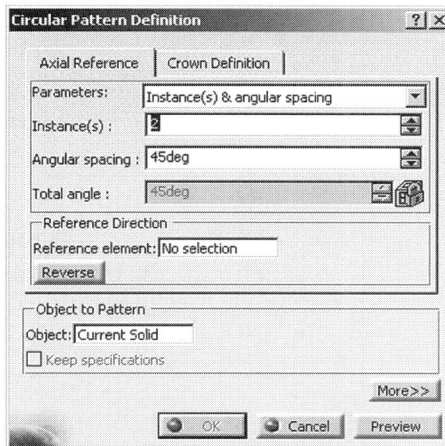

Fig. 4-137. Circular Pattern dialog box.

User Defined

User-defined patterns, created via the User Defined icon (shown in figure 4-138), allow for the duplication of features, groups of features, or an association of bodies to a user-defined location. These operations are performed from the User Pattern dialog box, shown in figure 4-139.

Fig. 4-138. User Defined icon.

Fig. 4-139. User Pattern dialog box.

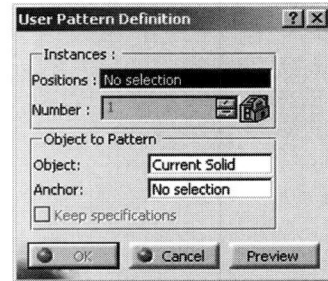

Patterning Tips

The following are tips to keep in mind when using the functions described in this section.

- Creating a pattern is a quick way to reproduce a feature.

- It is often more efficient to perform one operation on multiple features in a pattern than to specify features individually.

- A pattern is parametrically controlled. Therefore, you can modify a pattern by changing pattern parameters, such as the number of instances, spacing between instances, and original feature dimensions.

- Modifying patterns is more efficient than modifying individual features. In a pattern, when you change dimensions of the original feature, the system automatically updates the entire pattern.

Scaling

Fig. 4-140. Scaling icon.

This command, accessed via the Scaling icon (shown in figure 4-140), resizes a body via a user-entered scaling dimension.

Transform Feature Exercise: Bearing Block Hole Pattern

This exercise patterns a hole feature within the bearing block model. Perform the following steps.

Fig. 4-141. Open icon.

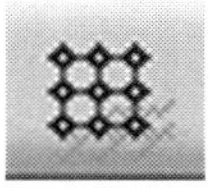

1   Retrieve the bearing block part (via the Open icon, shown in figure 4-141) from the previous exercise.

2   Click on the Rectangular Pattern icon (shown in figure 4-142), from the Dress-Up toolbar. This accesses a dialog box consisting of two tabs, labeled First Direction and Second Direction.

3   Within the First Direction tab, accept the default parameter setting of Instance(s) & Spacing.

4   Enter a value of *100* (mm) for the spacing and a value of *2* for the number of instances.

Fig. 4-142. Rectangular Pattern icon.

5   Select the hole feature as the Object to Pattern and the top surface of the model as the Reference Element direction. These selections are shown in figure 4-143.

6   Duplicate the same settings and values for the Second Direction tab. The model should look like that shown in figure 4-144.

7   Experiment by changing some of the values and options within each tab to get a feel for how the model reacts to different combinations of settings.

Boolean Operations

Boolean operations within CATIA V5 are modeling methods that make use of two bodies that overlap and therefore share part of the same space. Boolean operations are a way of adding or removing material between multiple bodies within the part design model.

In Boolean union, the geometry of the overlapping body is eliminated, and a single body is created from the two overlapping bodies using all of the exposed surface area. Union is generally used to merge bodies that are most easily built from component bodies modeled separately.

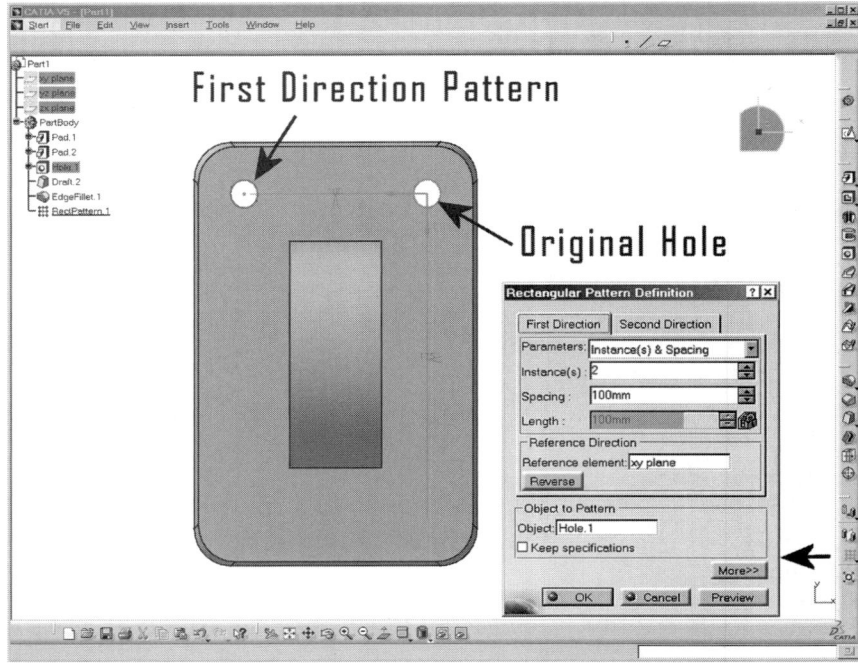

Fig. 4-143.
Object to
Pattern and
Reference
Element
specifications.

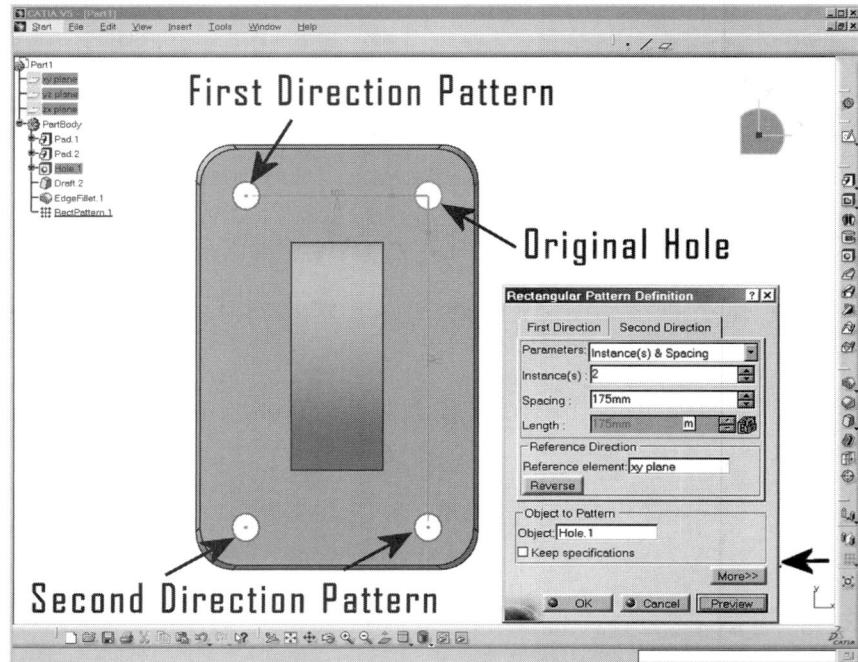

Fig. 4-144.
Hole feature.

Boolean subtraction is used to sculpt out the overlapping volume from one body or the other body. After the operation, one body is left, minus its overlapping region with the other body. Boolean intersection preserves the overlapping body only, eliminating the rest of both bodies.

Fig. 4-145. Boolean Operation toolbar.

CATIA V5 offers a variety of ways to perform Boolean operations within the part design model. These operations, explored in the sections that follow, are performed from the Boolean Operation toolbar, shown in figure 4-145.

Assemble

This command, accessed via the Assemble icon (shown in figure 4-146), assembles two selected bodies together. During this operation, the first body's material is assembled from the second body.

Fig. 4-146. Assemble icon.

Add

This command, accessed via the Add icon (shown in figure 4-147), adds the first selected body's material to the selected second body.

Fig. 4-147. Add icon.

Remove

This command, accessed via the Remove icon (shown in figure 4-148), removes the first selected body's material from the selected second body.

Fig. 4-148. Remove icon.

Intersect

This command, accessed via the Intersect icon (shown in figure 4-149), intersects the first selected body with the second selected body.

Fig. 4-149. Intersect icon.

*Fig. 4-150.
Union Trim
icon.*

*Fig. 4-151.
Remove
Lump icon.*

*Fig. 4-152. 3D
Constraints icon.*

Union Trim

This command, accessed via the Union Trim icon (shown in figure 4-150), combines the first selected body with the second selected body.

Remove Lump

This command, accessed via the Remove Lump icon (shown in figure 4-151), removes "as a lump sum" the first selected body from the second selected body.

NOTE: *Chapters 5 and 6 cover in more depth the application and use of Boolean operations and proper body organization.*

3D Constraints

This command, accessed via the 3D Constraints icon (shown in figure 4-152), allows for the application of 3D constraints to individual part features. CATIA V5 supports 3D constraints such as distance, length, angle, fix/unfix, tangency, coincidence, parallel, and perpendicular. Depending on how a feature was created, 3D constraints placed on these features may react in two ways: as references or as constraints. Examples of 3D constraint options are shown in figure 4-153.

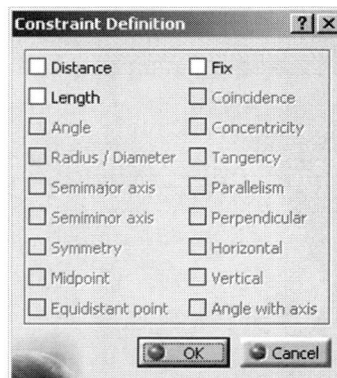

Fig. 4-153. 3D constraint options.

Summary

CATIA V5 has a well-rounded set of sketch-based and dress-up features. Several tools are also available for manipulating these features and structuring the design model. This hybrid approach to modeling offers great flexibility in the design process. At the end of this chapter you should have a basic understanding of the following.

- Reference elements
- Geometry-based features
- Transform features
- Boolean operations

Review Questions

1 Name the various types of reference elements within CATIA V5.

2 How do these reference elements assist in the creation of 2D sketched profiles and 3D solid model features?

3 Where are reference elements typically stored in the configuration tree?

4 How is a reference element point different from a sketched point?

5 What is the difference between sketcher-based features and dress-up features?

6 Name five sketcher-based features and their respective uses.

7 Name four dress-up features and their respective uses.

8 List four limiting options available for a pad feature.

9 What are the four fillet-creation options?

10 List four types of Boolean operation.

CHAPTER 5
MODEL ORGANIZATION

Introduction

GOOD MODEL ORGANIZATION IS ESSENTIAL for the management of large models and for model reuse by multiple users. This chapter outlines the method of organizing features and bodies as functional groups of the design model. These techniques may seem cumbersome for smaller design models, but they are essential for the management of change within large models. It is up to the user how far these techniques need to be employed in any given scenario, based on the user's internal design processes and methodologies.

Model organization is extremely important for building a robust design within CATIA V5. The key is to build flexibility into the design for feature manipulation, downstream design changes, and assembly operations. This chapter examines modeling methods and best practices focused on bodies and features within bodies. The grouping of features, bodies, and open bodies is the most important aspect of model management.

NOTE: *The main purpose of this chapter is to demonstrate various rules and techniques for model organization of solid model features, with less emphasis on open bodies. Open bodies are discussed in the context of supporting the creation of solid model features. Advanced methods of open body organization, related to advanced surfacing methods, are similar but are without the Boolean assembly operations used for body interaction.*

Objectives

This chapter explores the following.

- The specification tree
- Body types
- Modeling practices
- Body organization
- Feature organization

Specification Tree Fundamentals

The specification tree in CATIA V5 is a container of all information related to your model. It is also the diagram of how that model was constructed. An important concept in CATIA V5 is the organization of this data within the specification tree. In addition to being a hierarchical representation of the design, the specification tree displays all types of design specifications, such as knowledge rules and checks, material characteristics, hyperlinked documents, and more. It is a track record of your design intent, as well as a specification-oriented representation of the design.

Upon opening any new CATPart document in CATIA V5, you will always receive the same default specification tree, containing the same five items. At the top of the tree, shown in figure 5-1, is *Part1*; followed by the three default X, Y, and Z planes; and then the default Part Body.

Part Name

P*art1* is the default name for the part number. *Part1* properties may be modified by pressing and holding down MB3, and then selecting the Properties icon. The name of the part may be changed within the Product tab, under Part Number, of the dialog box, shown in figure 5-2. The new part name will be accepted when the dialog box is saved and exited.

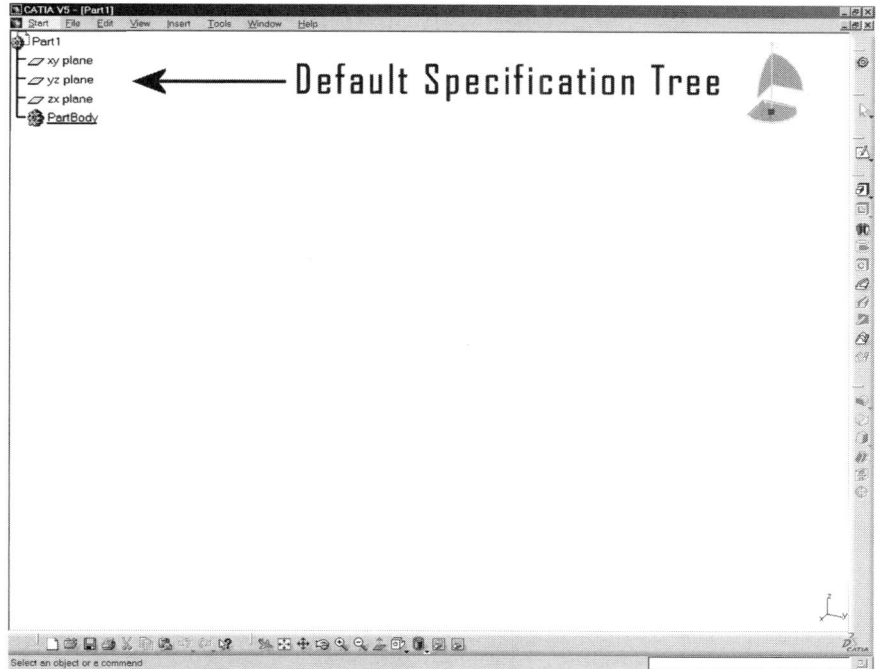

Fig. 5-1. Default specification tree.

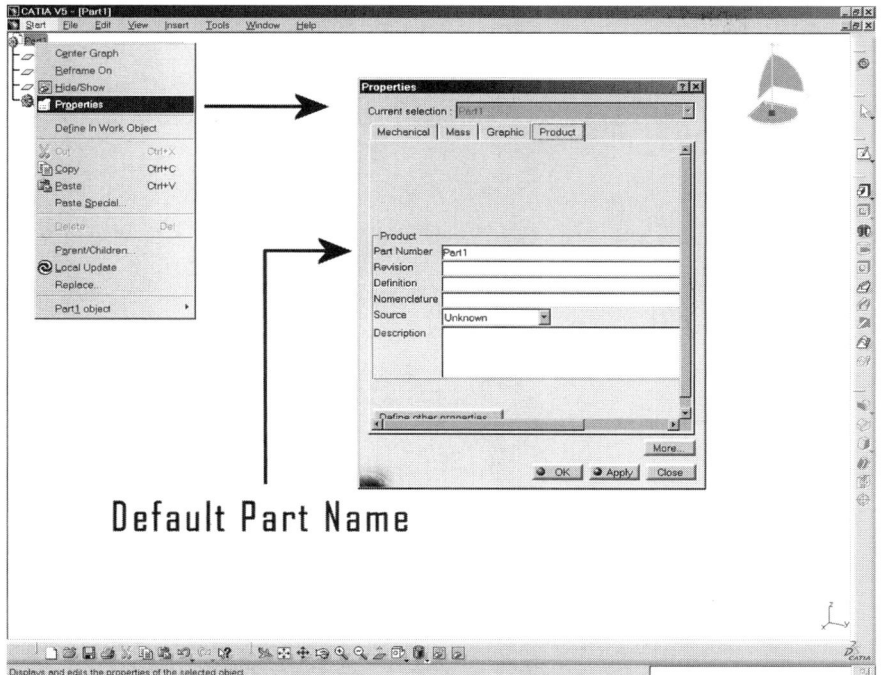

Fig. 5-2. Default part name.

Default Datum Planes

The default X, Y, and Z planes are fixed and cannot be deleted. Because these planes are fixed it is generally good practice not to constrain any geometry to these elements. The preferred method would be to create duplicate planes on top of the three default datum planes. This provides more flexibility in the model, because geometry is constrained to elements that are more modifiable. In doing this you will note that these new planes (see figure 5-3) are automatically placed in a new open body. Open bodies are discussed in detail later in the chapter.

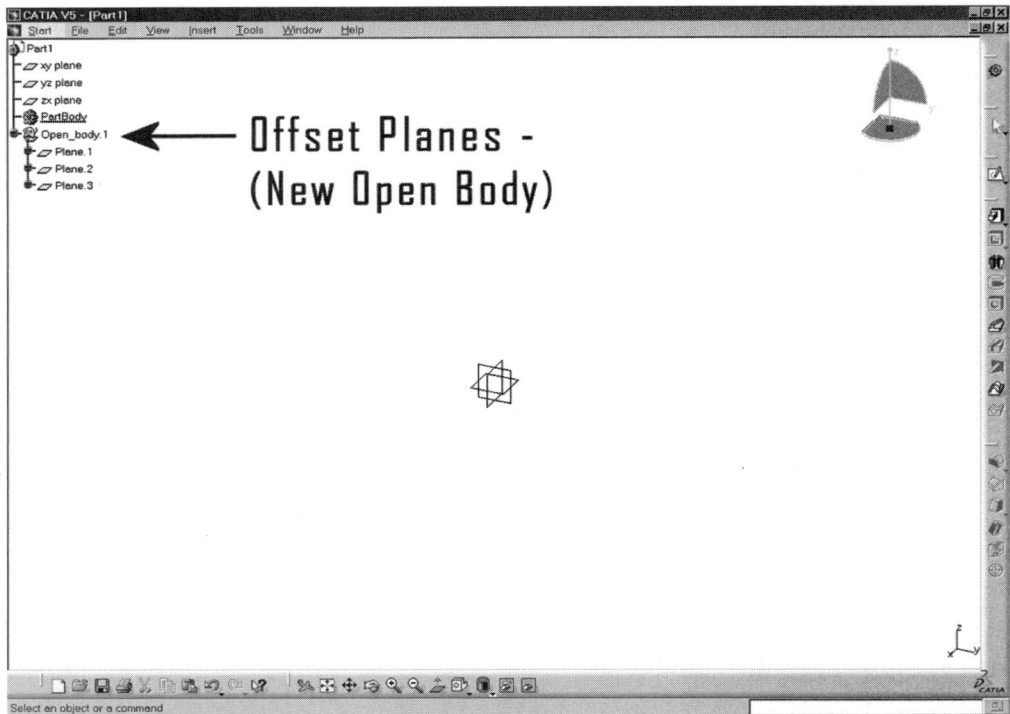

Fig. 5-3. Default datum planes.

NOTE: *The creation of new datum planes is subjective and depends on the size and complexity of the model.*

TIP: *No-show the default datum planes so that they are not accidentally selected.*

Default Part Body

The default part body is a permanent body created upon opening a new CATPart. This body, an example of which is shown in figure 5-4, is the root of the specification tree and cannot be manipulated. It should define the result of all operations and body assemblies. Therefore, it is recommended that this body remain an empty body until completion of all elements of the model. When the design is complete, all model bodies should be assembled within the default part body as the final operations of the part design process.

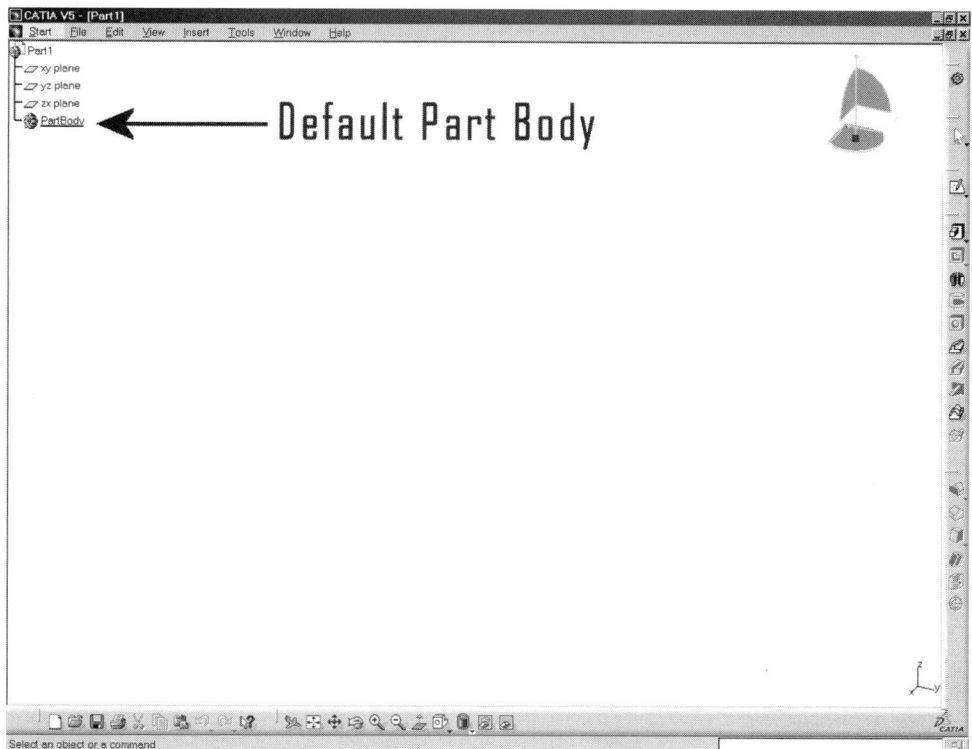

Fig. 5-4. Default part body.

It is also recommended that the name of the default part body be changed to more accurately and meaningfully reflect the part. This is important to note because all mass properties, analysis, and drawing feature links are located only in this body.

Body Types

CATIA V5 works via a "multi-body" principle. There are two types of bodies CATIA V5 uses, referred to as bodies and open bodies. These body types are the containers for all geometry element types within CATIA V5.

Body

*Fig. 5-5.
Body icon.*

A body within CATIA V5, created via the Body icon (shown in figure 5-5), houses only solid geometry. It is very similar to the default part body, except that this body can be manipulated. In this body type, and example of which is shown in figure 5-6, you will find standard features such as pads, pockets, and holes and the sketches that support these features. You may also find dress-up features, such as fillets, drafts, and thickness parameters. You can add a body to a model using the Insert Body command, located in the main menu. A body may be inserted within the Part Design workbench only. The icon for this body varies, depending on the type of the first feature created (i.e., positive or negative geometry).

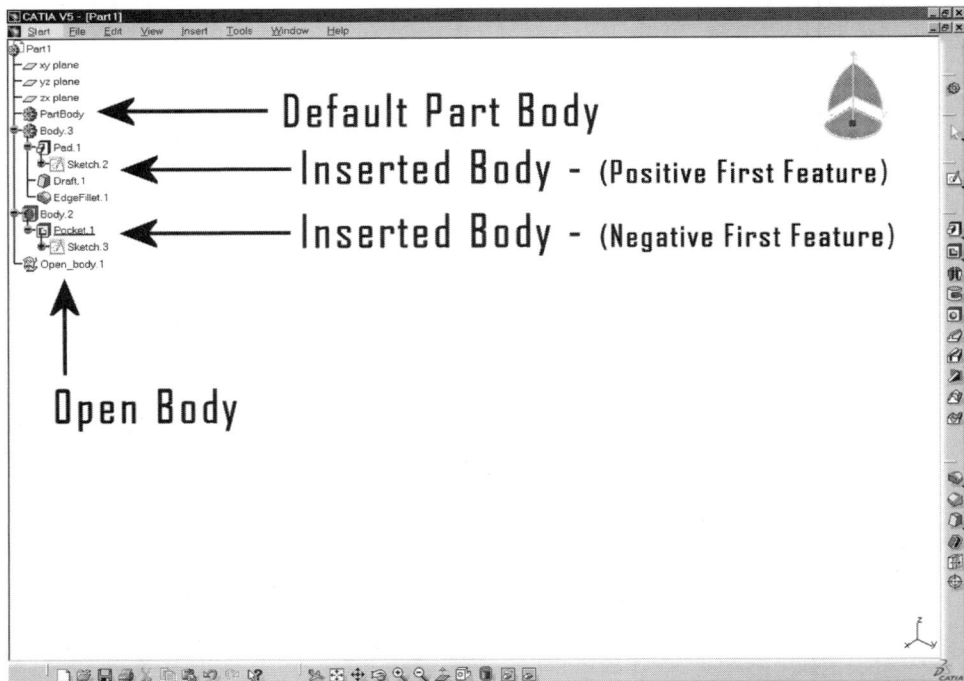

*Fig. 5-6.
Part body.*

Open Body

Fig. 5-7.
Open Body
icon.

An open body, created via the Open Body icon (shown in figure 5-7), contains all geometry used for part creation that is not solid geometry. An example of this is construction geometry such as points, lines, and planes. All surface geometry types are also stored within an Open Body. These features can be referenced and constrained during the creation of part geometry to help establish and maintain associativity within the document. An open body, an example of which is shown in figure 5-8, may be inserted into the model using the Insert Body command located in the main menu. An open body is automatically inserted into a model upon creation of a reference element. An open body may also be inserted "blank," using the Insert Open Body command.

Fig. 5-8.
Open
body.

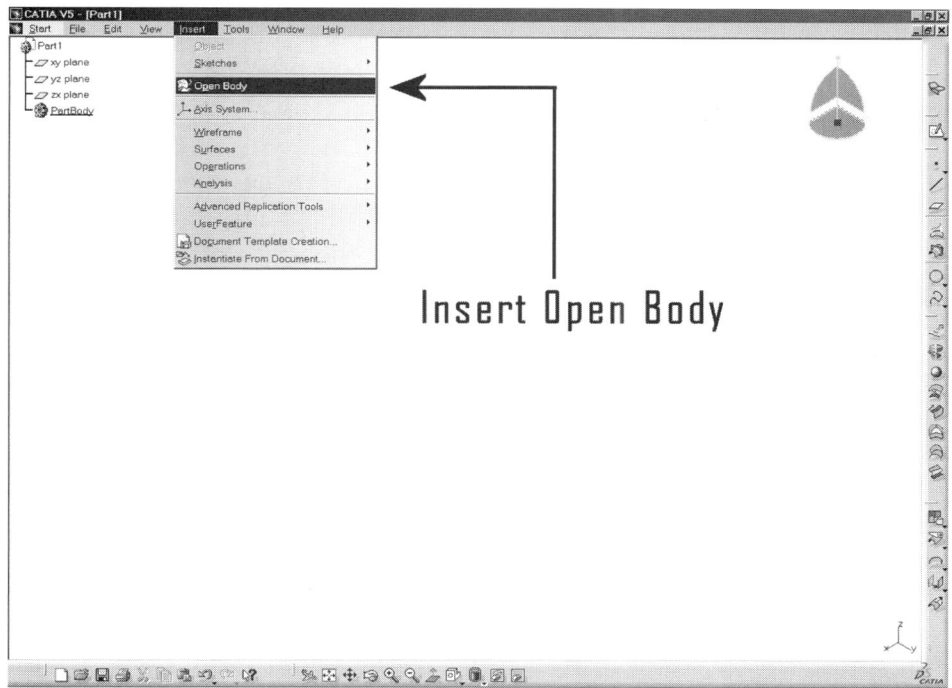

Good Modeling Practices

Organization of geometry within bodies and open bodies is extremely important to maintaining the integrity of the part design. Good modeling practices are essential in building a robust model that handles constant change during the product development process. The sections that follow demonstrate the need for good basic organizational techniques for both bodies and features. These techniques maximize model performance, reliability, and stability during the part creation process and help ensure that these aspects exist should the part need to be modified.

- *Performance:* Performance is measured as software speed or processing time. This can be categorized as follows: (A) model update speed of the software, (B) software response speed when creating geometry, (C) overall update process time required (human and software), and (D) time required for tasks using the software.

- *Reliability:* Reliability is a measure of confidence in software performance level. That is, the dependability of the software to repeatedly produce the same, correct results.

- *Stability:* Stability is measured by how often the software locks up or crashes. All software crashes, but the frequency determines the stability of the software. Stability is one measure of software quality. The goal is to utilize good practices and techniques to reduce the stress on the software, resulting in more stability.

Body Organization

Functional features should be created and contained within their own respective bodies. One of the first things to consider regarding body organization is the type of part in which you will be working. Take a few minutes to visualize how this might work. Think about how the part model might be constructed and identify the basic functional features of the model. As mentioned earlier, it is good practice to always rename the body to some functional name. This makes it easier to locate important features within the specification tree. These bodies will eventually interact with one another, through the use of Boolean operations, to create the final part design.

Boolean Operation Recommendations

When interacting with a body, the Assemble command should be your primary means of performing operations. The Assemble command represents a "neutral" operation and the Boolean operation result will rely on the positive or negative signs associated with the bodies.

Body Organization Example

The following design exercise is an example that demonstrates the thought process of creating a part and organizing its features within the proper bodies. The functional features identified for the part are a base, a clearance feature, and a mounting boss. This sample is not a step-by-step exercise but a demonstration of the thought process involved in organizing a new part model.

1 Start a new CATIA V5 session and create the part design, shown in figure 5-9.

Fig. 5-9.
New part.

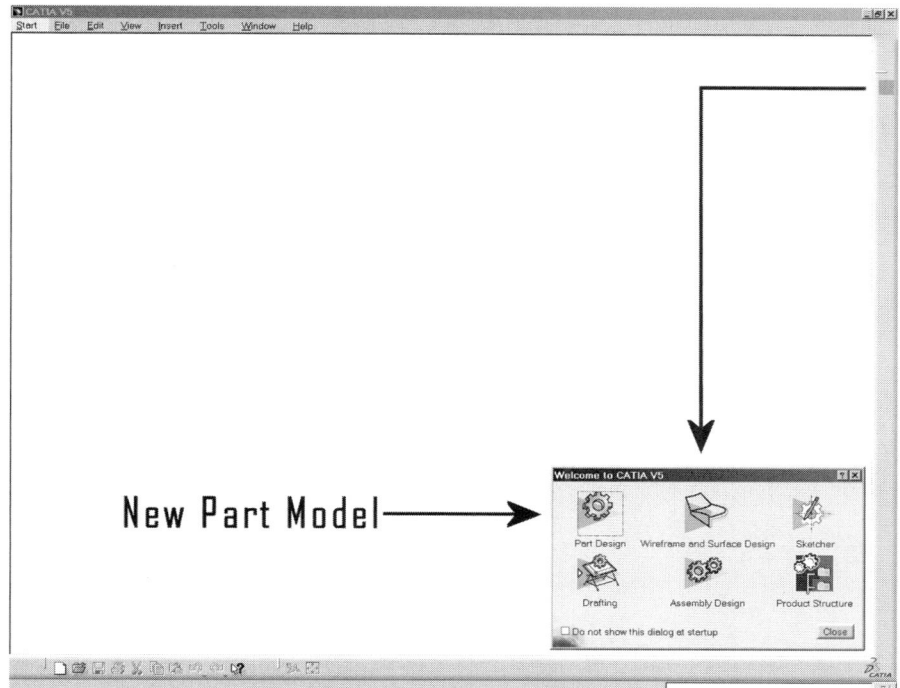

2   Renamed the part *#12345678* by highlighting the default part name (*Part1*) and modifying its properties by clicking MB3 (see figure 5-10).

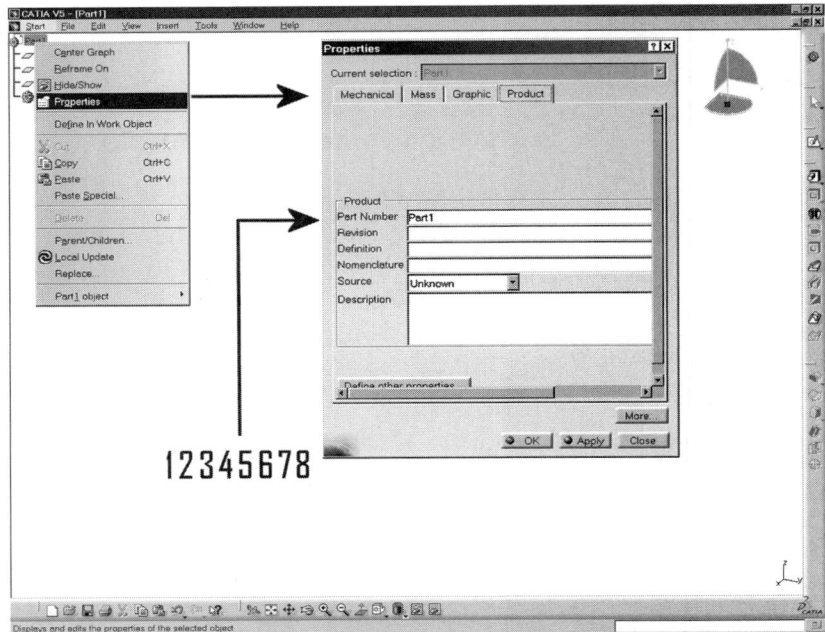

*Fig. 5-10.
Renamed
CAT part.*

3   Create duplicate planes from the original default planes and specify these as "no-show." Rename the three new planes and use these as the primary reference elements (i.e., the open body) for the construction of the part.

4   Rename the new open body (created when the three new reference construction planes were created) *Construction Features,* as shown in figure 5-11.

5   Rename the original default part body (CAT part) to reflect what it will represent. In this case, name it *Total Casting,* as shown in figure 5-12.

6   Insert a new body in the specification tree and name it *Body Casting Features.*

This body will eventually represent the Boolean assembly of the main three functional (casting) feature bodies, shown in figure 5-13.

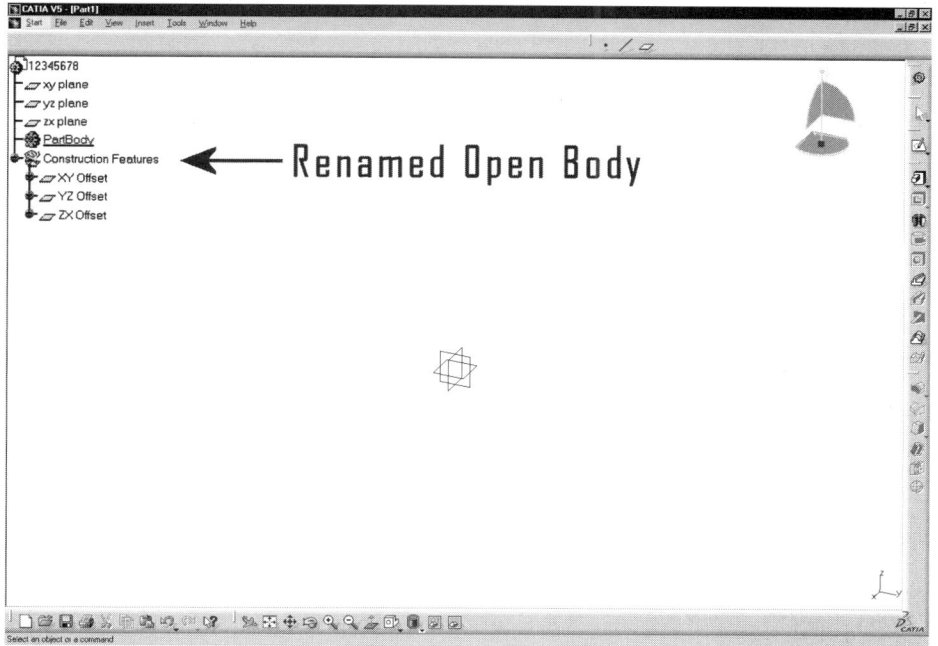

Fig. 5-11. Renamed open body.

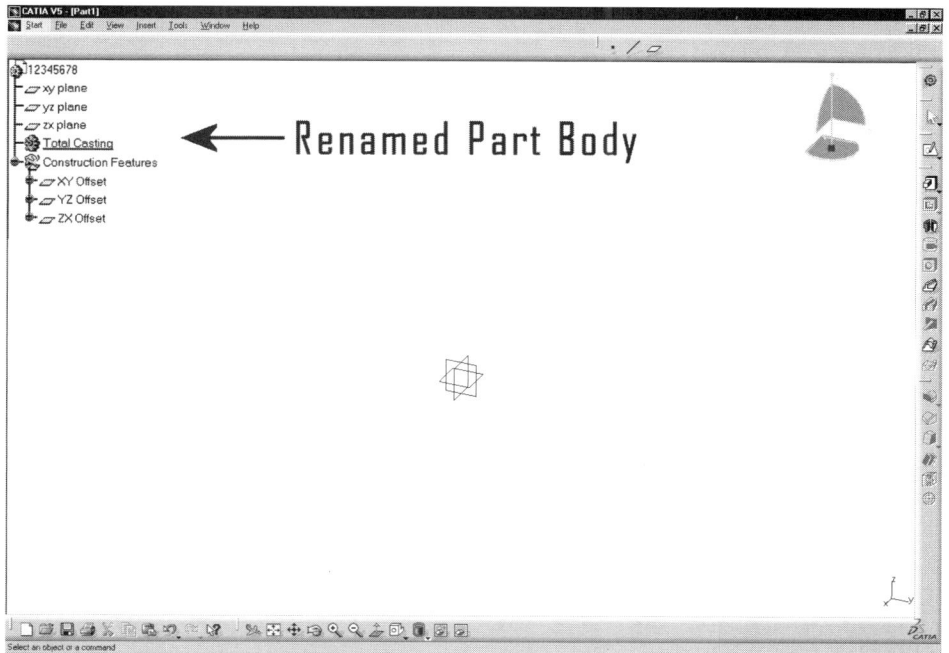

Fig. 5-12. Renamed original default part body.

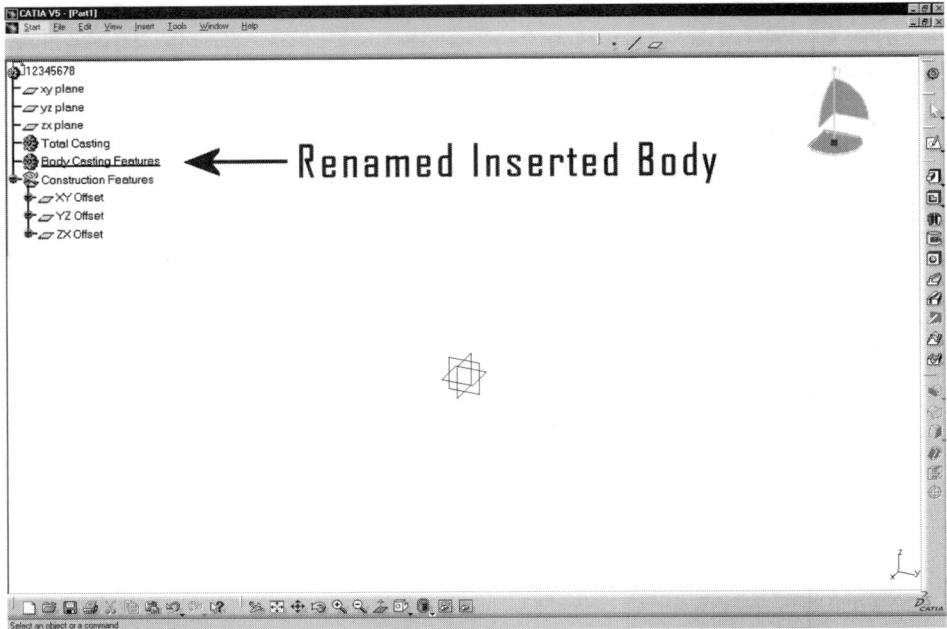

Fig. 5-13. Body casting features.

7 Create three new bodies and name them *Base, Bosses,* and *Clearance Feature.*

The next steps in this process would be to create the appropriate geometry within each functional body. After all geometry is completed, the functional bodies would be assembled into the *Body Casting Features* body. Figure 5-14 shows all of the bodies assembled into the *Body Casting Features* body.

Note that the icon for the *Clearance Feature* body (created in another part of the geometry creation process) is visually different than the icon for the *Body Casting Features* body. This is because a pocket (removal of material) would have been the first feature created in the *Clearance Feature* body. The geometry of the pocket feature represents a negative property and therefore the body that contains it (Clearance Feature) is displayed differently from the all-positive *Body Casting Features* body.

NOTE: *An empty body cannot be assembled.*

Finally, the *Body Casting Features* body would be assembled into the *Total Casting* body. The final specification tree would look like that shown in figure 5-15.

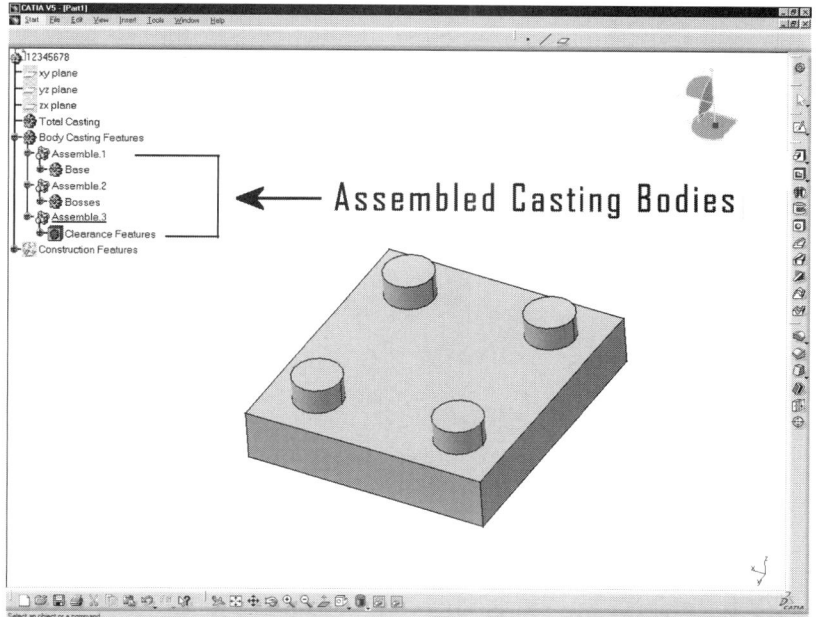

Fig. 5-14.
Body Casting
Features *body
consisting of
all assembled
functional
bodies.*

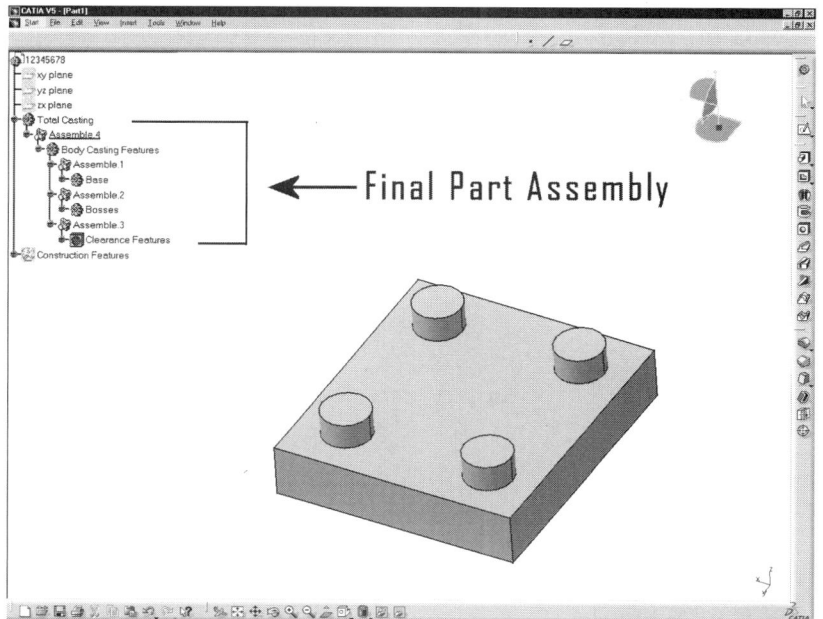

Fig. 5-15.
Body Casting
Features *body
with* Total
Casting *body
assembled
within it.*

Feature Organization

An equally important aspect of good model organization is how features are arranged within a particular body. Previously you learned that the functional features of a part should be organized as respective bodies. These bodies may also contain dress-up features associated with respective functional features.

The goal is to establish the proper parent/child relationships within these bodies. Toward the end, you want dress-up features as close as possible to the design features they affect. Think of the features within a body as a family. It is easier to work with these features when they are contained within a group (family) represented by a body. It makes it much easier to copy and paste, make changes, and reorder features when they exist in the same body.

Feature Organization Rules of Thumb

The sections that follow describe a few rules of thumb you should employ to achieve modeling consistency during the geometry creation process.

Draft/Thickness/Fillet Rule

This rule has to do with dress-up feature order. If the created design feature is intended to have draft, create the draft first. If the methodology you are following uses thickness on this feature, the thickness feature should be created next (prior to any fillets). If the design model requires fillets, all fillet features should be created after the draft and thickness features.

First Fillets Rule

All fillet features of a particular design feature should reside in that feature's body. These fillets become children of that single parent design feature. Fillets that are part of the relationship established between intersecting design features (i.e., children of these features) will be located in the design tree immediately after the assembly of bodies representing the intersecting design features.

Second Fillets Rule

This rule of thumb relates to the filleting order for a single design feature. The thought is to first place all fillets that add extra material to the design feature, and to place all fillets that remove material from the design feature last. This order applies when both types of fillets are needed to finish the design feature before any assembly is done.

Benefits of a Feature Organization Strategy

The following are benefits of the feature organization strategy dealt with in this chapter.

- This "best practices" approach, well-known to the experienced CATIA V4 user, helps new users make the transition from V4 to V5.

- Maximized performance and stability during, and facilitation of, model changes.

- Ease of establishing relationships between and among features.

- Modifications such as duplication, copy and paste, and reorder are more easily performed.

- Better interaction with outside documents when using cut, copy, and paste functions.

- Selected bodies within the tree can interact with other documents.

- Facilitation of patterning operations.

- Specification tree organizational structure is more logical.

Feature Organization Example

The following exercise is an example intended to provide you with a better understanding of how to organize bodies and features within a part model. The part used in the exercise consists of two design features. This exercise is a demonstration of the thought process involved in organizing a new part model.

1   Start a new CATIA V5 session and create the new part design shown in figure 5-16.

2   Insert two bodies in the model and name the first body *Design Features 1* (shown in figure 5-17) and the second *Design Features 2*.

Fig. 5-16. New part design.

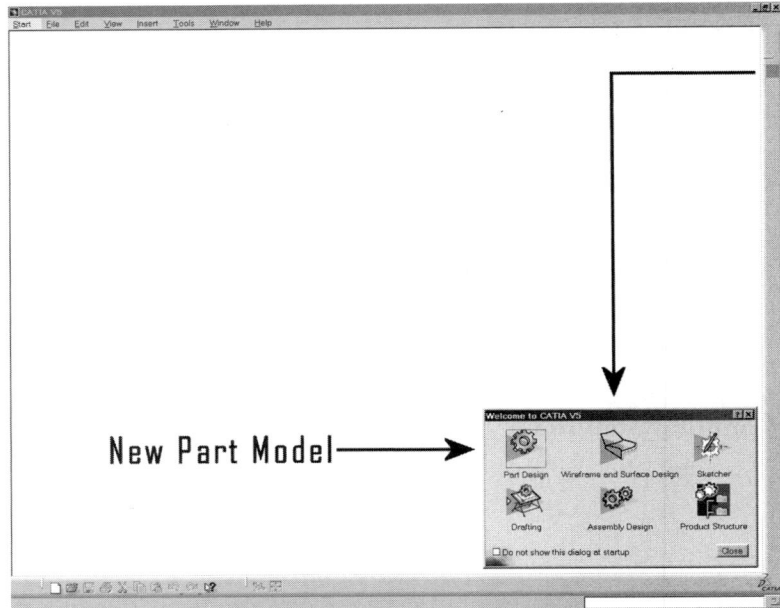

New Part Model

Fig. 5-17. Design Feature 1 *body*.

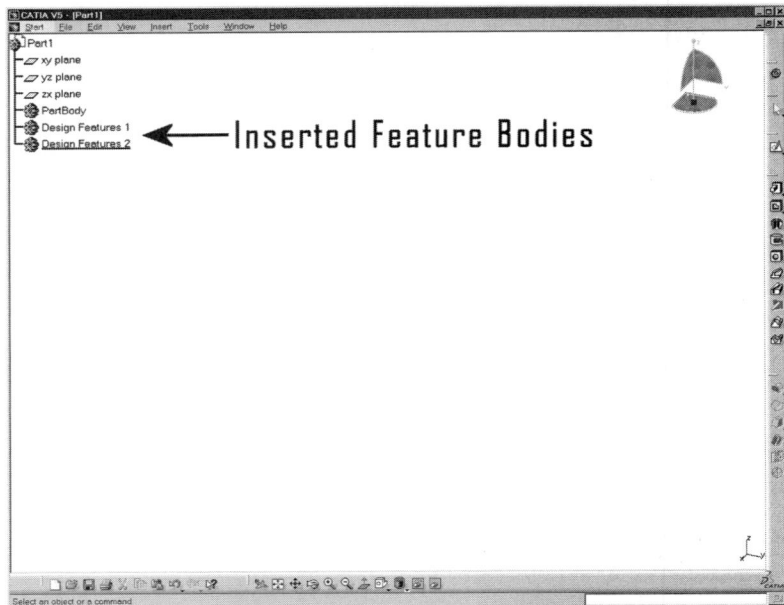

Inserted Feature Bodies

3   Create a rectangular pad that resides within the *Design Features 1* body, as shown in figure 5-18.

4   Add draft to the pad feature, as shown in figure 5-19.

5   Add thickness to the drafted pad feature, as shown in figure 5-20.

Fig. 5-18.
Pad feature.

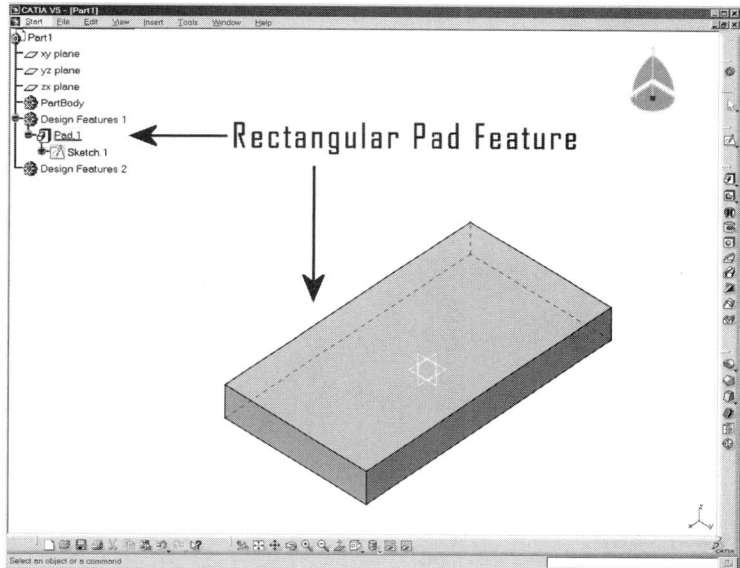

Fig. 5-19.
Draft feature.

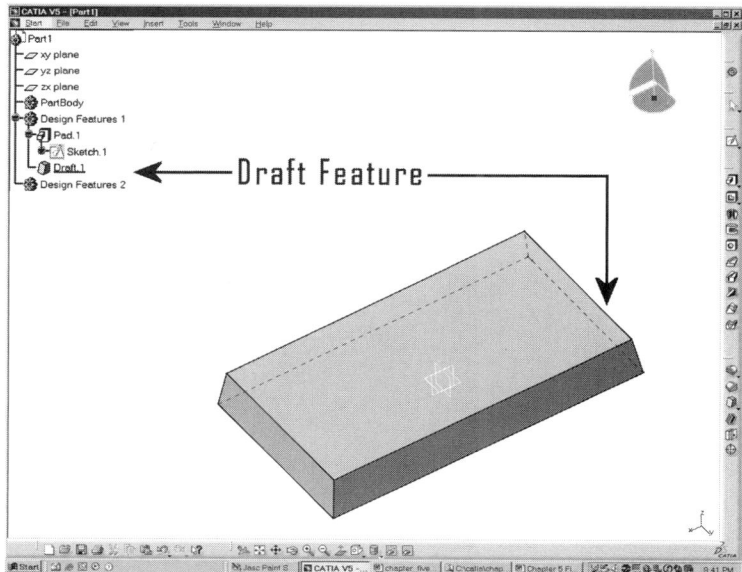

6   Finish the design with the appropriate fillets. (Considering
     the shape of this feature, it is recommended that you fillet the
     shorter vertical edges first.) Fillet the top surface edges. Fig-
     ure 5-21 shows the fillets in place.

Fig. 5-20.
Thickness
feature.

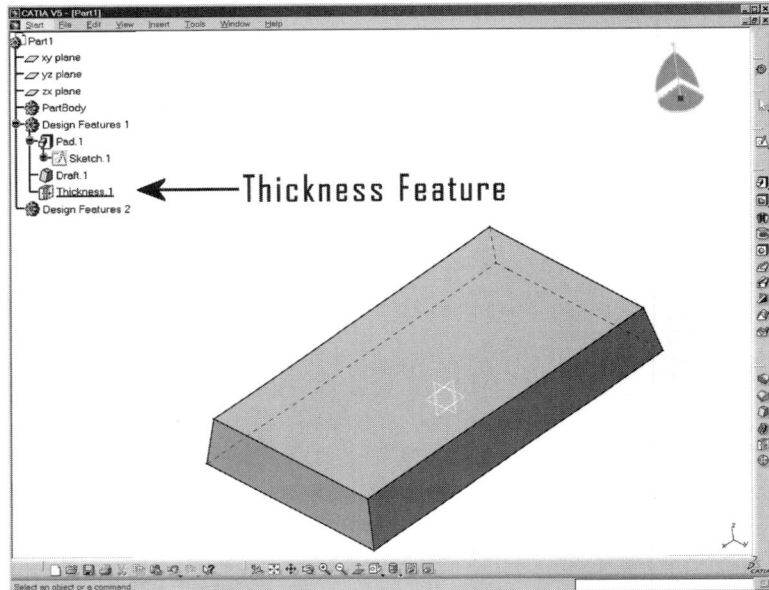

Fig. 5-21.
Fillets in
place.

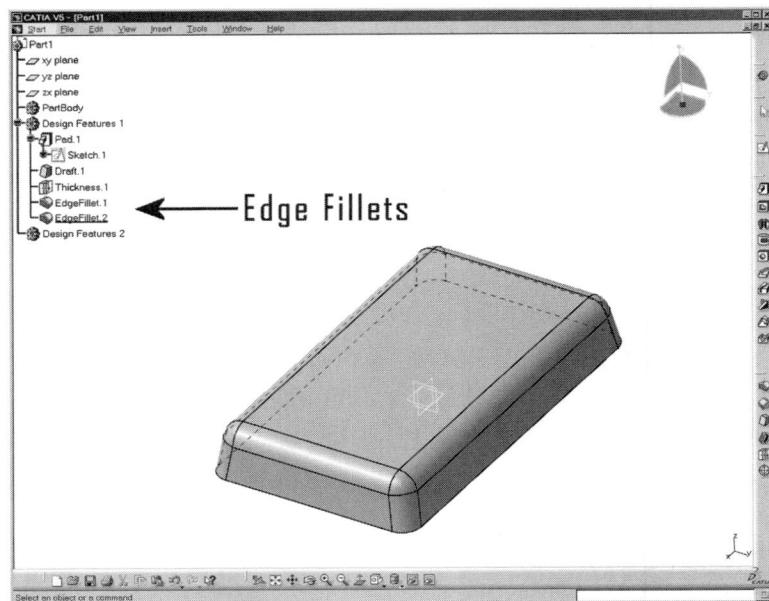

7 Create a circular (round) feature located within *Design Feature 2*. Add to this feature draft similar to that applied to the rectangular feature in *Design Feature 1*. Figure 5-22 shows the round features as propagated.

8 Using the Boolean assembly command, assemble the *Design Feature 1* and *Design Feature 2* bodies to the default part body.

9 Create a fillet feature between the two assembled bodies.

The result is shown in figure 5-23.

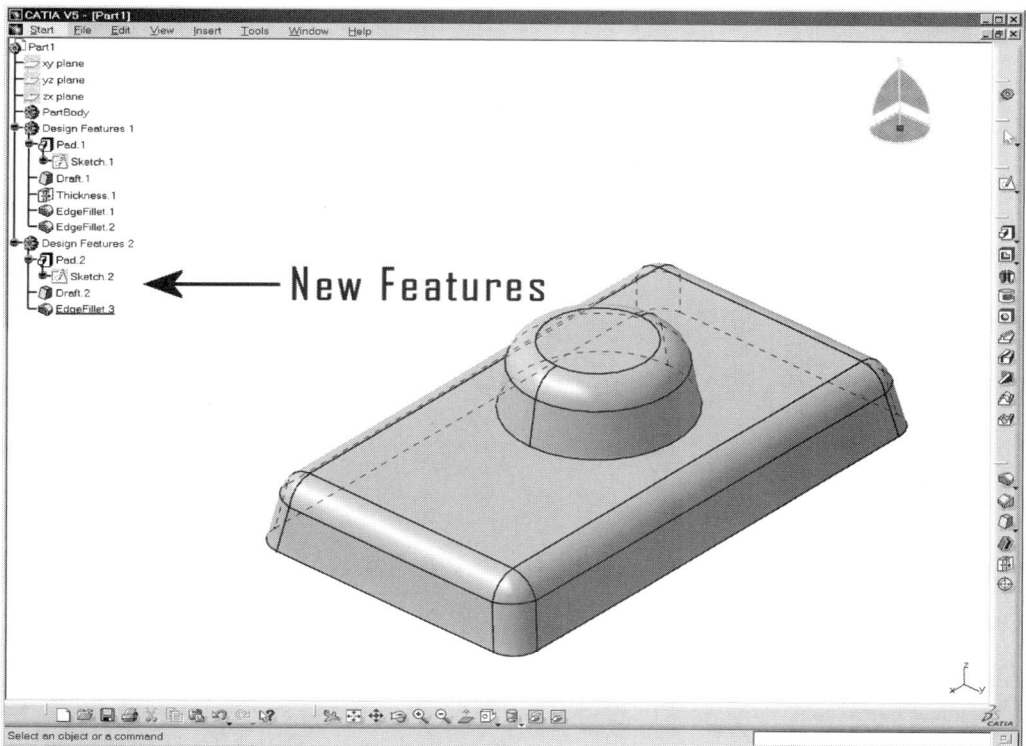

Fig. 5-22. Round features in place.

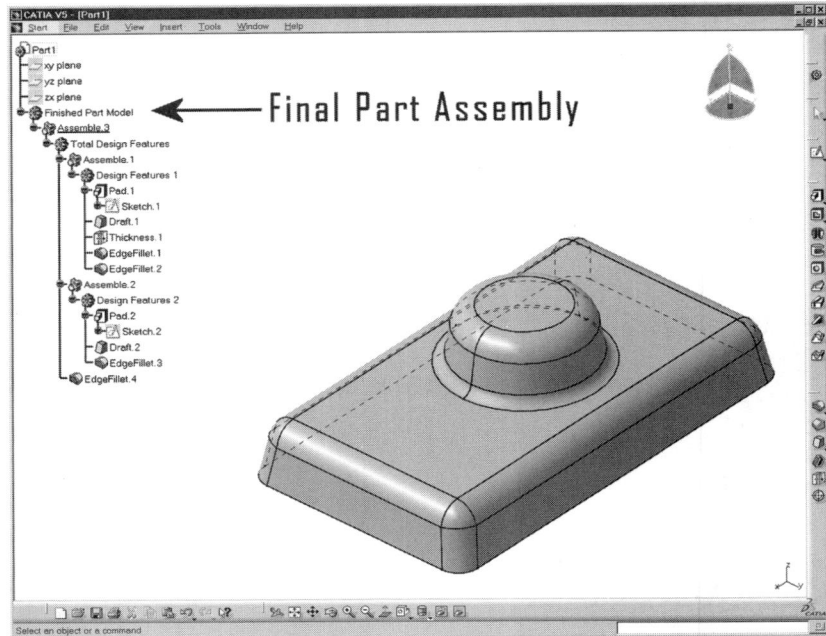

*Fig. 5-23.
Final
assembly.*

A Note on Organization

Body organization is the key to building a robust design. Parts can become complicated and disorganized very quickly. Mistakes will be made in terms of organization, but CATIA V5 provides the tools required to reorganize a model. For example, you have the ability to reorder and redefine features within bodies. Such techniques are explored in Chapter 6. However, it is generally best to flex and review a model during the initial creation process rather than to wait and reorganize an untested mature model.

Part Design Exercises

This section consists of a series of part design exercises. These exercises incorporate the principles and design practices discussed in this chapter.

Part Design Exercise 1: Shifter Boot

This exercise involves the creation of part bodies that contain unique geometric features. These bodies interact with each other by way of Boolean operations such as assembly and removal. The methods involved in this exercise represent a use of CATIA V5 that arrives at a stable and modifiable part. The shifter boot part involved, shown in figure 5-24, is very simple. Simple parts generally allow you a free hand in terms of design options and techniques used. The approach used in this example illustrates successful modeling techniques used in much larger, more complex, parts. This exercise incorporates the following features and functions.

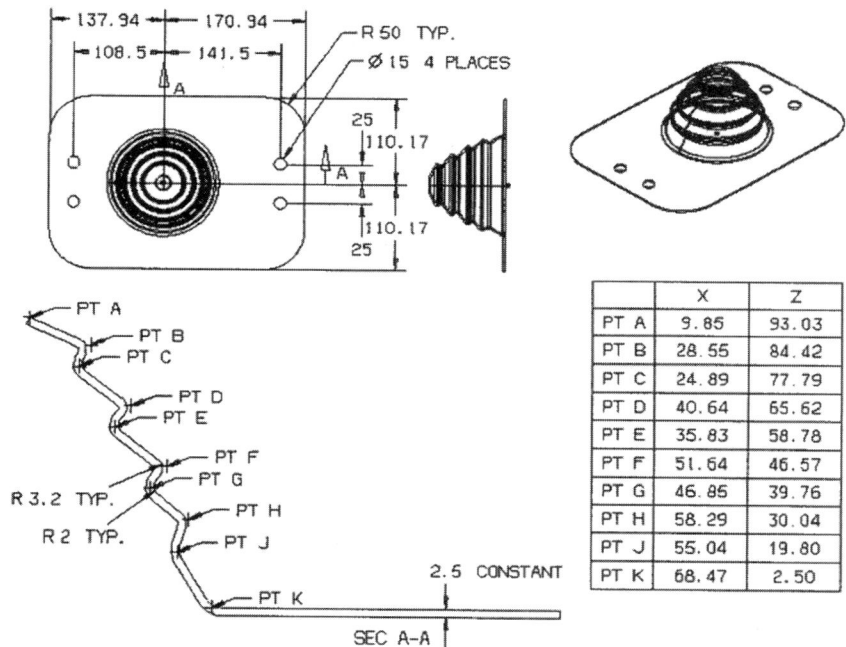

Fig. 5-24. Shifter boot part.

	X	Z
PT A	9.85	93.03
PT B	28.55	84.42
PT C	24.89	77.79
PT D	40.64	65.62
PT E	35.83	58.78
PT F	51.64	46.57
PT G	46.85	39.76
PT H	58.29	30.04
PT J	55.04	19.80
PT K	68.47	2.50

- Creation of reference elements such as points, lines, and planes

- Use of part bodies and associated organizational methods toward development of the tree structure

- Use of the Sketcher to create profiles, points, and construction elements

- Development of pads and shafts from sketch profiles

- Use of simple formulas for establishing relationships between and among elements and operations

- Adding a shell feature

- Assembly and union/trim functions as Boolean operations.

1 Create a new CATPart and rename the default part body *Boot.*

NOTE: *This is done by clicking on the Part Body icon with MB3 and modifying the properties within the Form menu.*

2 Create a new point with the coordinates 0,0,0. This represents the construction elements, shown in figure 5-25.

3 Rename the default open body *Construction Elements* by clicking on the Open Body icon with MB3 and modifying the properties.

4 Using Point-Direction, create a new line element, shown in figure 5-26. Rename the new line element *Centerline* by clicking on the Open Body icon with MB3 and modifying the Properties menu.

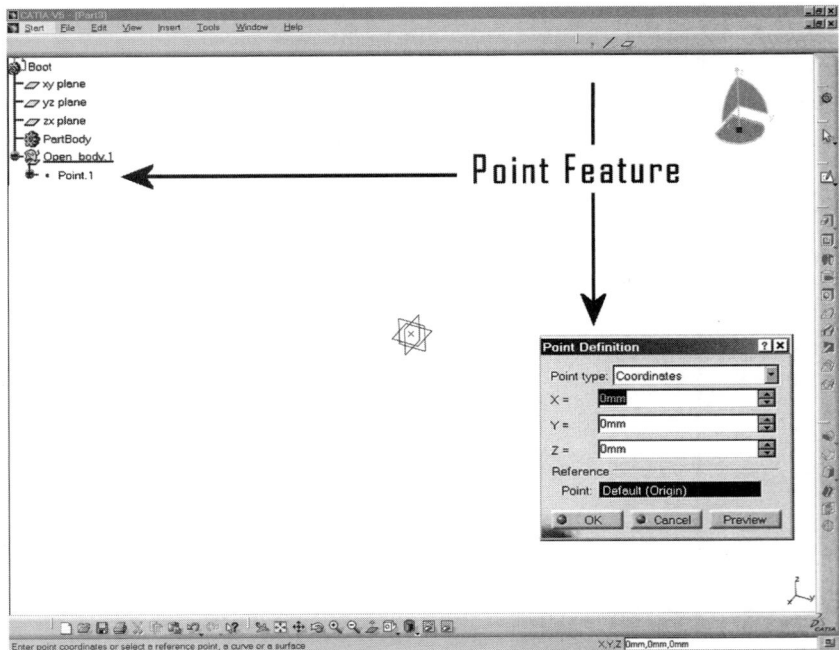

Fig. 5-25. Construction elements.

5 Insert a new body and name it *Boot Sleeve*, as shown in figure 5-27.

6 Create a new sketched profile within the *Boot Sleeve* body. Use the ZX plane as the Sketcher reference plane.

7 Click on the Axis icon on the Profile toolbar. Select on screen the two endpoints of the axis, shown in figure 5-28.

Fig. 5-26.
Line element.

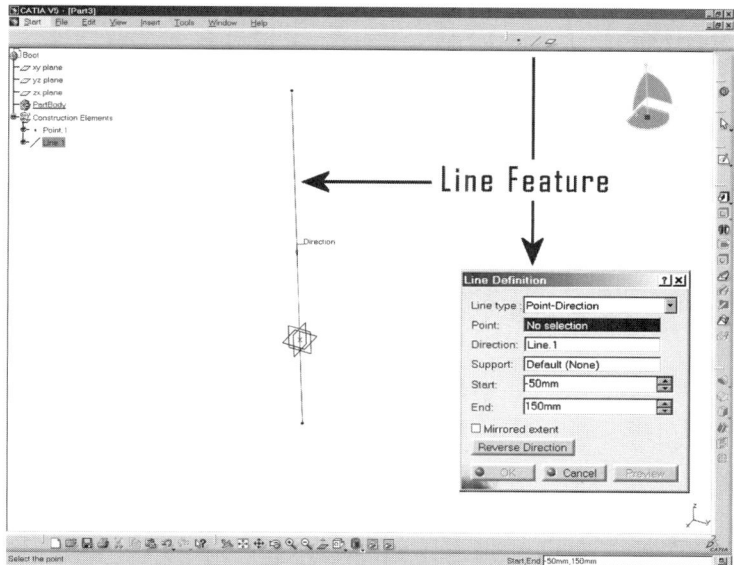

Fig. 5-27.
Boot Sleeve
body.

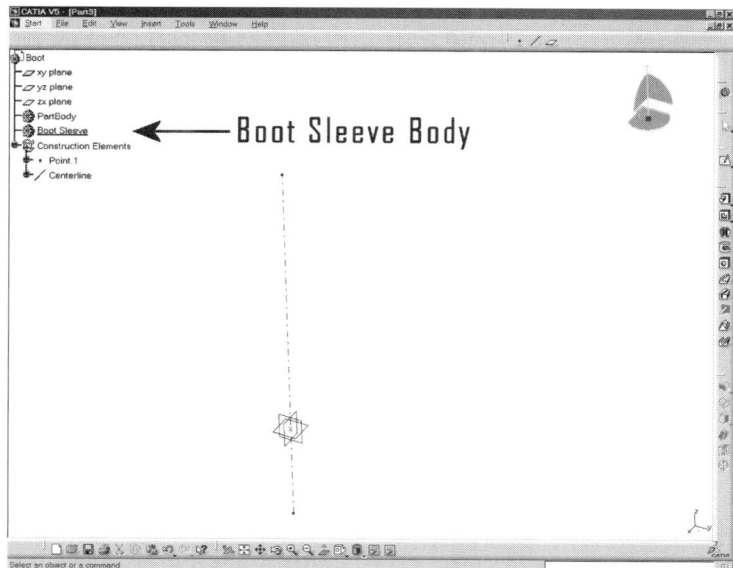

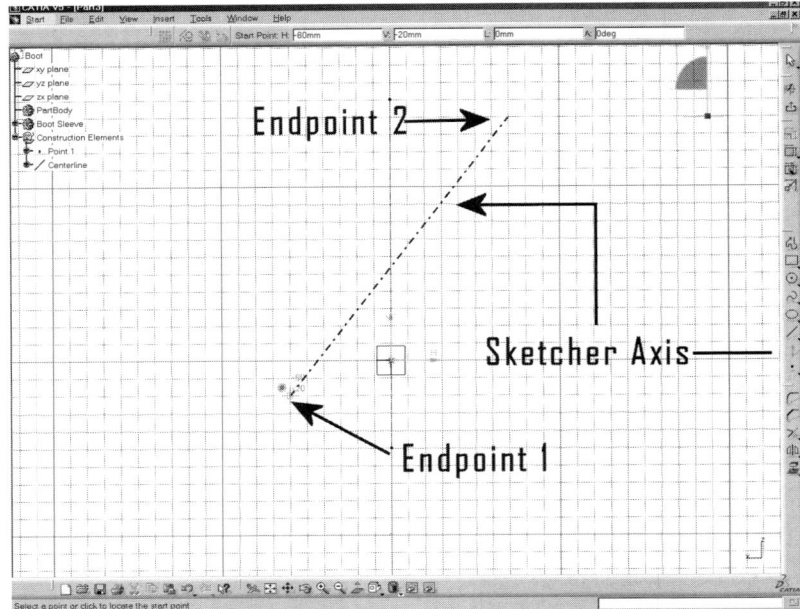

*Fig. 5-28.
Axis
endpoints.*

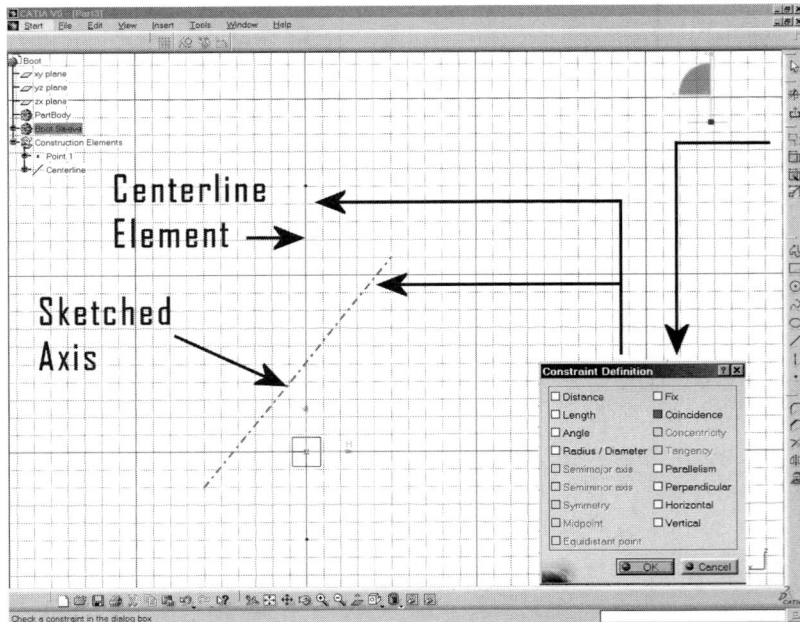

*Fig. 5-29.
Coincident
constraint.*

8 Select Constraints Defined in Dialog Box and create a coincident constraint between the sketched axis and the previously created centerline element, as shown in figure 5-29.

9 Click on the Construction Elements icon on the Sketch Tools toolbar. (Making this selection allows for elements to be created, viewed, and referenced in Sketcher mode, but not seen or accessed outside Sketcher mode.) Click on the Point icon and type in a value of *9.85* (mm) for H and *93.03* (mm) for V for the point shown in figure 5-28.

In the model, a blue dot-and-cross symbol will temporarily indicate the position of the point, as shown in figure 5-30.

Fig. 5-30.
Temporary
position of the
point indicated.

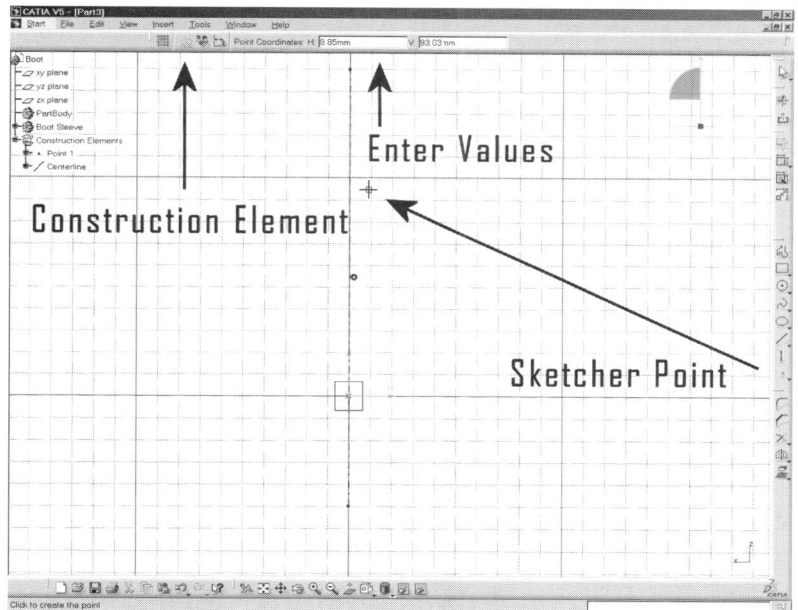

10 Create the remaining nine points by entering the values shown on the point chart, shown in figure 5-31.

11 Constrain the points by clicking on the Auto-Constrain icon and then selecting the XZ and YZ planes.

The resulting constrained points, shown in figure 5-32, need to have their associated locational values modified so that the dimension lines they represent display more clearly on screen.

12 Modify the point (dimension) locations by selecting their associated dimension lines and dragging them to create a more organized display, as shown in figure 5-33.

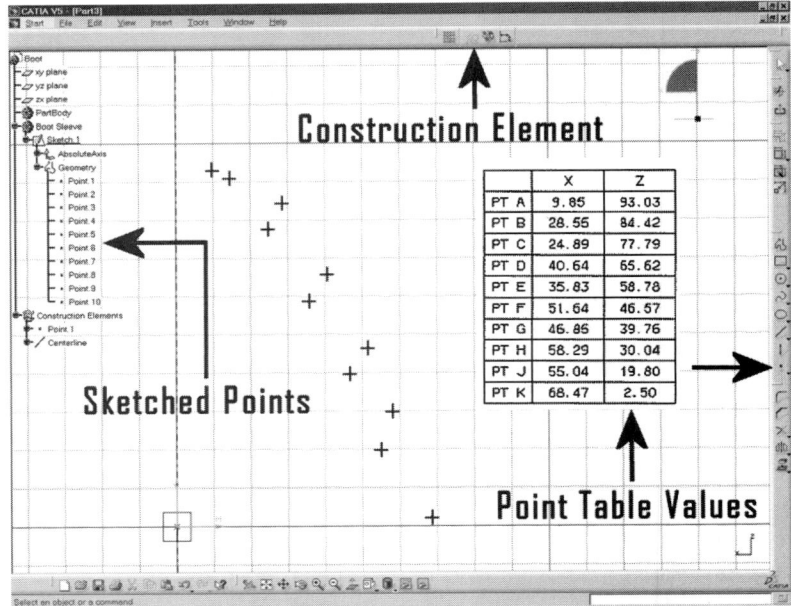

Fig. 5-31.
Point chart.

	X	Z
PT A	9.85	93.03
PT B	28.55	84.42
PT C	24.89	77.79
PT D	40.64	65.62
PT E	35.83	58.78
PT F	51.64	46.57
PT G	46.85	39.76
PT H	58.29	30.04
PT J	55.04	19.80
PT K	68.47	2.50

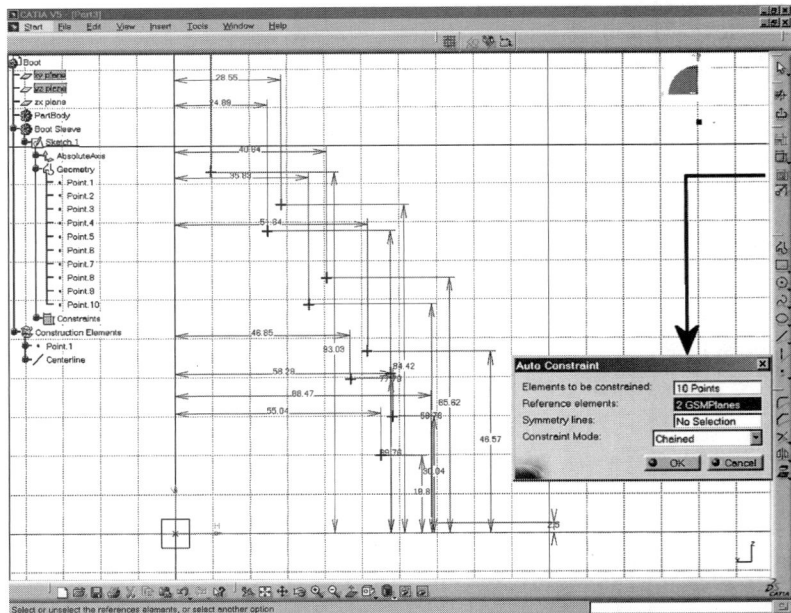

Fig. 5-32.
Constrained
points.

13 Click on the Profile icon and "connect the dots" (points) by creating a top line and a bottom line (respectively in each case from one endpoint to the other coincident to the axis line).

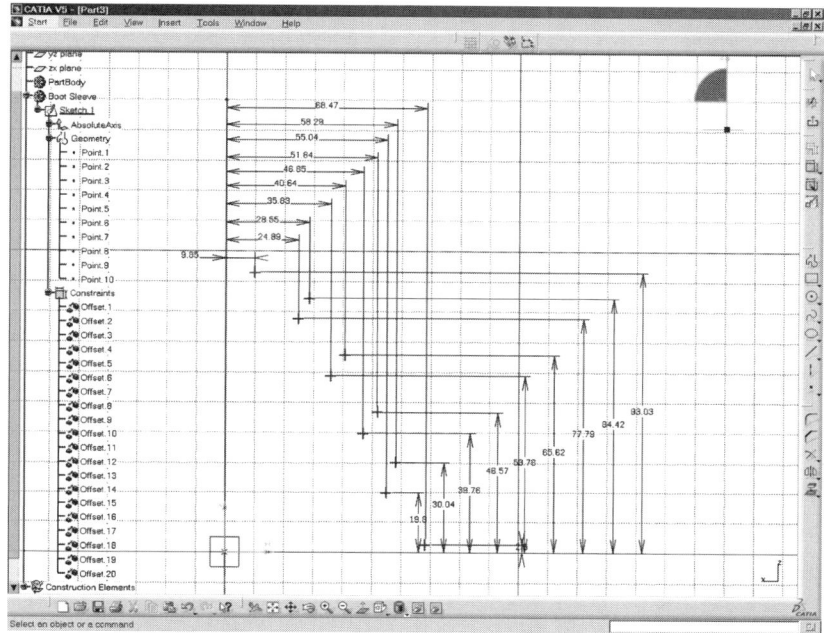

Fig. 5-33.
Relocated
dimension
lines and
points.

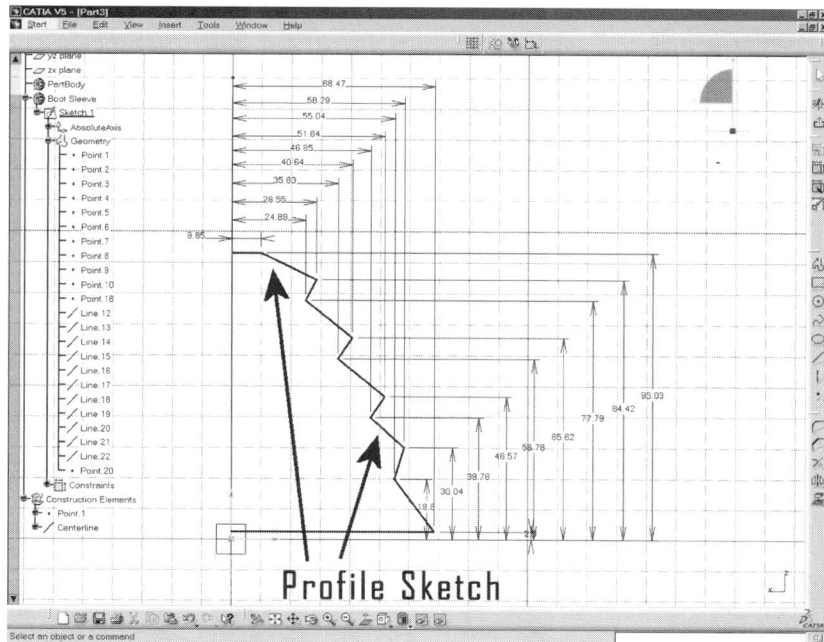

Fig. 5-34.
Sketched
profile closed
to the axis.

(The result is a profile closed to the axis, as shown in figure 5-34.) Exit the Sketcher.

14 Create a new feature by clicking on the Shaft icon and then selecting the previously sketched profile. Select the centerline element as the axis and rotate the shaft 360 degrees. The shaft feature is shown in figure 5-35.

15 Create an inside edge fillet, shown in figure 5-36, by clicking on the Edge Fillet icon and then on the five inner edges of the shaft. Enter a radius value of 2 mm. Click on OK.

16 Create an outside edge fillet, shown in figure 5-37, by clicking on the Edge Fillet icon and then on the five inner edges of the shaft. Enter a radius value of 3.3 mm. Click on OK to complete the shaft features.

17 Create a simple formula to maintain the associativity between elements.

A material thickness offset formula will create a link (relationship) between the boot feature and the base feature. The formula will be stored in the specification tree and referenced throughout the part creation process.

Fig. 5-35.
Shaft feature.

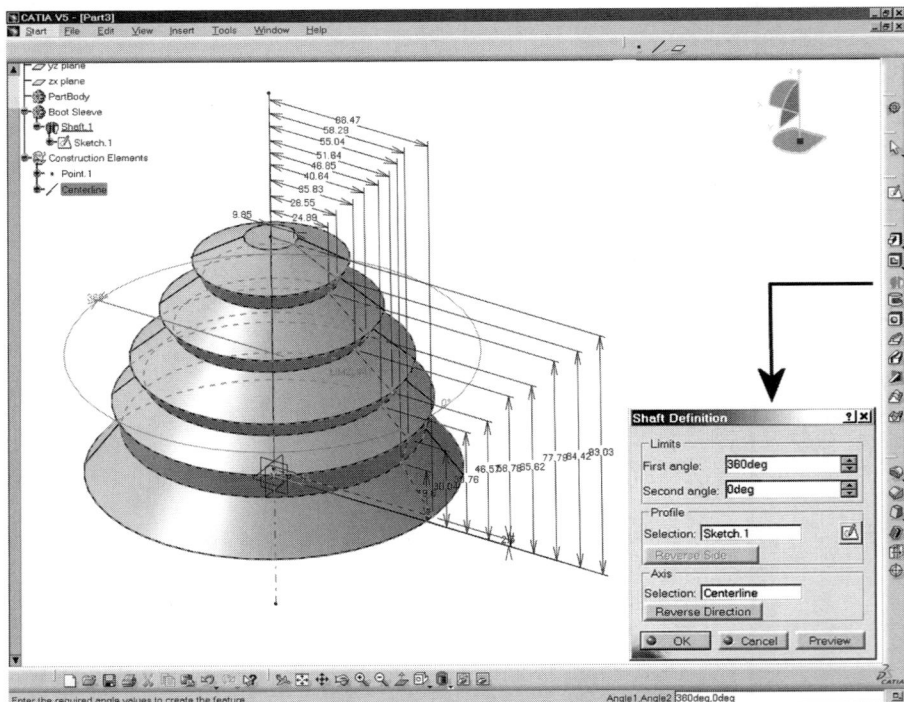

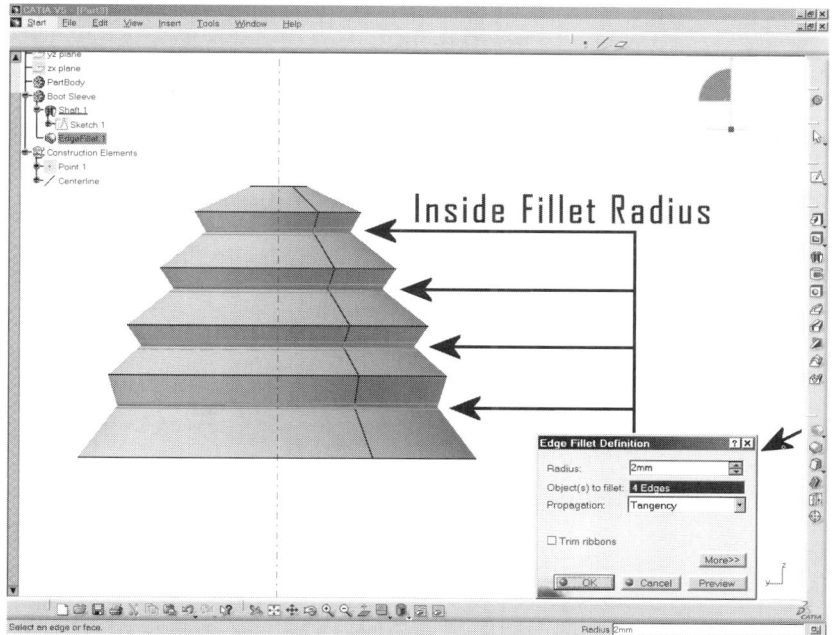

Fig. 5-36.
Inside edge
fillet.

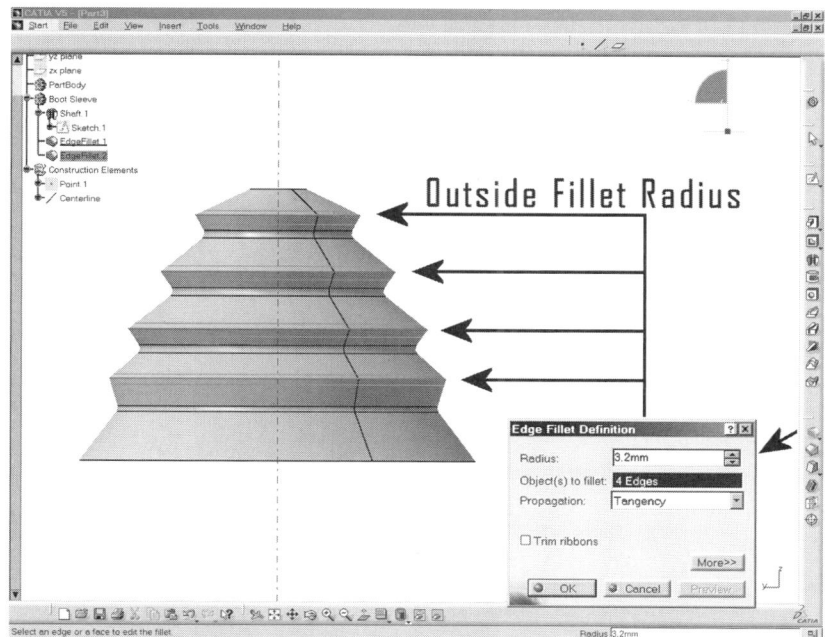

Fig. 5-37.
Outside
edge fillet.

18 Click on the Formula icon (fx). In the Formulas pop-up window, establish a length parameter in the Single Value option of the New Parameter of Type drop-down box. Click on the

Add Formula button to assign the new parameter and initiate the Formula editor. The length parameter is shown in figure 5-38.

19 In the Formula editor, shown in figure 5-39, name of the current parameter *Material Thickness* and enter a value of 2.5 mm. Click on OK to complete the formula.

Fig. 5-38.
Length
parameter.

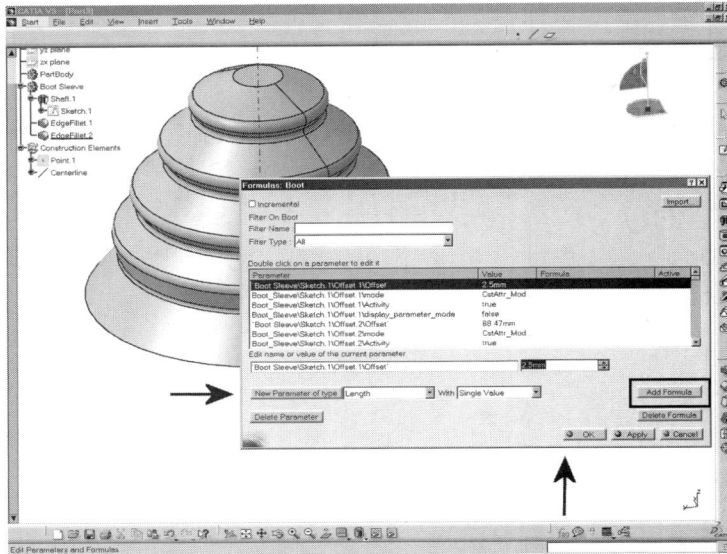

Enter New Value

Fig. 5-39.
Material thickness
parameter
established in the
Formula editor.

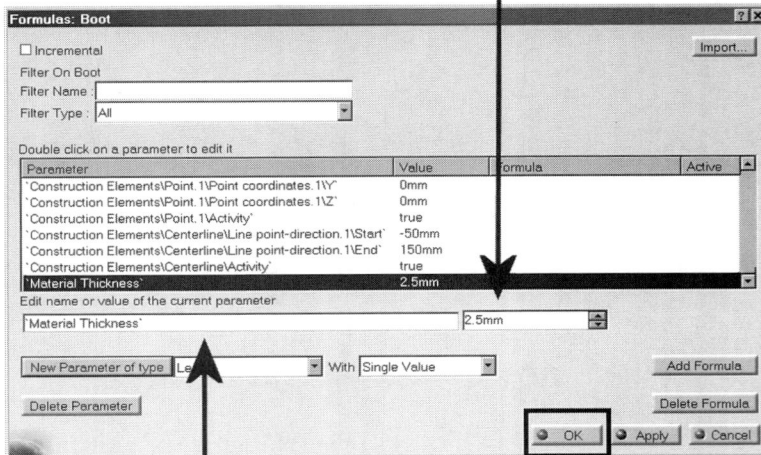

Enter New Name

The formula resides in the *Parameters* folder of the formula tree, as shown in figure 5-40. It can be accessed throughout the creation process and used unlimited times. When the value is changed, all associated elements will update to reflect the change. These associative links can be terminated, and unassociated values added if required.

Fig. 5-40.
Formula
tree.

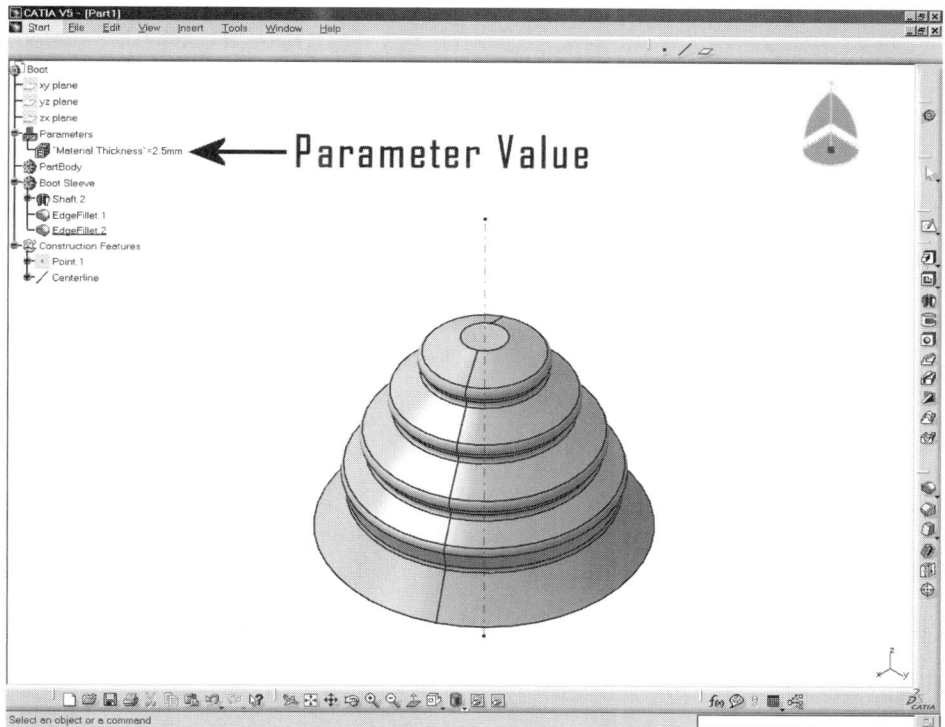

20 Click on the Shell icon and type an equals sign (=) in the Default Material Thickness area and select *Material Thickness Formula* from the specification tree. The formula will be selected as the input for the shell operation default thickness. The shell feature is shown in figure 5-41.

NOTE: *The formula value is now entered into the Default Material Thickness area. The value is de-highlighted to indicate that it is linked to an outside input. The Relations folder in the formula tree stores the links to the associated elements.*

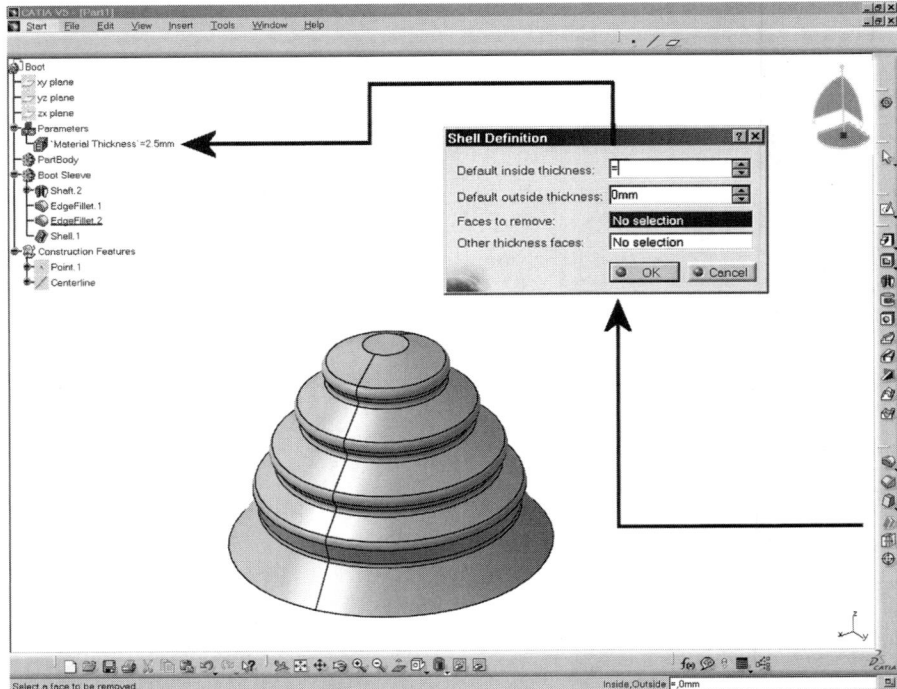

Fig. 5-41.
Shell feature.

21 Select the bottom face as the face to be removed during the shell operation, as shown in figure 5-42. Click on OK to complete the operation.

NOTE: *All faces not selected will be offset the given thickness. The shell operator, like the thickness operator, will produce a "normal" offset from the remaining faces. The shell feature is complete, with the bottom face removed and the material thickness parameter applied, as shown in figure 5-43.*

22 Insert a new part body and name it *Shaft Opening*. This body will be defined and removed to create the top opening of the boot feature. The opening is a separate (hole) feature associated with another part that if modified will affect only the opening.

23 Click on the Hole icon and select the circular top face, to which the hole will be applied. (The hole will automatically center itself on the selected face.) Select the Blind Hole option and enter the values shown in figure 5-44. Click on OK.

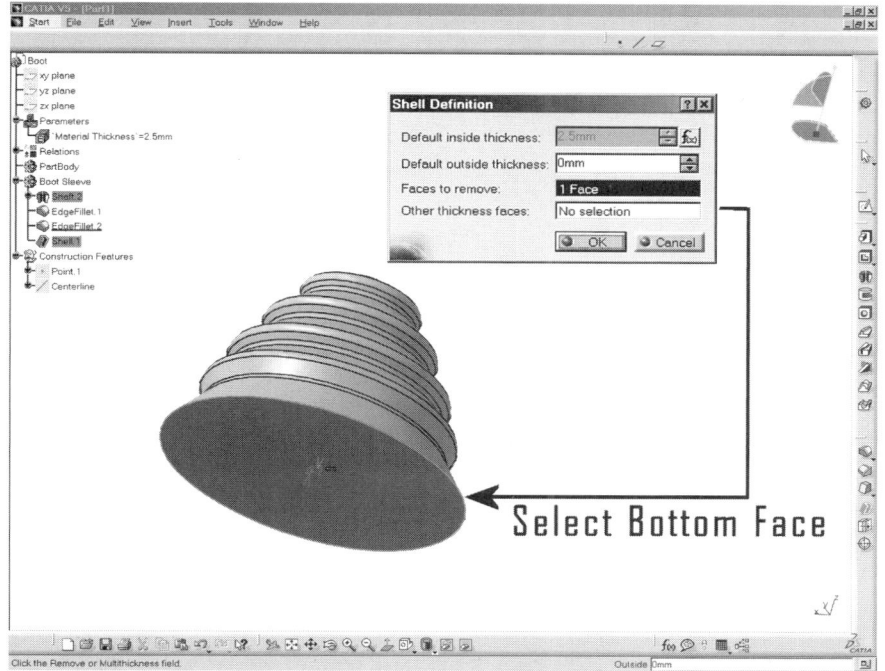

Fig. 5-42.
Normal offset
of the shell
feature.

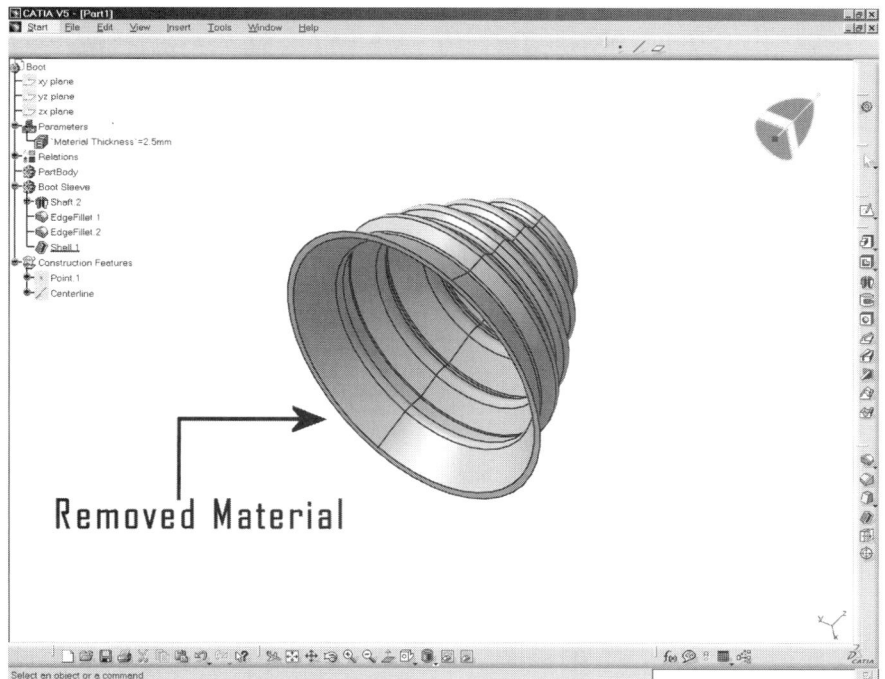

Fig. 5-43.
Completed
shell feature.

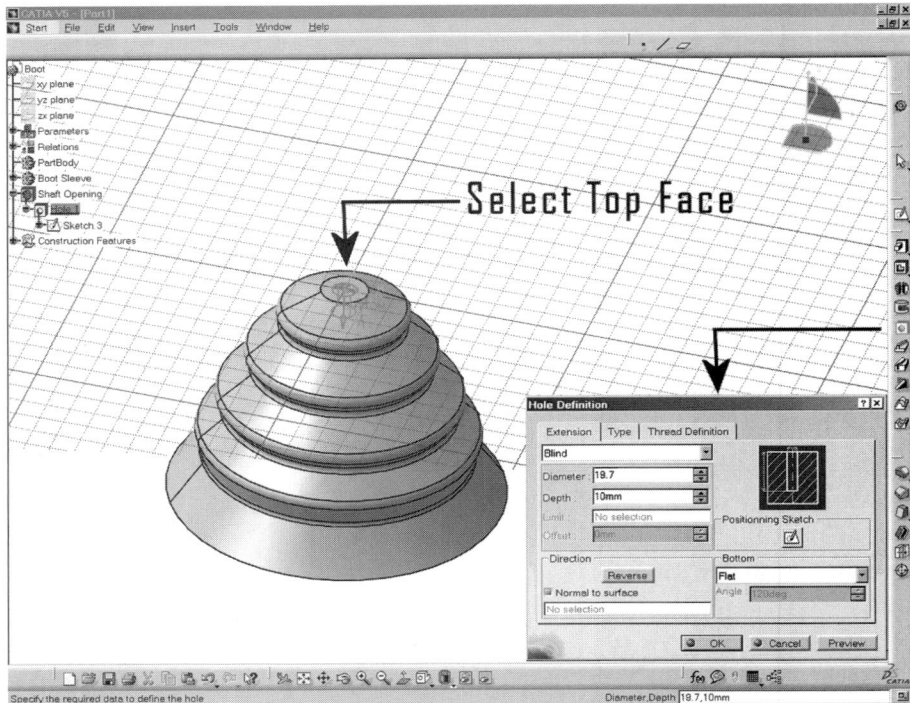

Fig. 5-44. Values for the settings under the Blind Hold option.

NOTE: *The body symbol for this hole feature is a negative due to the fact it is the first feature in the body. The next step is to assemble the two bodies to obtain a finished result.*

24 Insert a new part body and name it *Boot Sleeve Complete.*

25 Begin a new Boolean operation by selecting the *Boot sleeve* body and, using MB3, selecting Object/Assemble, as shown in figure 5-45. Select Boot Sleeve in the Assemble box and Boot Sleeve complete in the After box. Click on OK to complete the operation.

26 Initiate a second Boolean operation by selecting the *Shaft Opening* body and, using MB3, selecting Object/Assemble. Select *Shaft Opening* and *Assemble.1*, as shown in figure 5-46. Click on OK to complete the operation.

NOTE: *Assembly is a "neutral" operation. The Boolean operation will be computed per the positive and negative signs associated with the bodies. In this case, the* Boot Sleeve Complete *and* Assemble.1 *part bodies*

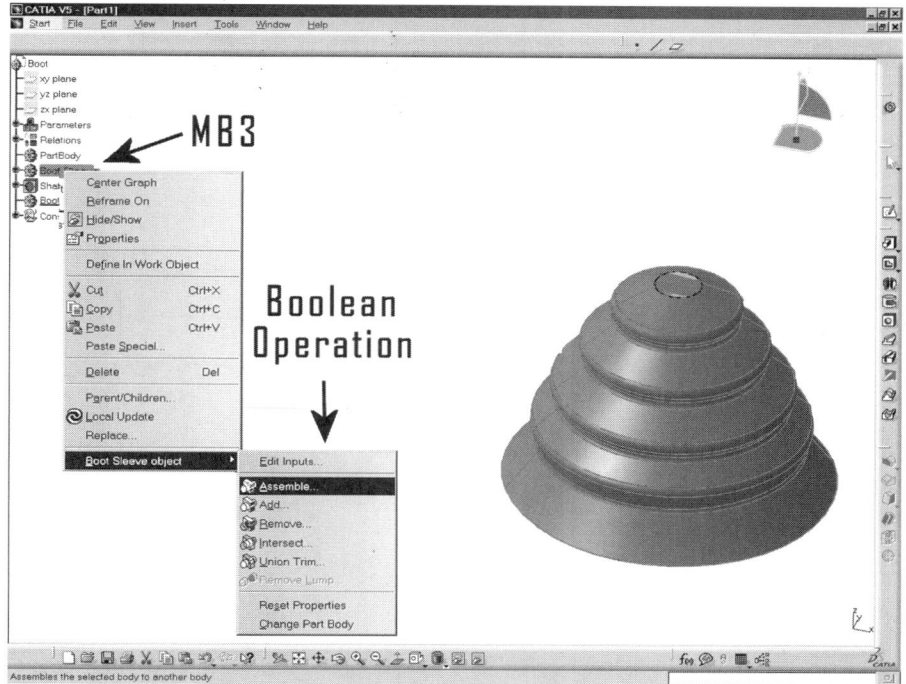

Fig. 5-45.
Boot Sleeve
assembly
operations.

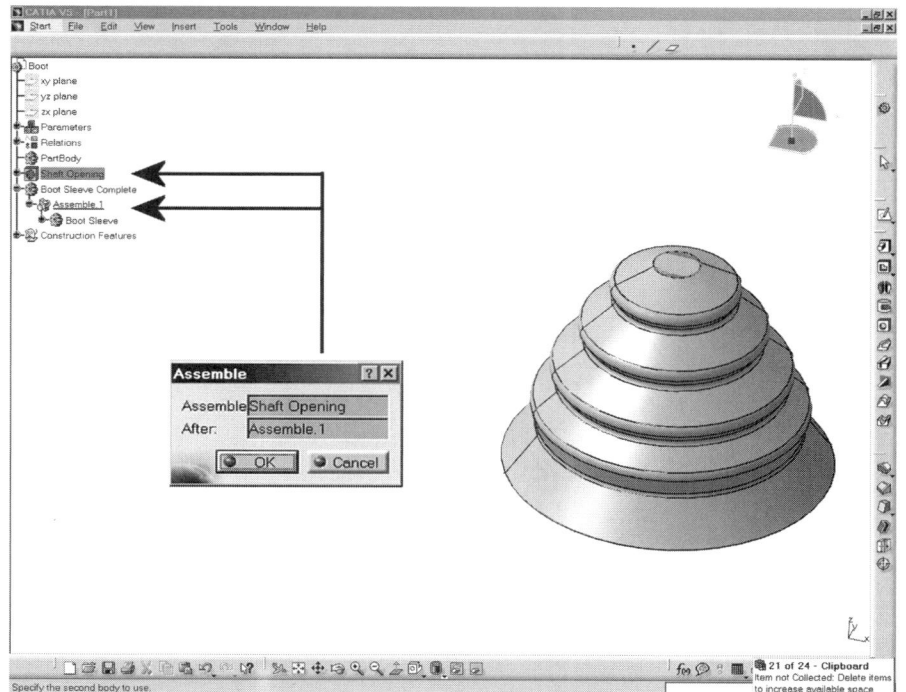

Fig. 5-46.
Boolean
assembly
operation.

are positive, and the Shaft Opening *body is negative. The hole will be removed as a result.*

Note that the resulting Boolean operation has taken place inside a third, empty body. Each body containing geometry maintains its identity within the tree and can be modified separately, as indicated in figure 5-47.

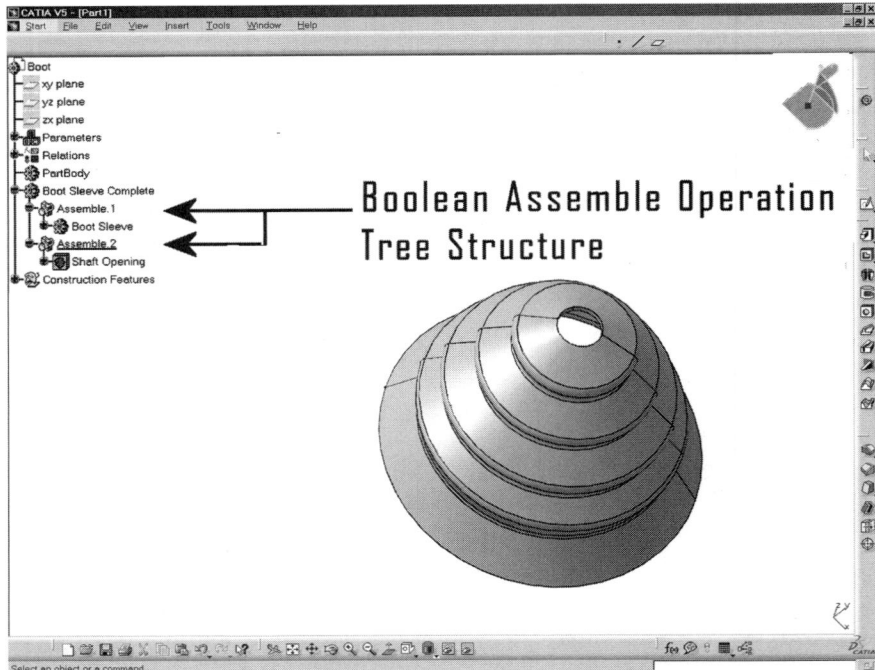

Fig. 5-47. Tree branches (body organization) indicating independence of bodies containing geometry.

27 Click on the Plane icon on the Reference Elements toolbar and select the XY plane. Enter *2.5* (mm) for the offset and move the plane away from the part to make selection easier later on. Click on OK to create the plane, shown in figure 5-48. *Plane.1* is created and is located in the specification tree under *Construction Elements.*

28 Create a new sketched profile by selecting *Plane.1* and then the Sketcher icon.

This sketch, shown in figure 5-49, will anchor the centers of the holes to be created on the base.

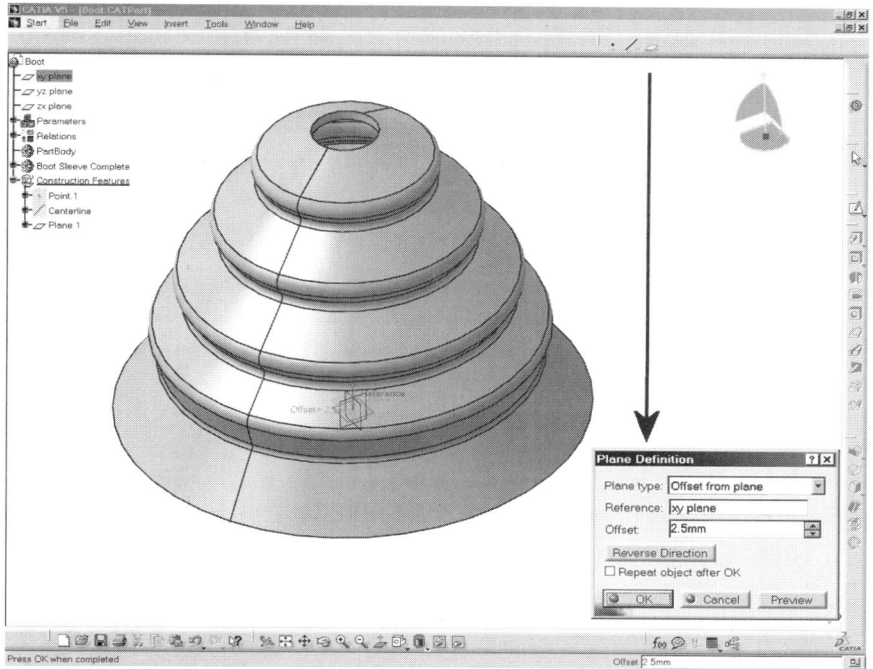

Fig. 5-48. Plane.1 in specification tree.

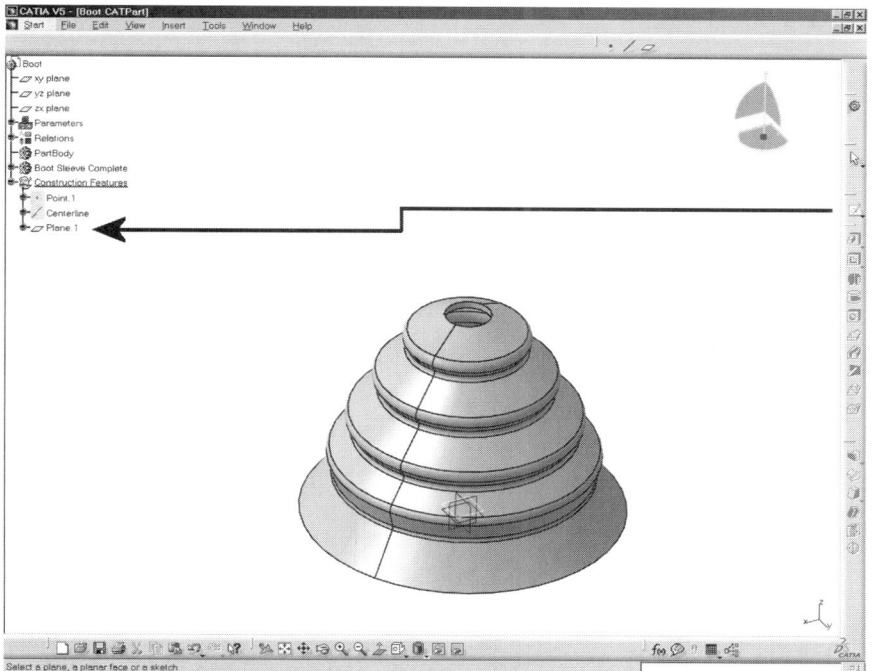

Fig. 5-49. Plane sketch.

29 Create four sketched points and constrain them to the planes, as shown in figure 5-50. Exit the Sketcher to complete the points.

30 Click on the Line icon. Select Normal to a Plane, and then select the point from the previous sketch. Enter *–50* (mm) for the start value and *75* (mm) for the end value, as shown in figure 5-51. Click on OK to create the line.

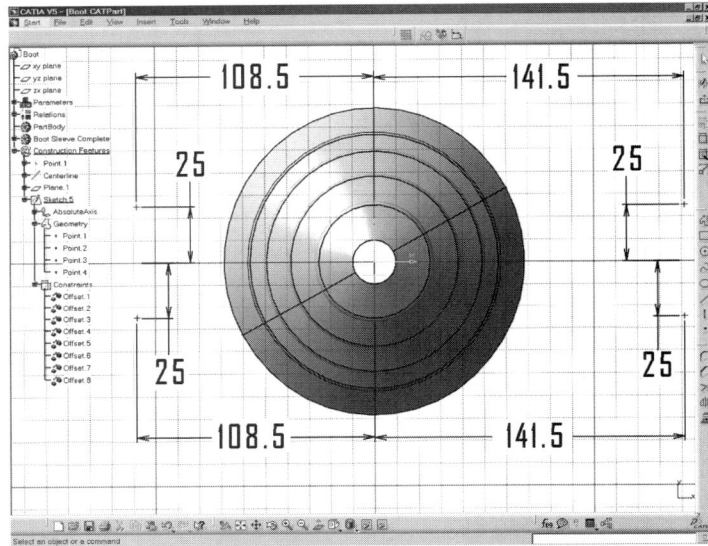

Fig. 5-50.
Points created and constrained.

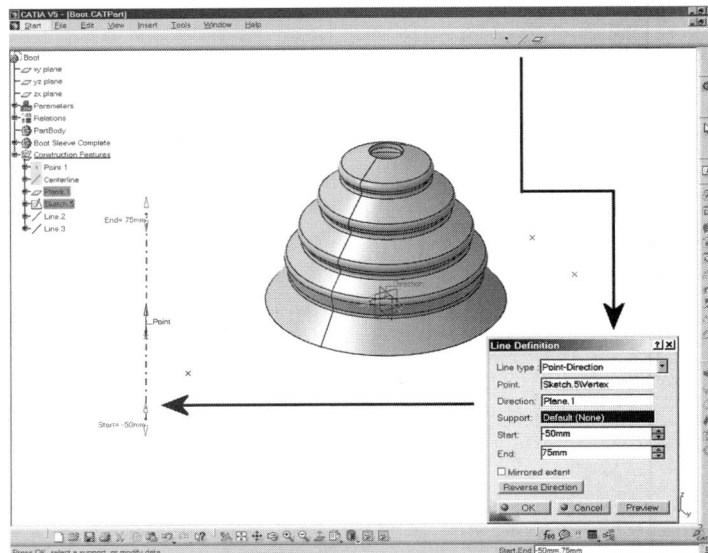

Fig. 5-51.
Line created normal to Plane.1.

31 Rename the line created in step 30 *Centerline A*. In the sketch, create and name three additional centerlines (B, C, and D), as shown in figure 5-52. Minimize the *Boot Sleeve Complete* to shorten the tree.

Fig. 5-52. Created and named centerlines.

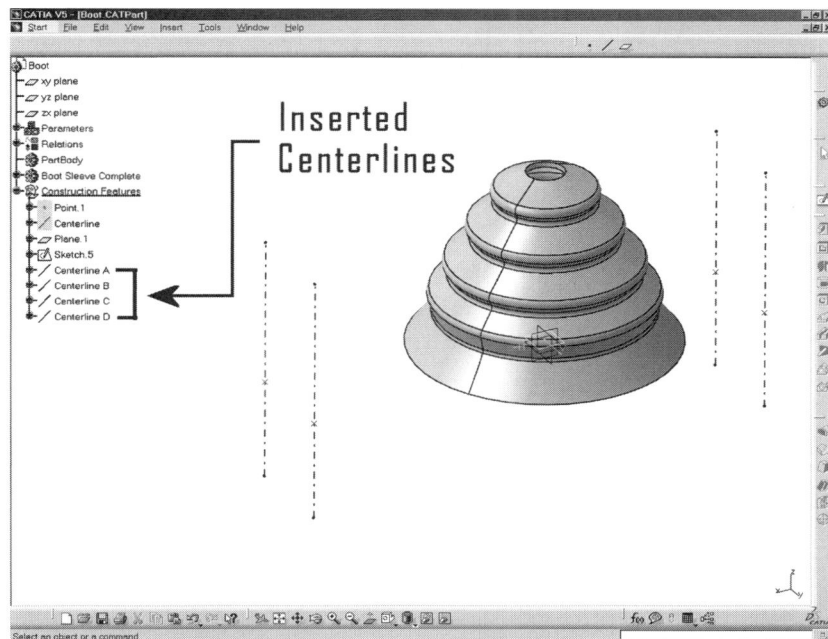

32 Insert a new part body and name it *Boot Base*. Create a new base profile sketch by selecting *Plane.1* and then clicking on the Sketcher icon. Constrain the base profile sketch as shown in figure 5-53. Exit the Sketcher.

33 Create a new base (pad) feature by clicking on the Pad icon and selecting the previously sketched profile. Reverse the direction of the offset. In the Length field, type an equals sign (=), and then select the Material Thickness parameter setting on the specification tree. Click on OK.

The pad feature is shown in figure 5-54.

NOTE: Formula.1 *and* Formula.2 *reference the same parameter. The shell and pad offset lengths are linked, so that a change in the material offset will affect both.*

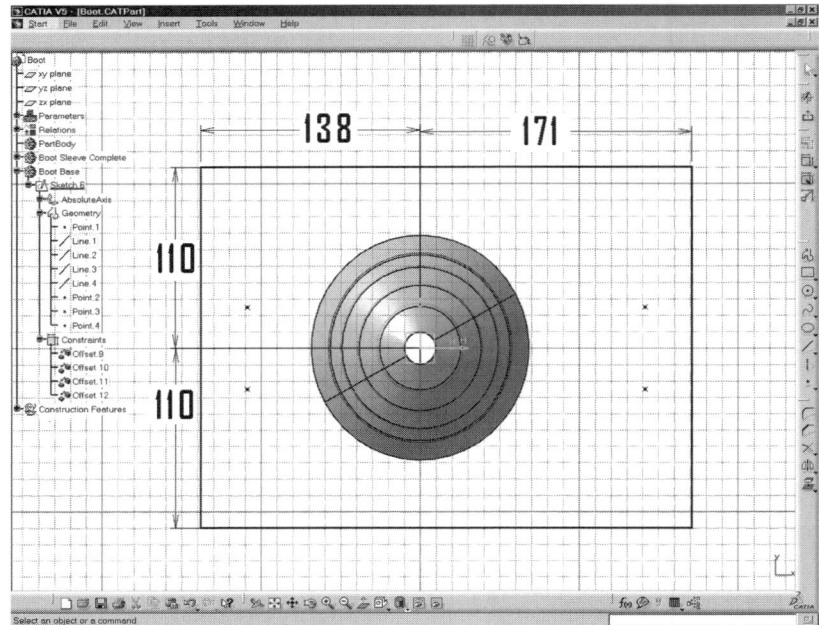

*Fig. 5-53.
Constrained
base profile
sketch.*

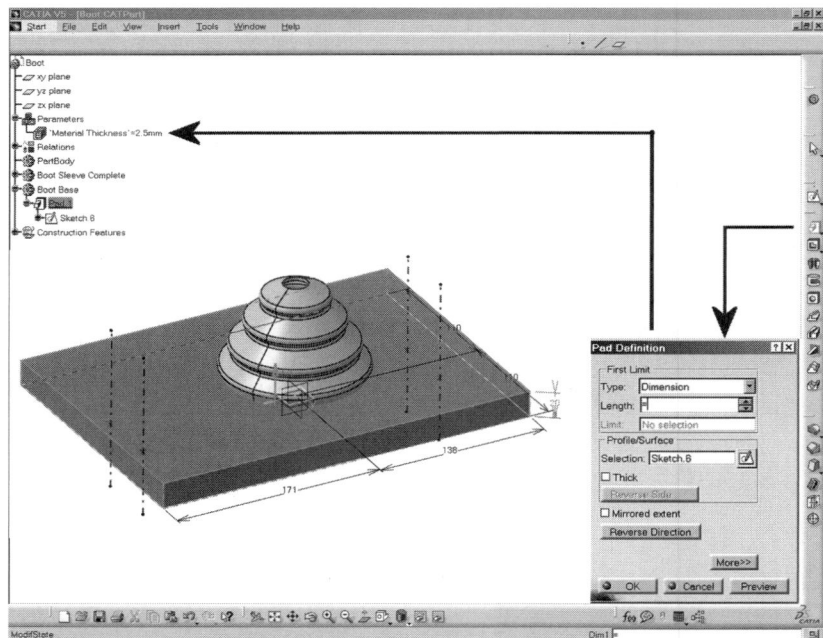

*Fig. 5-54.
Pad feature.*

34 Click on the Edge Fillet icon, and then select the outer four corners of the base. Enter *50* (mm) as the radius value. Click on OK to create the fillet, shown in figure 5-55.

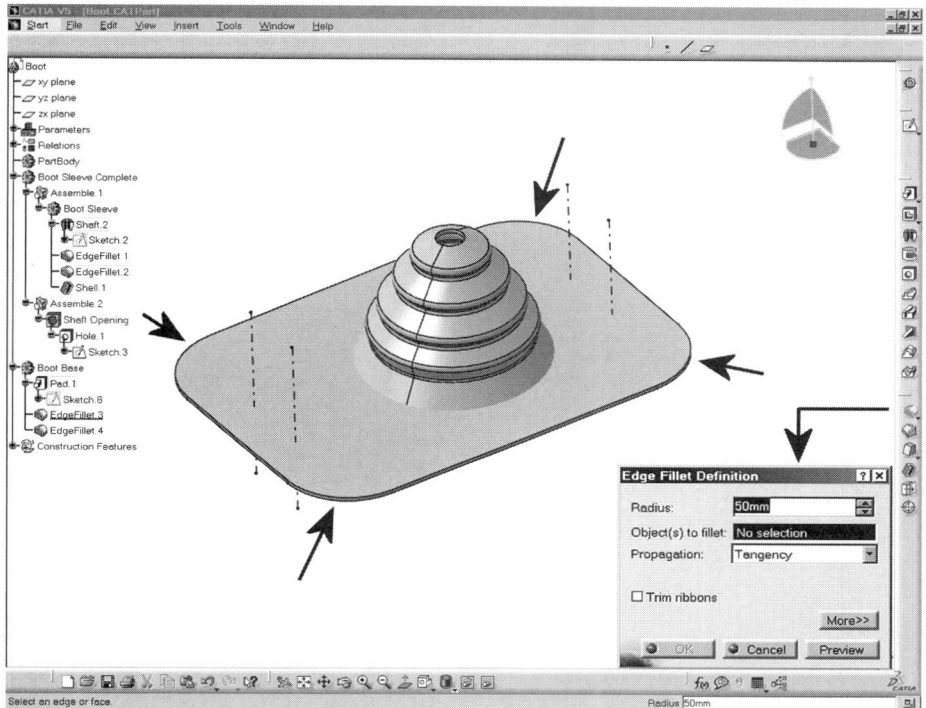

*Fig. 5-55.
Edge fillets.*

35 Insert a new part body and name it *Base Holes.*

36 Click on the Hole icon and create a hole feature by selecting *Plane.1* near centerline A. In the Hole Definition dialog box, create a 15-mm-diameter hole, as shown in figure 5-56. Click on the Sketcher icon, located in the Hole Definition dialog box, to enter the Sketcher workbench for the hole feature.

37 Create a coincident constraint by selecting the center of the hole and the axis of centerline A and then opening the constraints in the dialog box, as shown in figure 5-57. (Note that the hole automatically snaps to the centerline.) Click on OK to create the hole feature.

38 From the Transformation toolbar, select User Defined Pattern, and then select the previous *Hole* object as the object to be patterned. Select *Sketch.4* as the anchor. Click on OK to create the pattern, shown in figure 5-58.

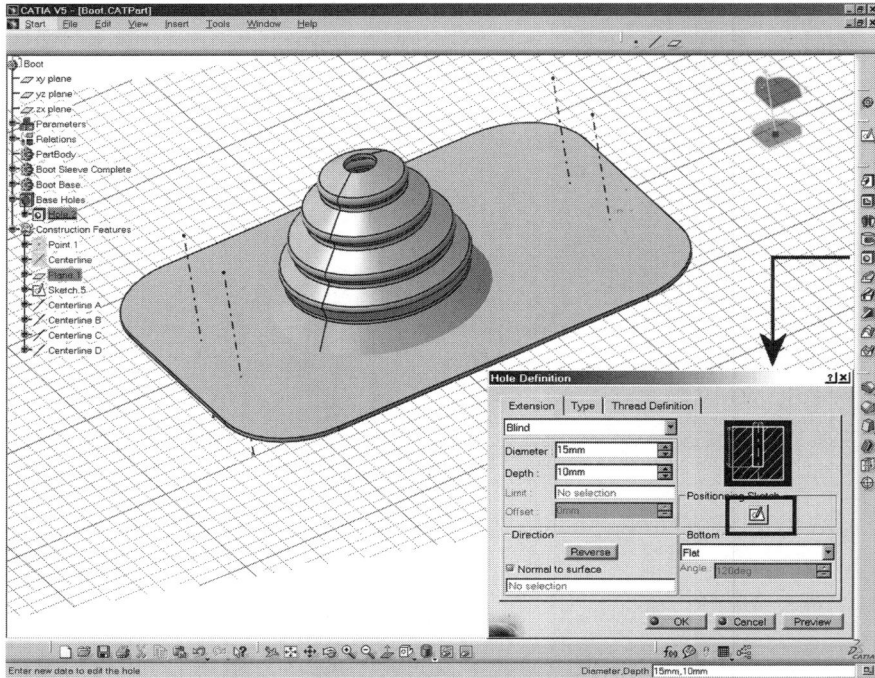

*Fig. 5-56.
Selections
made in the
Hole Definition
window.*

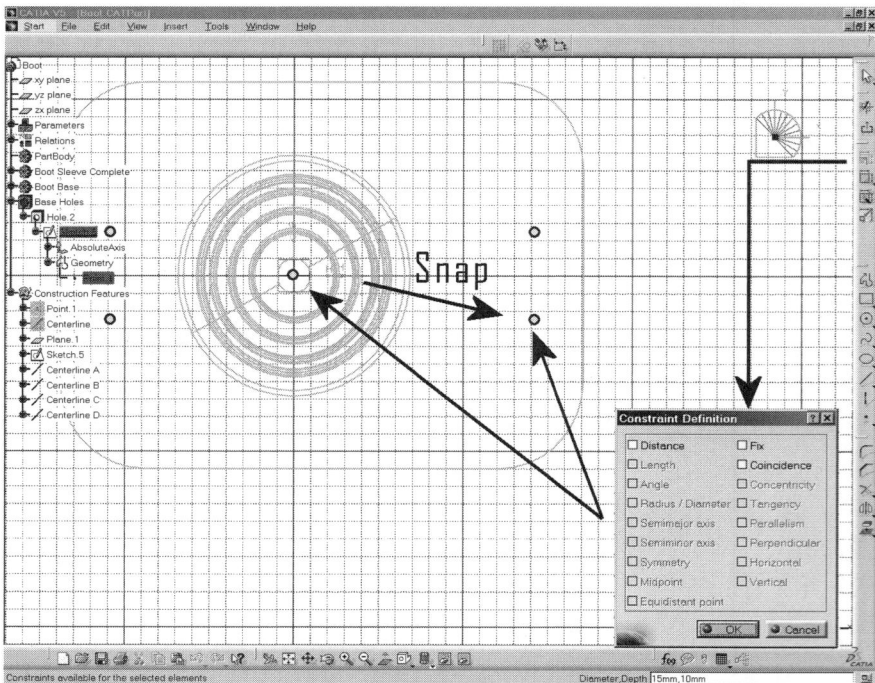

*Fig. 5-57.
Hole
Definition
window
settings.*

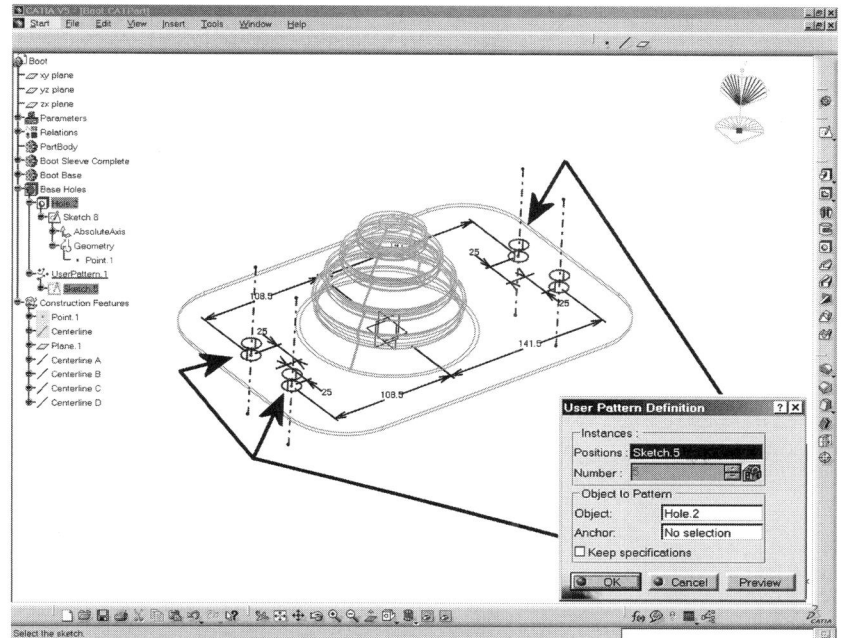

Fig. 5-58.
Holes located
at points
specified in
Sketch.4.

Note in figure 5-58 that the holes are located at the points specified in the sketch.

39 Insert a new part body and name it *Boot Base Complete*.

40 Using MB3, click on the *Base* part body, and then select Object/Assemble in the pull-down window. Select *Boot Base Complete*, and then click on OK to assemble. Using the same technique, assemble *Base Holes* and *Assemble.3*. The assembled bodies are shown in figure 5-59.

NOTE: *The main areas of the part have been defined. The associated, mutually exclusive Boolean interactions have been divided into two main bodies: Boot Base Complete and Boot Sleeve Complete. You now need to assemble the two main bodies, shown in figure 5-60, and compute a final body named Boot Complete.*

41 Insert a new part body and name it *Boot Complete*. This body will contain the information, arrived at by the final operations, that defines the part. The two main bodies will interact via a union/trim operation. For this to be successful, the boot feature must be extended through and beyond the base feature. To add thickness where it belongs in the tree, it is necessary to locate the applicable branch. To perform these operations, continue with the following steps.

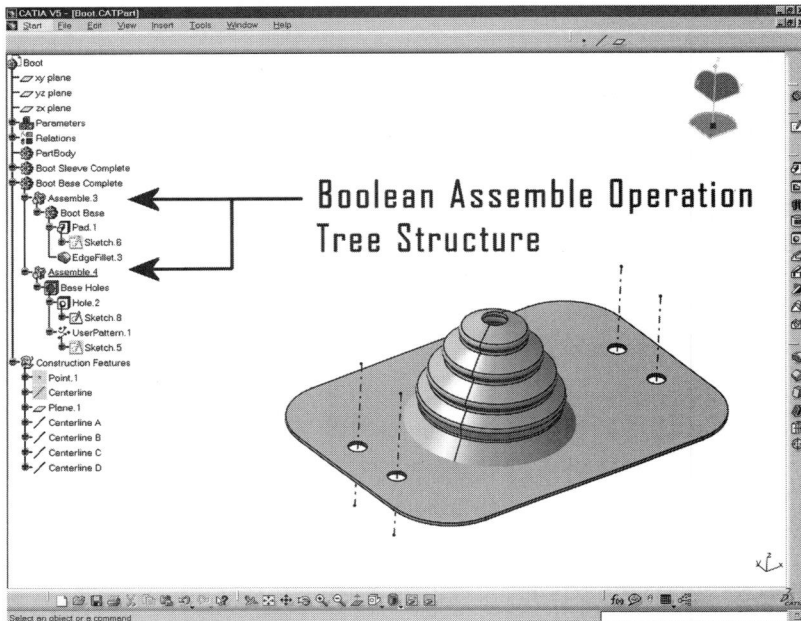

Fig. 5-59. Assembled bodies.

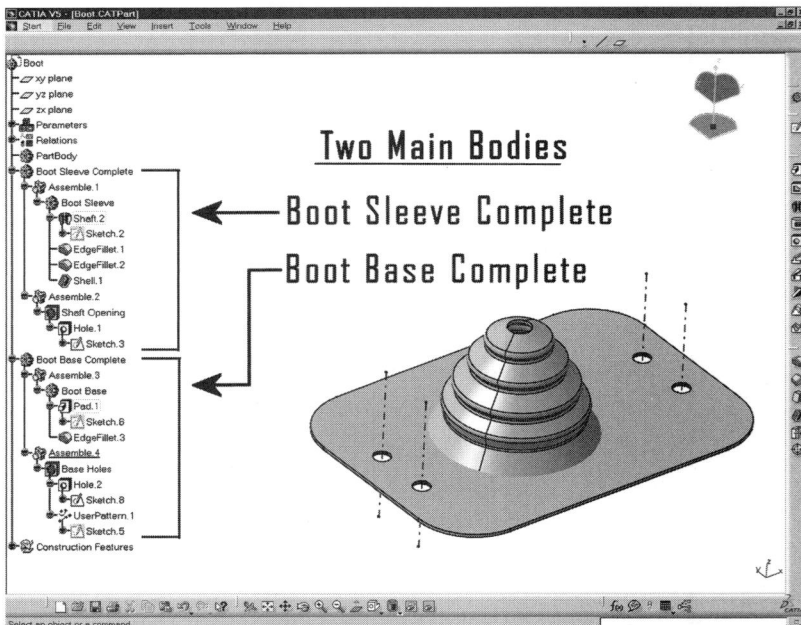

Fig. 5-60. Two main bodies to be assembled.

42 Change the active body to the *Boot Shape* body. Activate the Thickness icon and select the bottom edge of the feature to add thickness as shown in figure 5-61. Enter *10* (mm) for the Offset Thickness value and then click on OK to create the offset.

Fig. 5-61. Thickness feature.

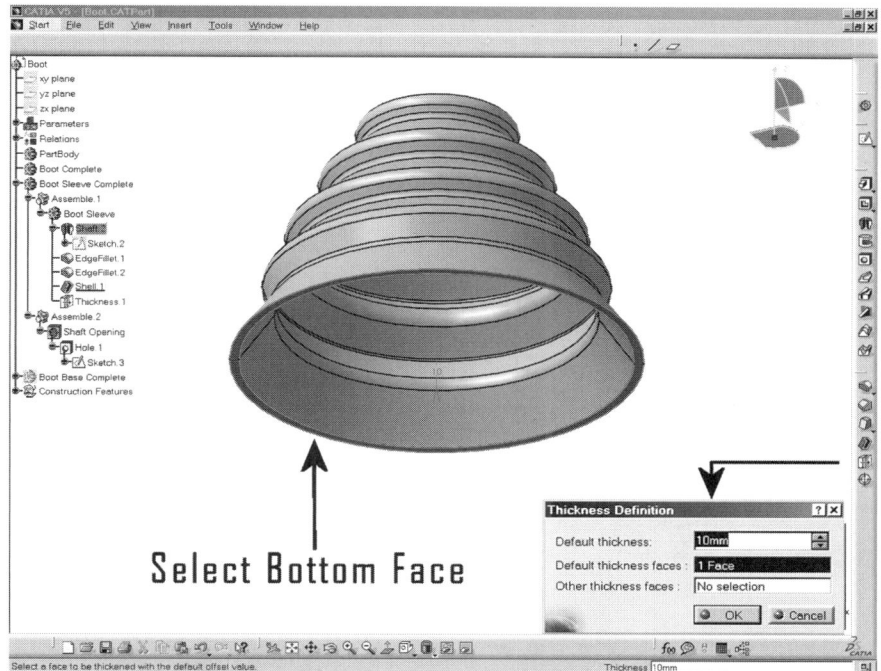

NOTE: *The thickness operation has extended the boot feature through the base feature. This is essential to the extension required for a successful union/trim operation.*

43 Perform the final assembly-and-trim operation by assembling the *Boot Sleeve and Boot Complete* bodies. Use the Union/Trim option when performing the Boolean operation.

44 Select the bottom thickness face and the round inside face of the boot feature. (The two faces selected, shown in figure 5-62, will turn magenta.) Click on OK to accept the operation, as shown in figure 5-63.

45 Create a fillet by selecting the resulting out intersection sharp edge and applying the Fillet icon. Enter a value of *2* (mm) for the feature, as shown in figure 5-64.

Fig. 5-62. Selected faces.

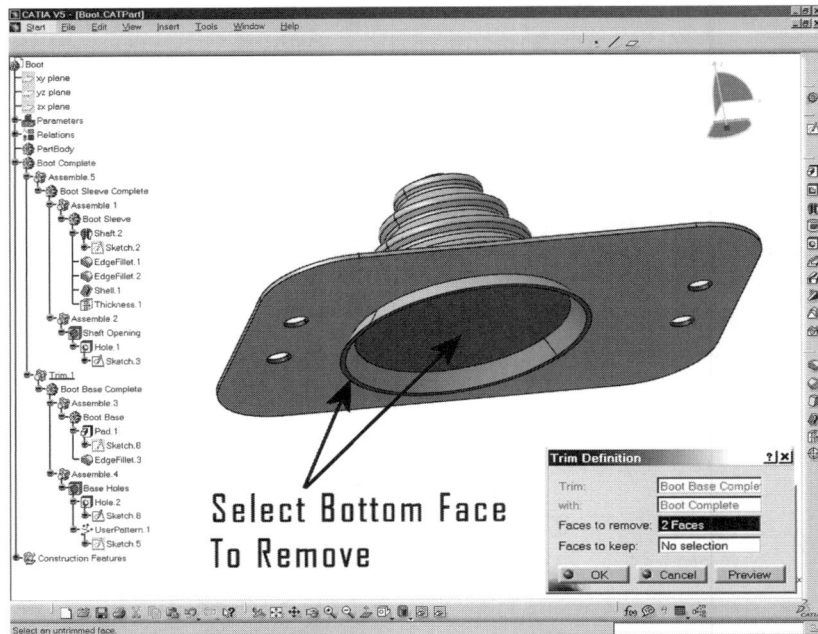

Fig. 5-63. Trimmed and removed faces.

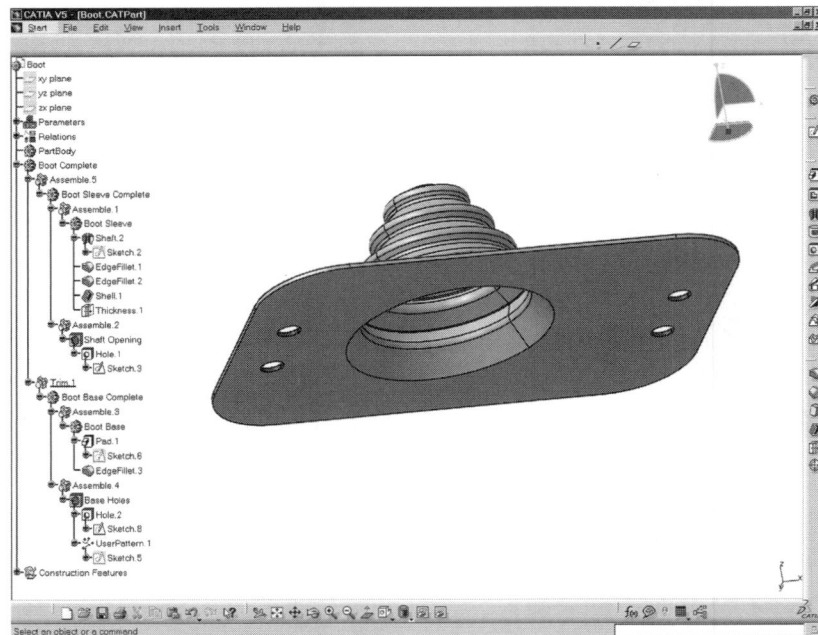

46 Create another fillet feature on the inside intersection edge at a value of 3.2 mm, as shown in figure 5-65.

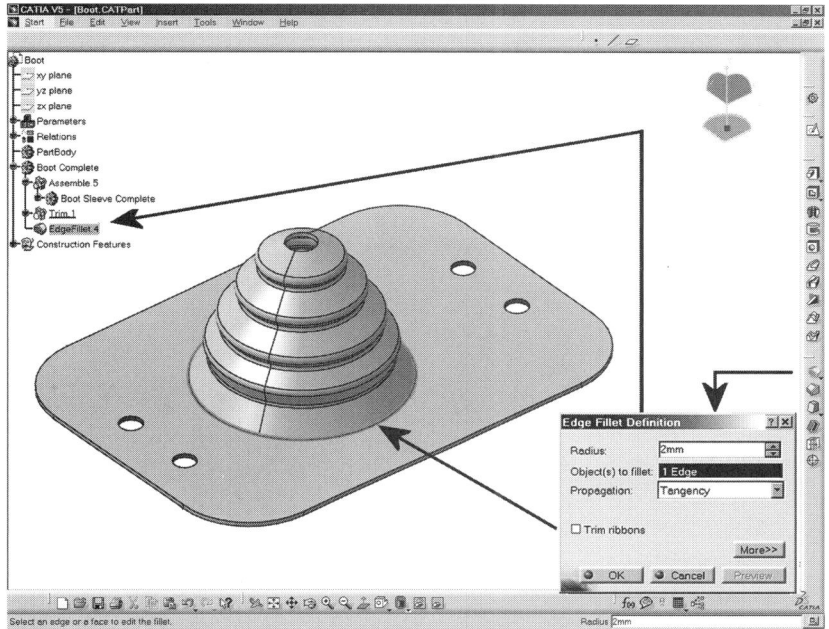

Fig. 5-64.
Intersection
fillet.

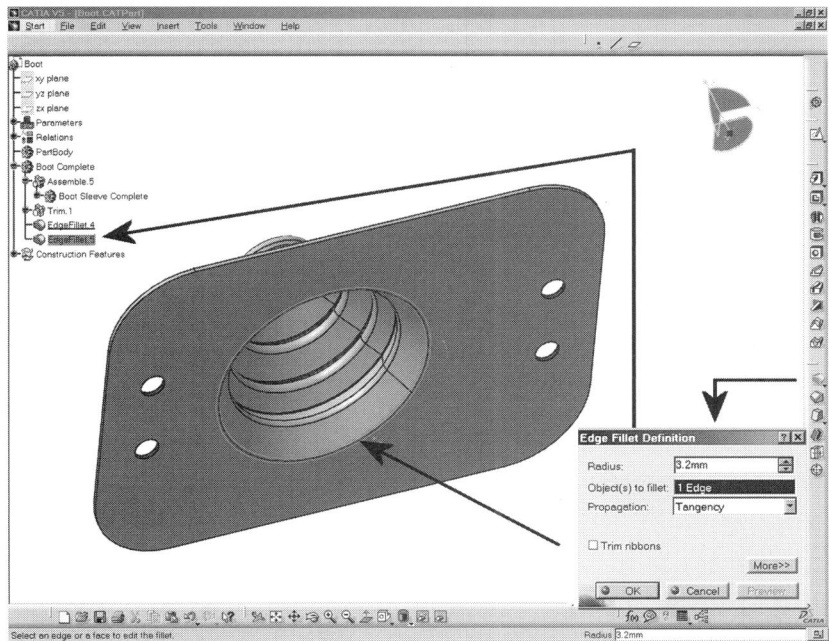

Fig. 5-65.
Fillets on
outer sharp
edge and
bottom
inside sharp
edge.

Both fillets, shown in figure 5-65, are dependent on the geometry
further up in the tree. They are by nature unstable, and may fail

to recompute upon modification. However, if they appear at the bottom of the tree they should be easy to recreate or rebuild if there is a failure.

47 Create the finished part by assembling the *Boot Complete Features* body and the default part body shown in figure 5-66.

Fig. 5-66.
Finished part.

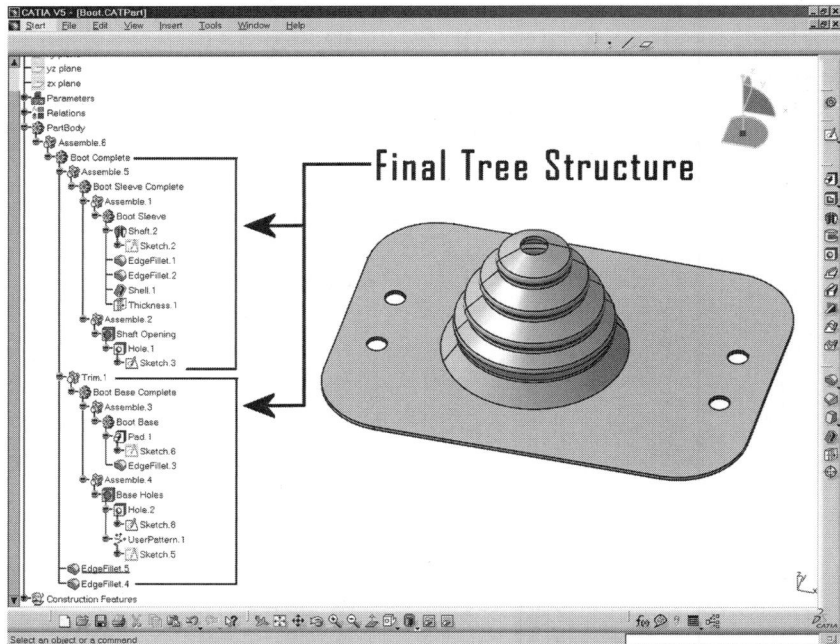

Part Design Exercise 2: Shifter Bracket

This exercise involves the part shown in figure 5-67. This exercise incorporates the following functions and applications.

- Customizing a toolbar by adding an icon
- Use of an open body sketch to help guide design work
- Extensive use of the Reference Elements toolbar
- Use of the Sketcher to define profiles
- Use and development of sketch-based and pocket features
- Use of the Shell feature to develop material thickness offset

- Use of the Hole feature to define the finished, machined part

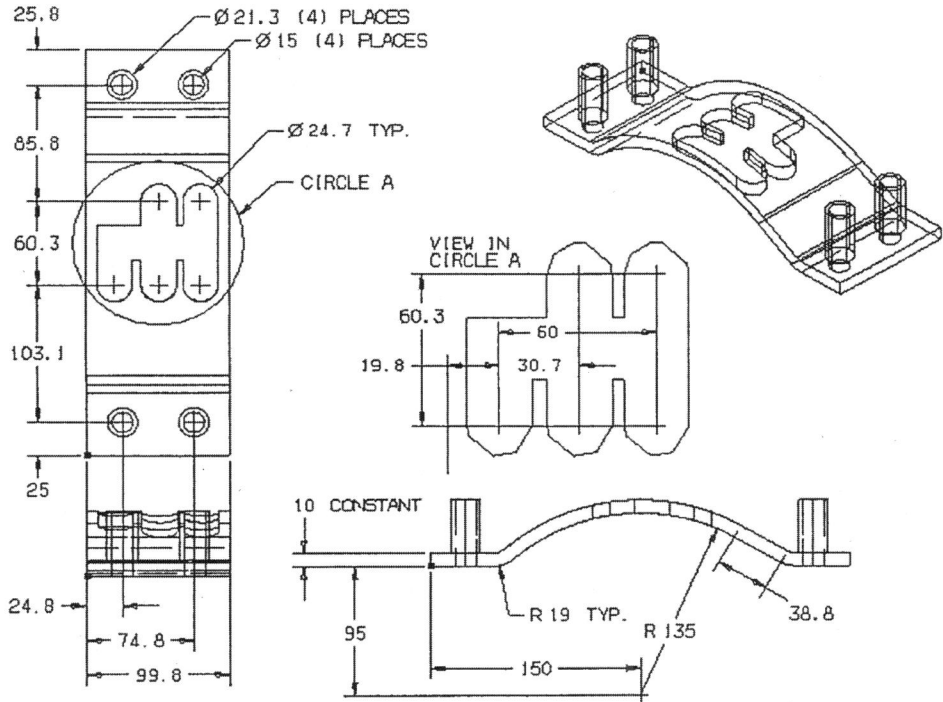

Fig. 5-67. Shifter bracket.

1   Create a new CATPart and rename the default part *Bracket Gear Selector.*

2   Insert a new open body within the model.

NOTE: *The Open Body icon is not by default made available in the Part Design workbench. To have the ability to insert a blank open body folder into the tree, the icon must be retrieved from a listing and inserted into a toolbar. Once the icon is located, it will be added to the Insert toolbar that currently holds the Insert Body option.*

3   Select the View Toolbar menu, located at the top of the window. Select the Customize option, shown in figure 5-68, to initiate the Customize menu.

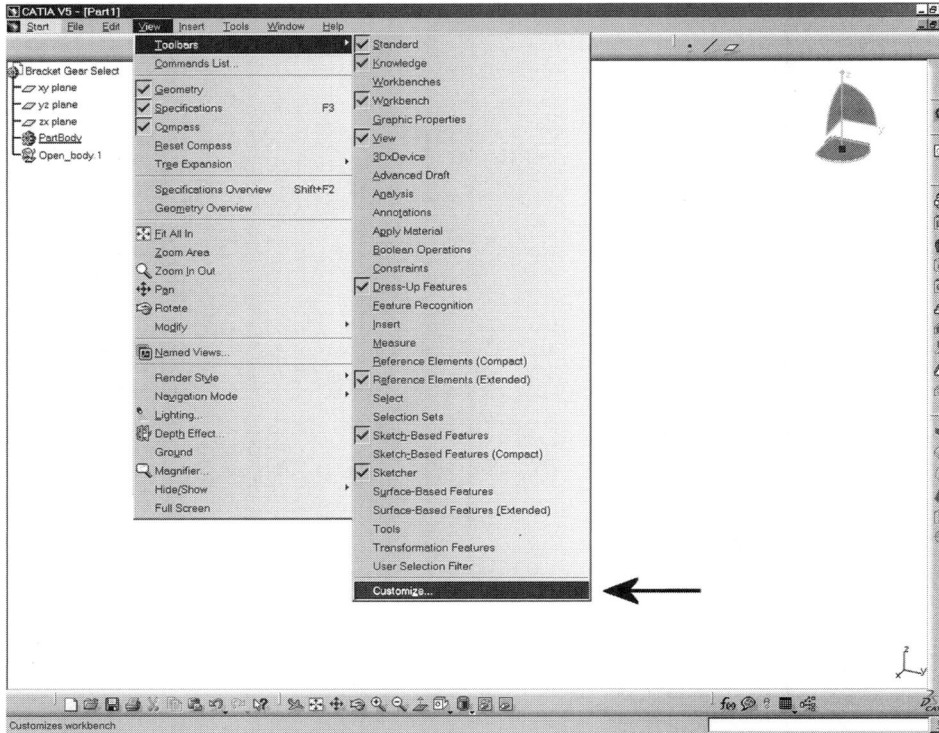

Fig. 5-68. Menu customization.

4 In the Customize menu, select the Toolbars tab, click on the Add Commands button, select Insert from the list, and click on the Open Body icon, shown in figure 5-69. Click on OK.

5 Click on the newly created Insert Open Body icon and create a new open body. Using MB3, access the Properties menu and name the new, open body *Construction Elements*. This open body, shown in figure 5-70, will contain the planes, points, and lines needed to anchor the basic geometry of the part.

6 Click on the Plane icon on the Reference Elements toolbar. Select Offset from Plane, and then select ZX as the reference plane. Set the offset value to 90 mm. Click on OK.

7 Create another offset plane feature 249.2 mm from the YZ plane.

8 Insert a new open body and name it *Locator Points*.

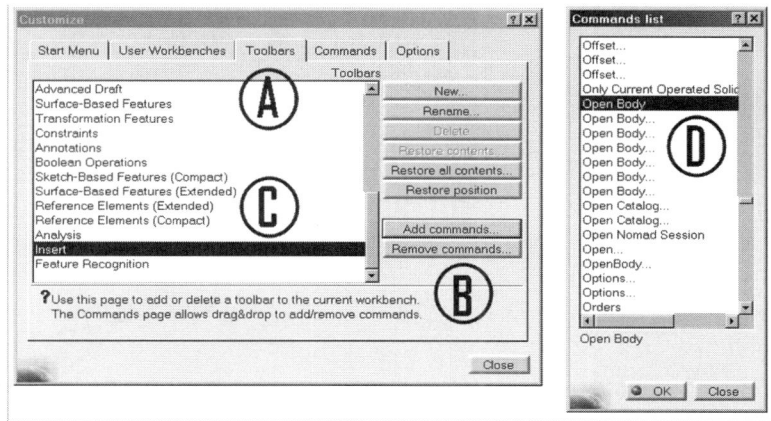

Fig. 5-69.
Open Body
icon.

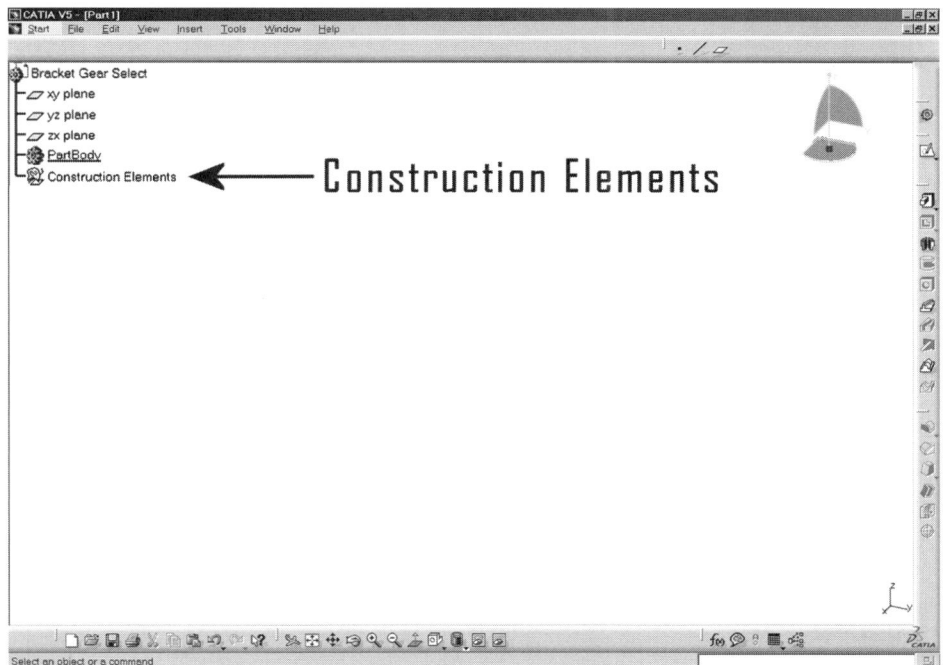

Fig. 5-70.
Construction
elements.

This open body will contain a sketch of points linked to the planes.

9 Create a new sketched profile by selecting the XY plane and clicking on the Sketcher icon.

10 Click on the Point icon on the Profile toolbar. Pick to indicate four points on the screen in a rectangular pattern, as shown in figure 5-71.

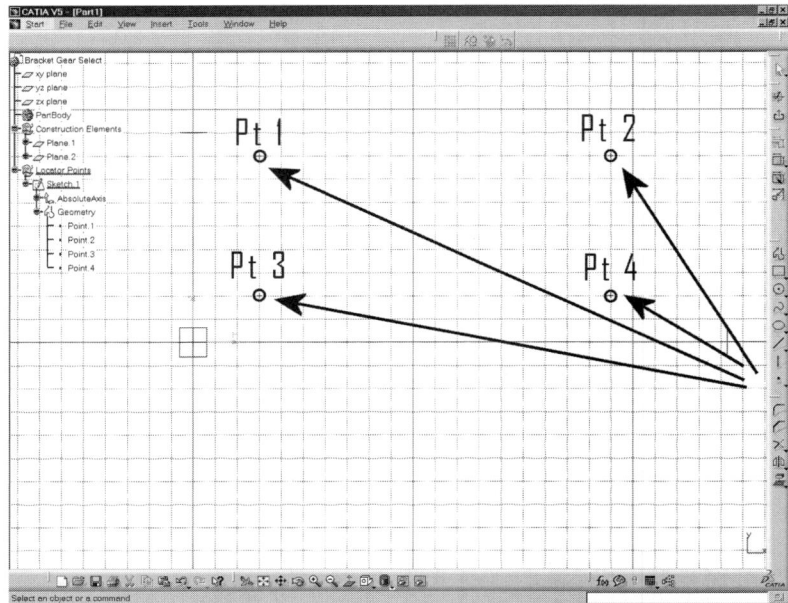

Fig. 5-71.
Four
element
points.

11 Constrain the top right-hand point to the corresponding top right-hand adjacent planes. Use the coincidence constraint located in the Constraints dialog box, as shown in figure 5-72.

12 Constrain the remaining three sketcher points to the corresponding planes' intersection, as shown in figure 5-73. The constraints are listed in the tree under *Sketch.1 / Constraints*. Exit the Sketcher.

13 Click on the Line icon on the Reference Elements toolbar. Select Point Direction, and select the XY plane as the direction. Enter a value of 75 (mm). Select the extent option Mirrored, to create a symmetrical line, as shown in figure 5-74. Click on OK.

14 Select the line feature, and using MB3 select the Properties option. Change the line's name to *Centerline A*, and the line type to the dash option 4.

15 Create three additional centerline features corresponding to the other three points. Name these centerlines B, C, and D, and change their line properties to dash option 4, as shown in figure 5-75.

16 Create a new sketched profile by selecting the XY plane and clicking on the Sketcher icon.

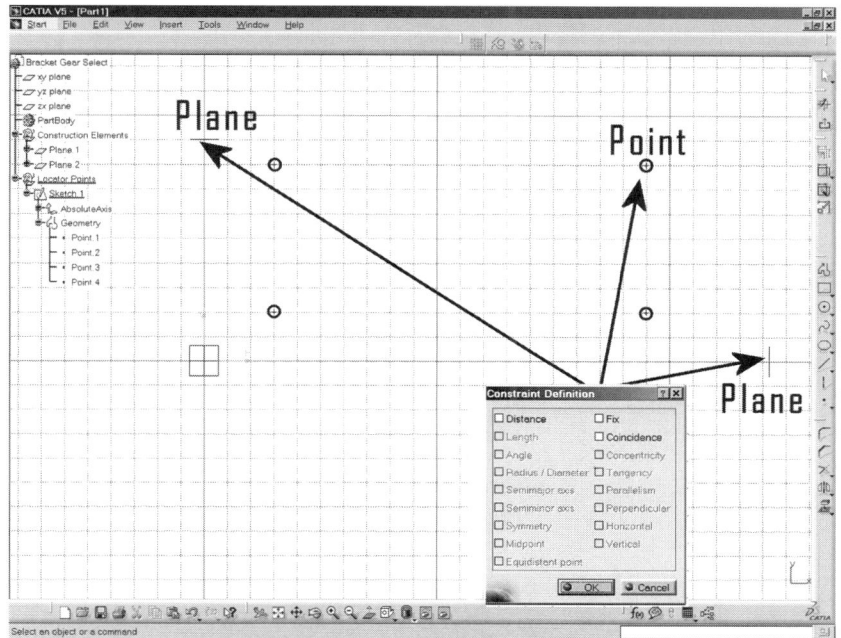

Fig. 5-72.
Point-to-plane
constraint.

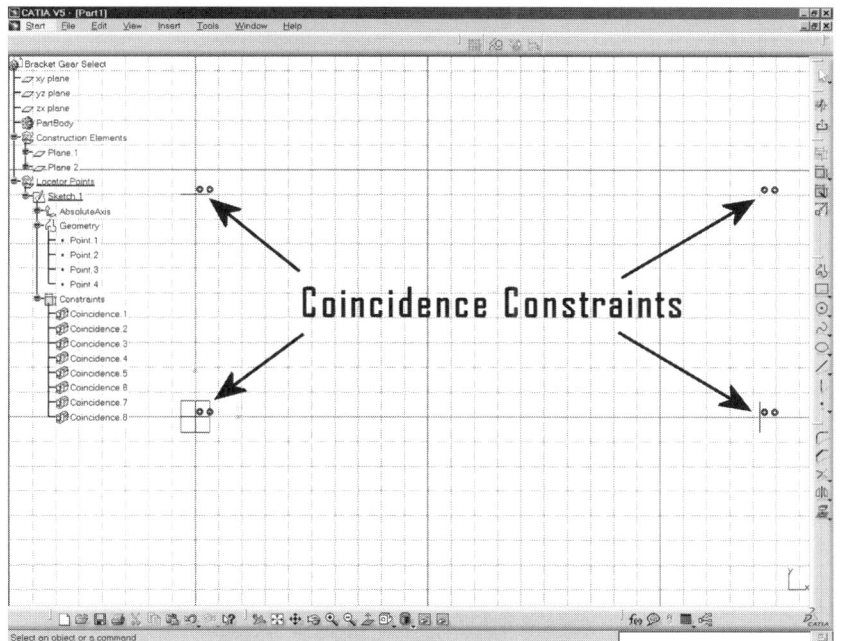

Fig. 5-73.
Point-to-plane
constraints.

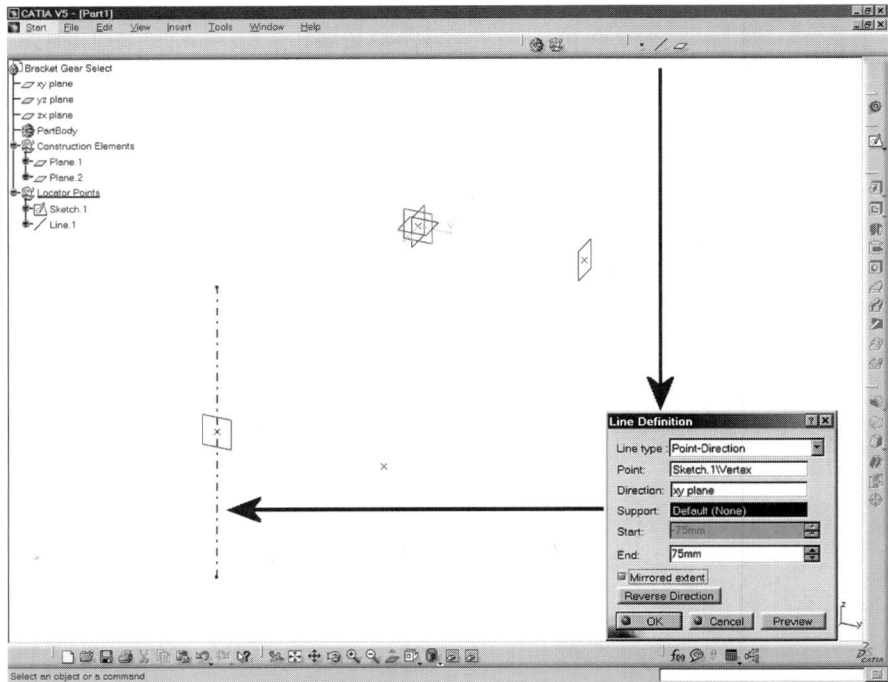

Fig. 5-74.
Line feature.

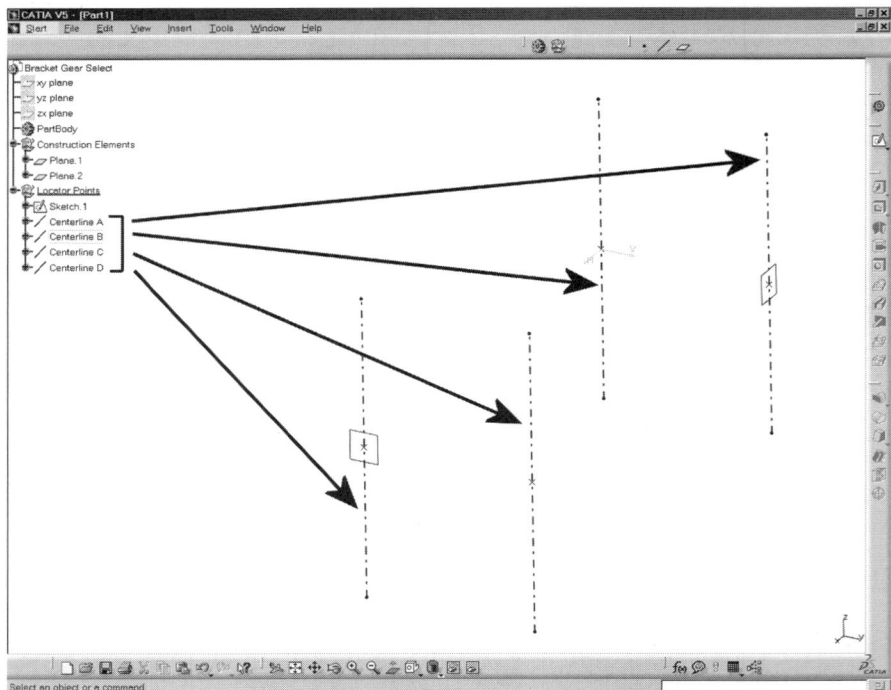

Fig. 5-75.
Additional
line
features.

17 Click on the Profile icon and create three line elements and a three-point arc element. Make sure the Automatic Geometric constraint option is selected. Create the three lines and arc approximately in the locations shown in figure 5-76.

18 Constrain the two horizontal lines coincident to the XY plane. Using the Trim option, trim the line and arc geometry as shown in figure 5-77.

19 Constrain the sketch with the values shown in figure 5-78.

20 Click on the Profile icon and create three additional lines to close off the top of the profile. (The height does not matter, as long as the horizontal line extends a fair distance above the lower profile.) Make sure all line elements are coincident at the endpoints, as shown in figure 5-79.

21 Click on the Pad icon and select the previous sketched profile (*Sketch.2*). Select More in the Pad Definition window to display more options for the feature. Enter a value of *25* (mm) for the first length, and a value of *75* (mm) for the second length, as shown in figure 5-80. Click on OK to create the pad.

Fig. 5-76. Sketched lines and three-point arc.

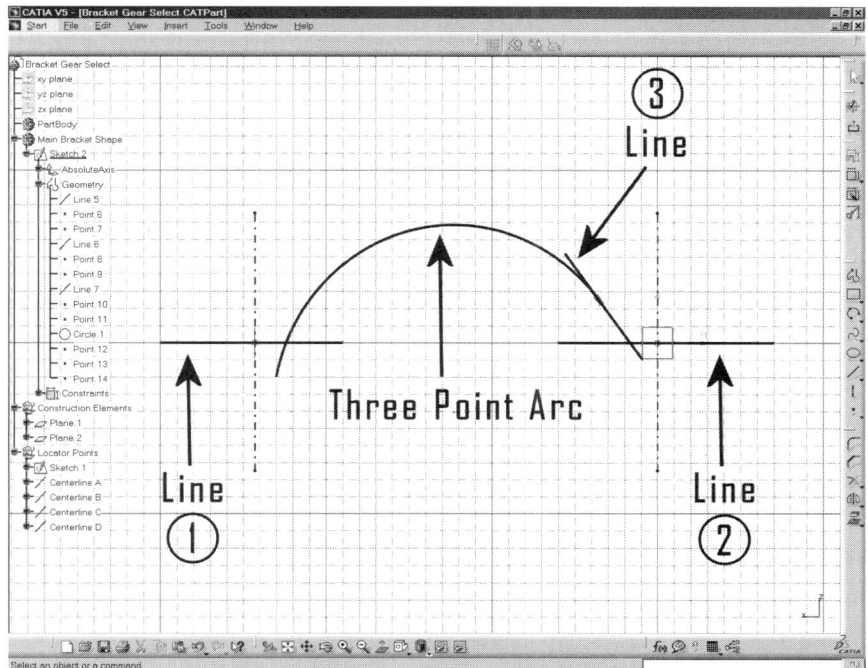

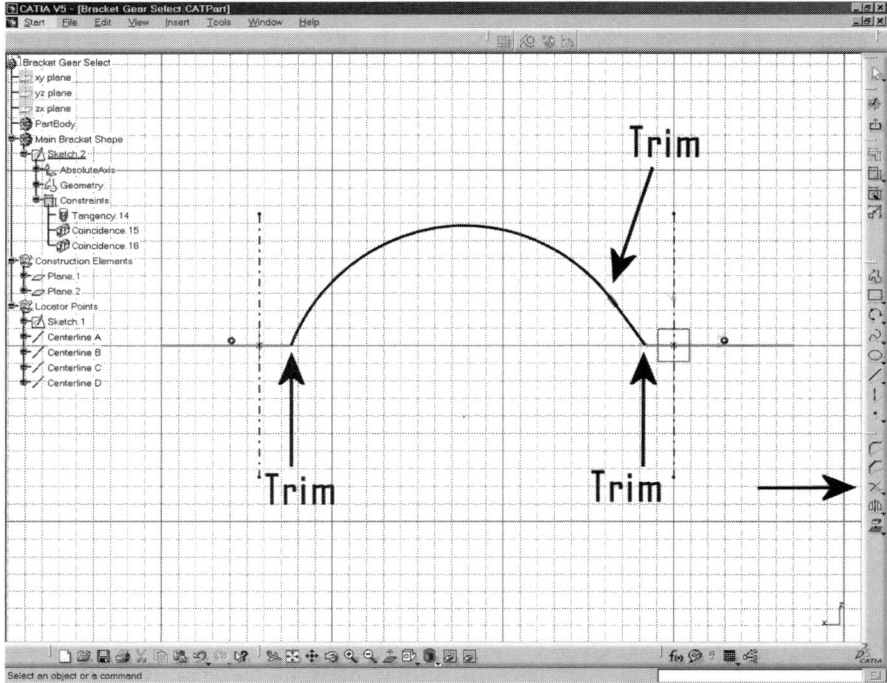

Fig. 5-77.
Constrained
lines and
curves.

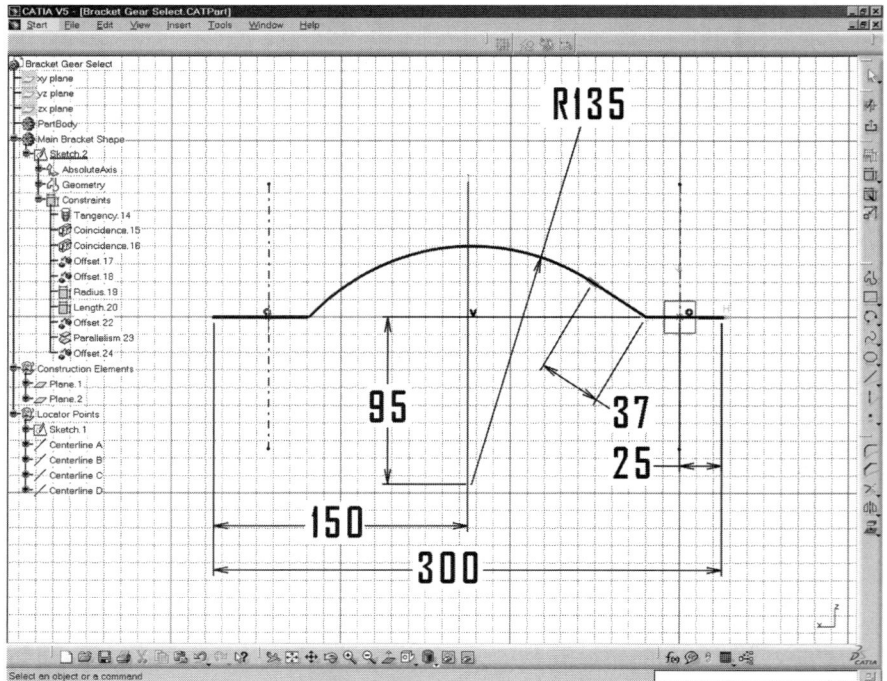

Fig. 5-78.
Sketch
direction
constraints.

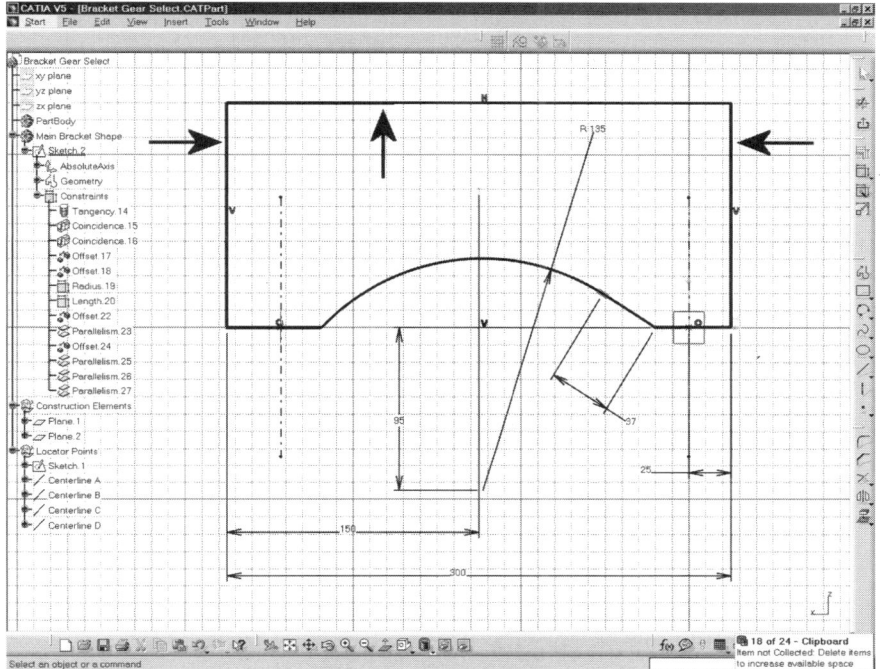

Fig. 5-79.
Sketch
dimensions.

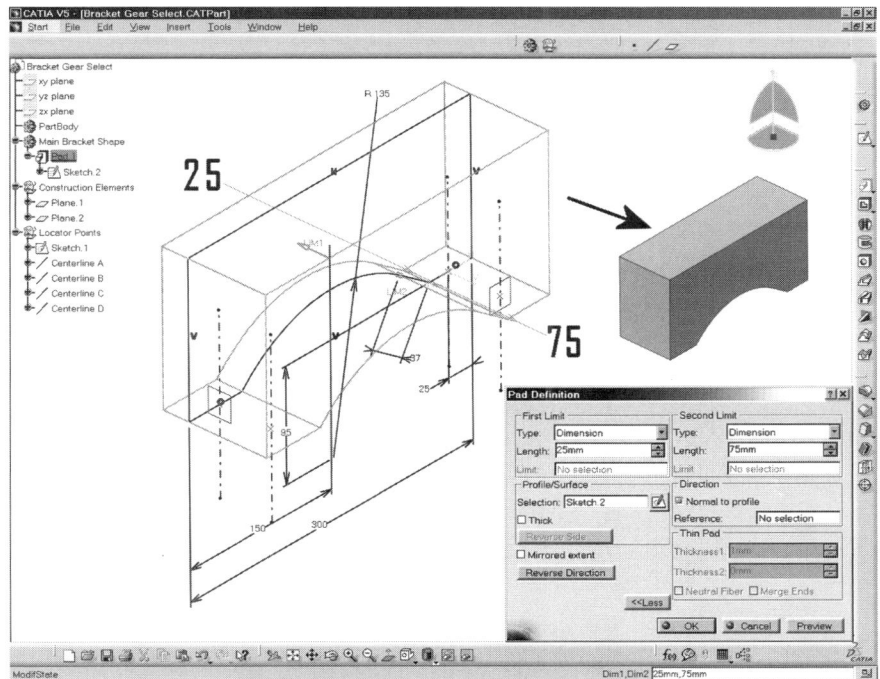

Fig. 5-80.
Pad feature
values.

22 Click on the Edge Fillet icon and select the two lower edges indicated in figure 5-81. Enter *19* (mm) for the radius value. Click on OK.

23 Click on the Shell icon and set the Default Inside Thickness option to 10 mm. Select for removal all faces except the main lower profile, shown in figure 5-82.

NOTE: *The shell operation should result in a material thickness offset of the lower profile and remove the other faces, as shown in figure 5-83.*

24 Click on the Edge Fillet icon and select the four outside corner edges. Enter *5* (mm) for the fillet value. Click on OK. The resultant outside edge fillets are shown in figure 5-84.

25 Create another 2-mm fillet on the outer top sharp edge, as shown in figure 5-85.

26 Create a new construction plane located within the *Construction Elements* body. Create an offset plane 75 mm from the XY

Fig. 5-81. Lower edge fillets.

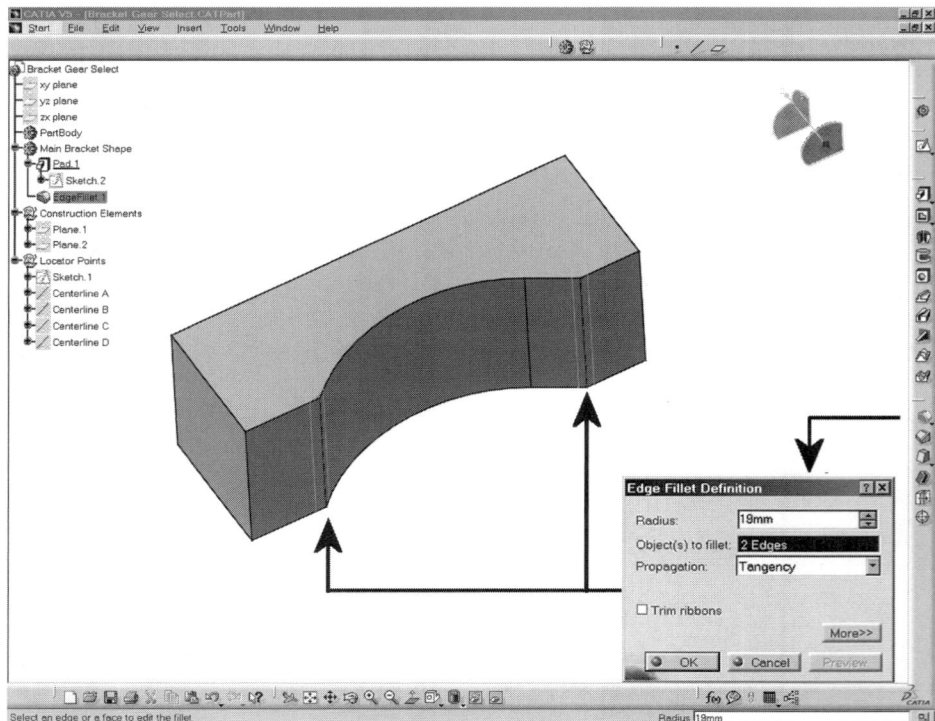

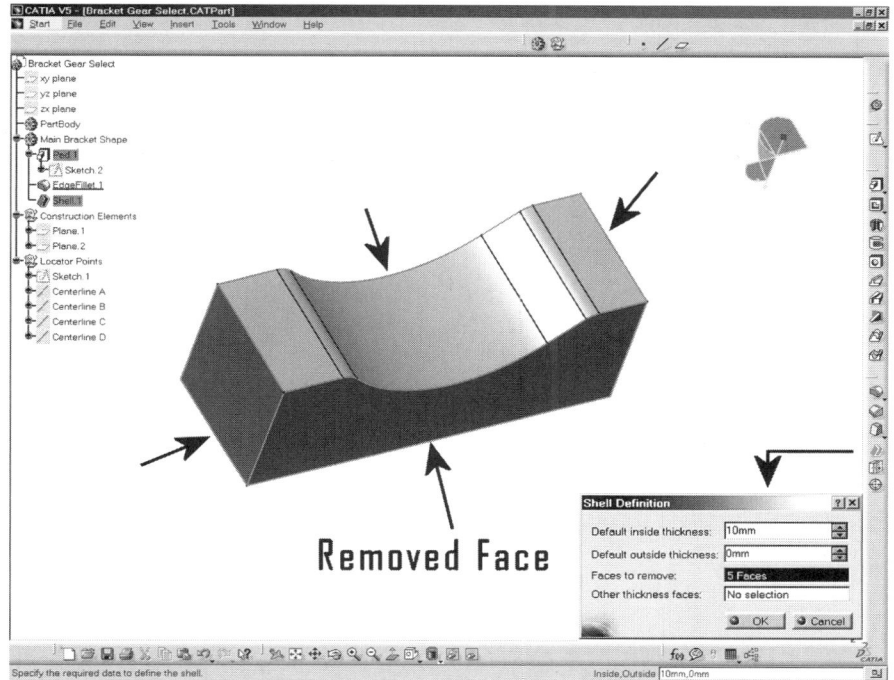

Fig. 5-82.
Shelling
parameters.

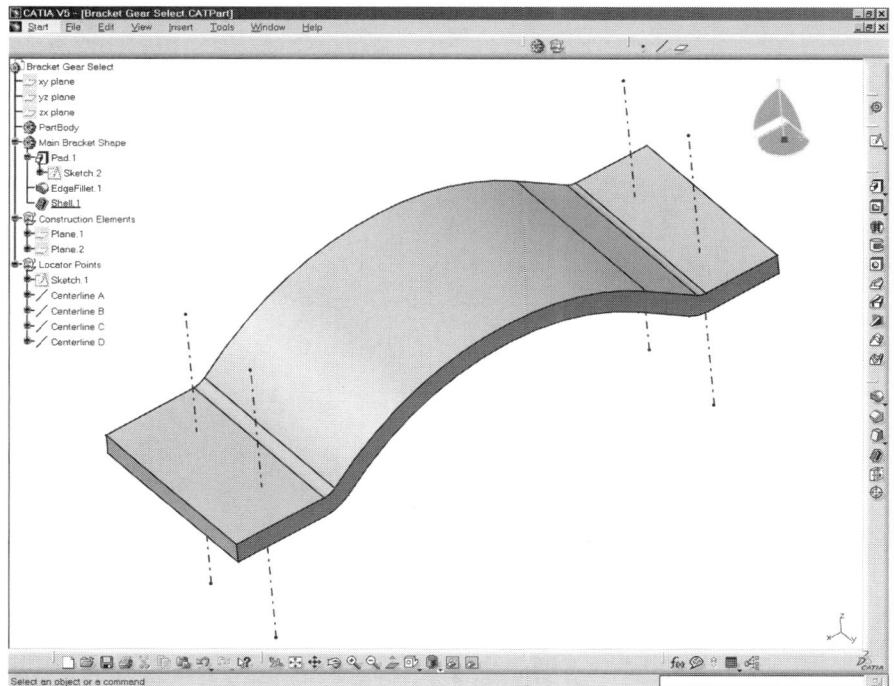

Fig. 5-83.
Shell
thickness
feature.

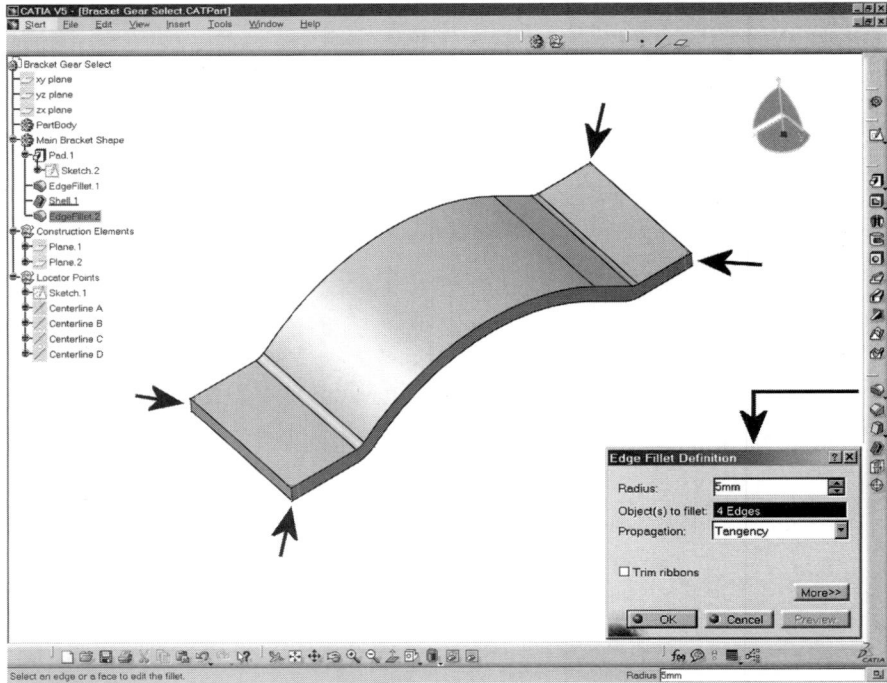

*Fig. 5-84.
Outside
edge fillets.*

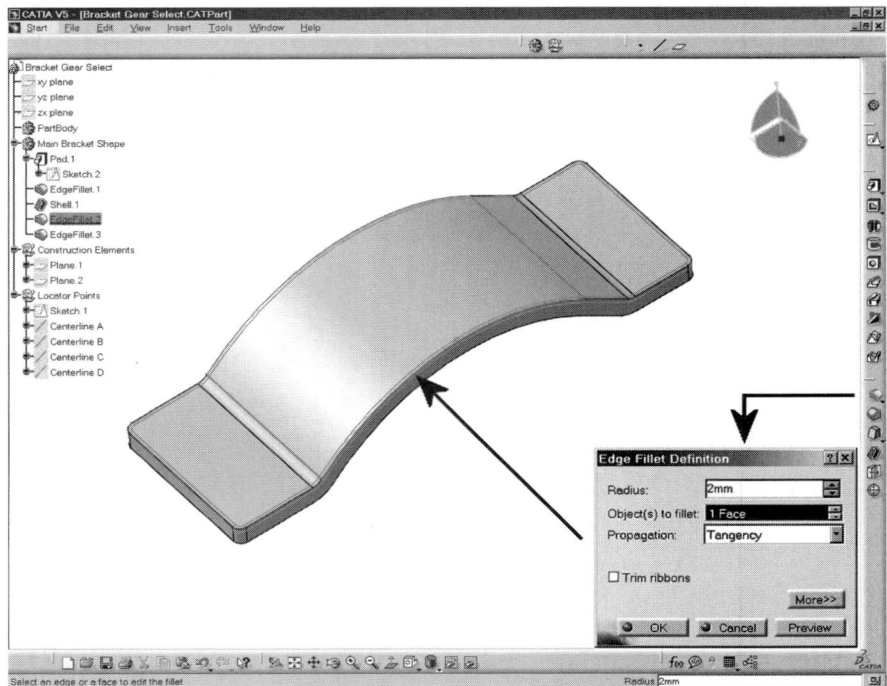

*Fig. 5-85.
Top edge
fillet.*

plane and name it *Slot Construction Plane*. This construction plane is shown in figure 5-86.

Fig. 5-86. Construction plane.

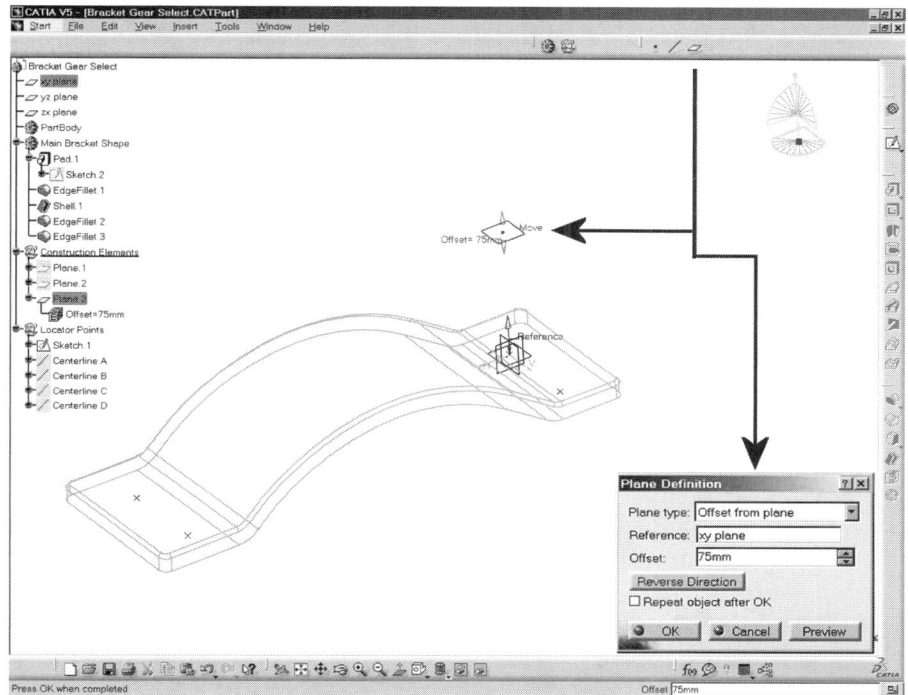

27 Insert a new part body and name it *Gear Slot*. Make sure this body is the active body.

NOTE: *An active body is identified by an underscore.*

28 Create a new sketched profile by selecting *Slot Construction Plane* and then clicking on the Sketcher icon. Create and constrain the profile sketch as shown in figure 5-87. Exit the Sketcher.

29 Click on the Pocket icon, and then select the previously sketched profile. Enter a value of *75* (mm). Click on OK. The resultant pocket feature is shown in figure 5-88.

30 Using MB3, click on the *Body Gear Slot* body and select Properties from the pull-down. Select the Graphic Tab option and change the fill color to Red. Click on OK.

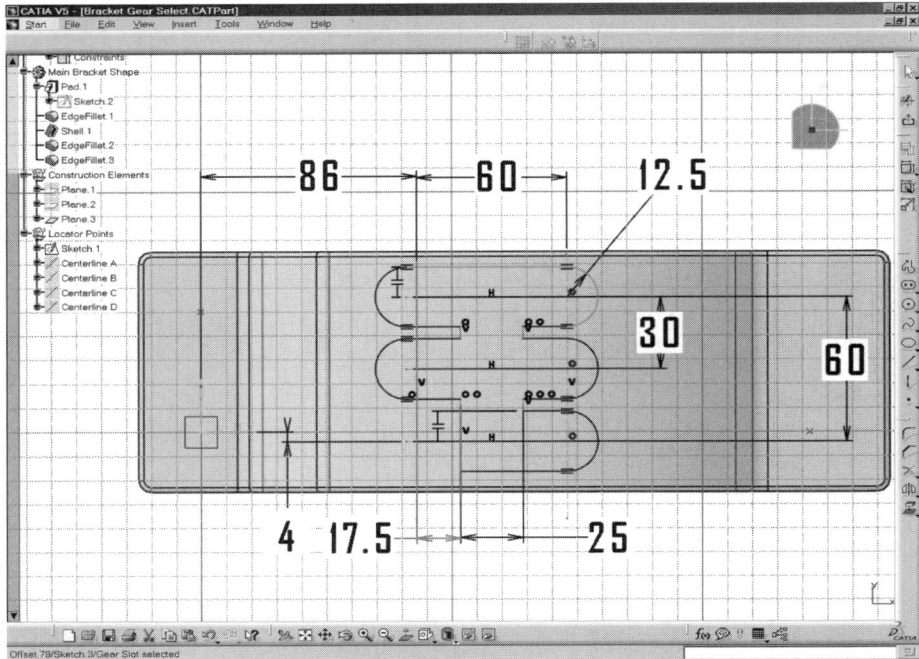

*Fig. 5-87.
Sketched
profile.*

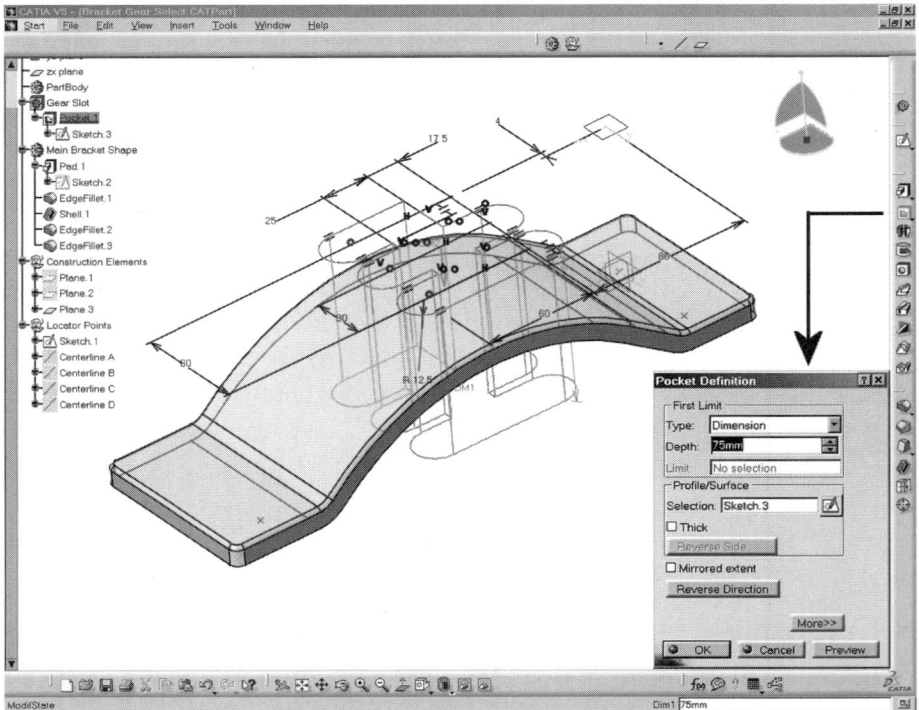

*Fig. 5-88.
Pocket
feature.*

31 Click on the Edge Fillet icon and select all of the eight vertical sharp edges on the pocket. Enter a value of *2* (mm). Click on OK.

32 Insert a new part body and name it *Gear Bracket & Slot*.

33 Using MB3, click on *Main Bracket Shape* and select Object/ Assemble in the pull-down window. Using MB3, select *Gear Bracket & Slot*. The resultant assembly is shown in figure 5-89.

Fig. 5-89. Main bracket and gear bracket body assembly.

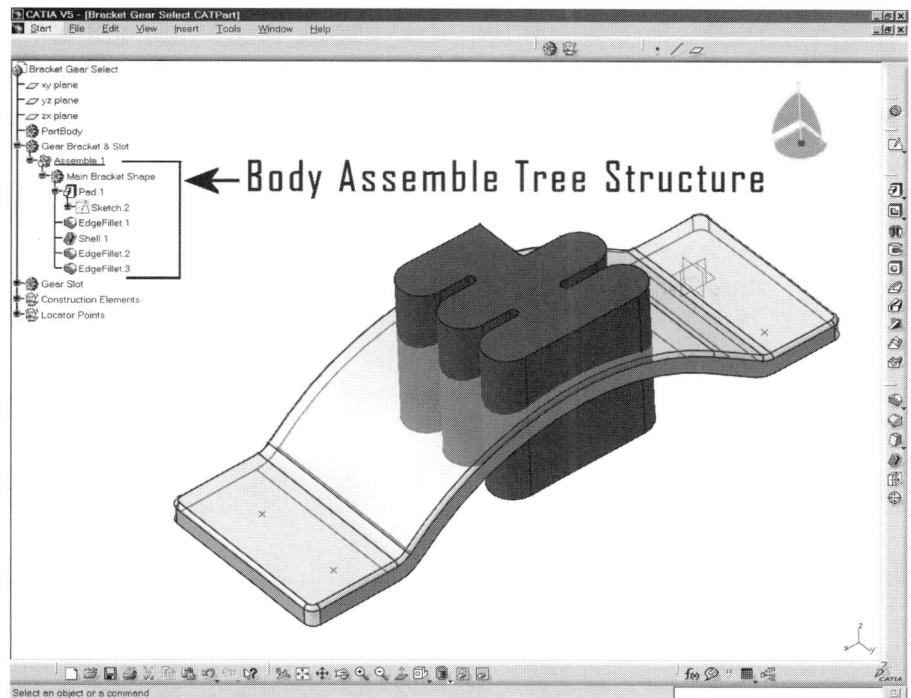

34 Using MB3, click on *Gear Slot* and then select Object/Assemble in the pull-down window. Select *Assemble.1* and then click on OK. The resultant assembly is shown in figure 5-90.

35 Add a 1-mm edge fillet to the resulting intersecting sharp edge. Select an edge as the start position and CATIA V5 will automatically propagate the fillet around the edge. The edge fillet is shown in figure 5-91.

NOTE: *This propagation is due to the continuous tangency of the edge.*

36 Make the *Construction Elements* body active, click on the Plane icon, and select Offset from Plane. Select the YZ plane and enter an offset value of *50* (mm). Name the plane *Boss Construction Plane*, as shown in figure 5-92.

Fig. 5-90.
Gear slot
assembly.

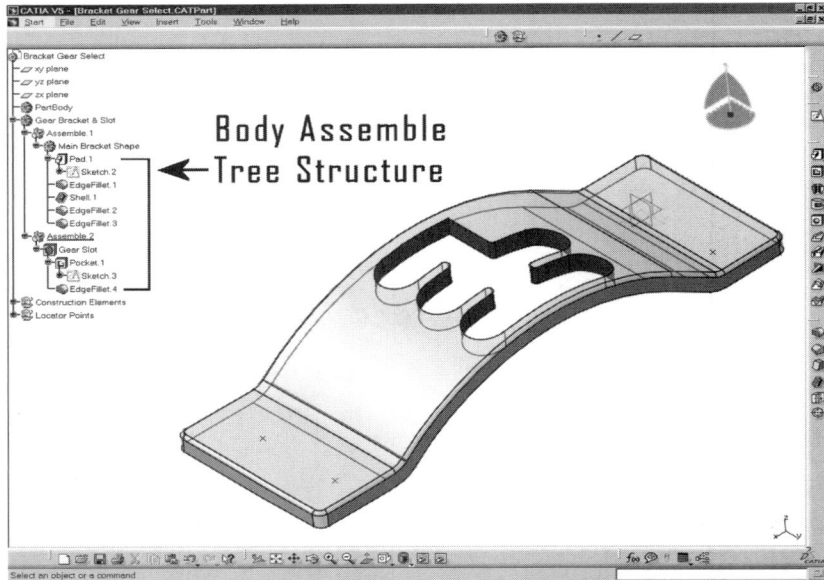

Fig. 5-91.
Edge fillet.

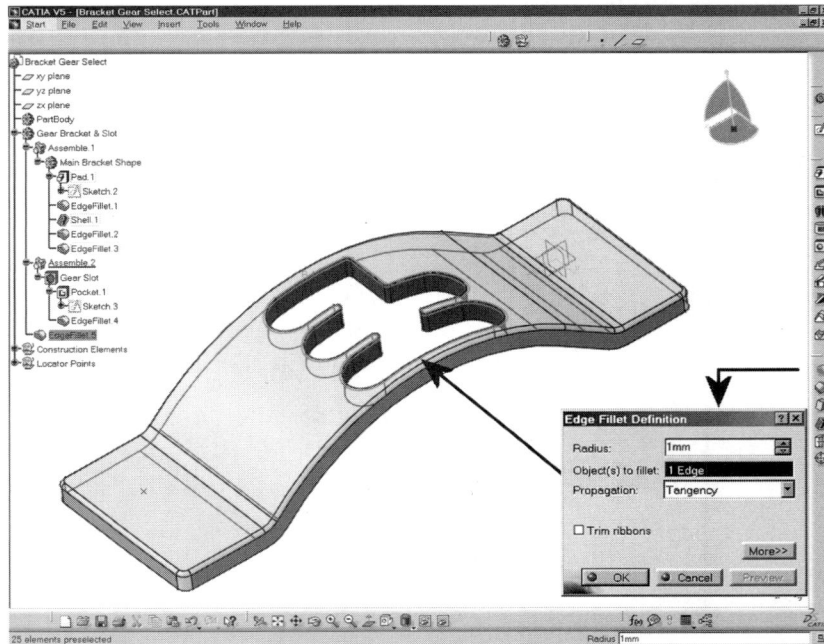

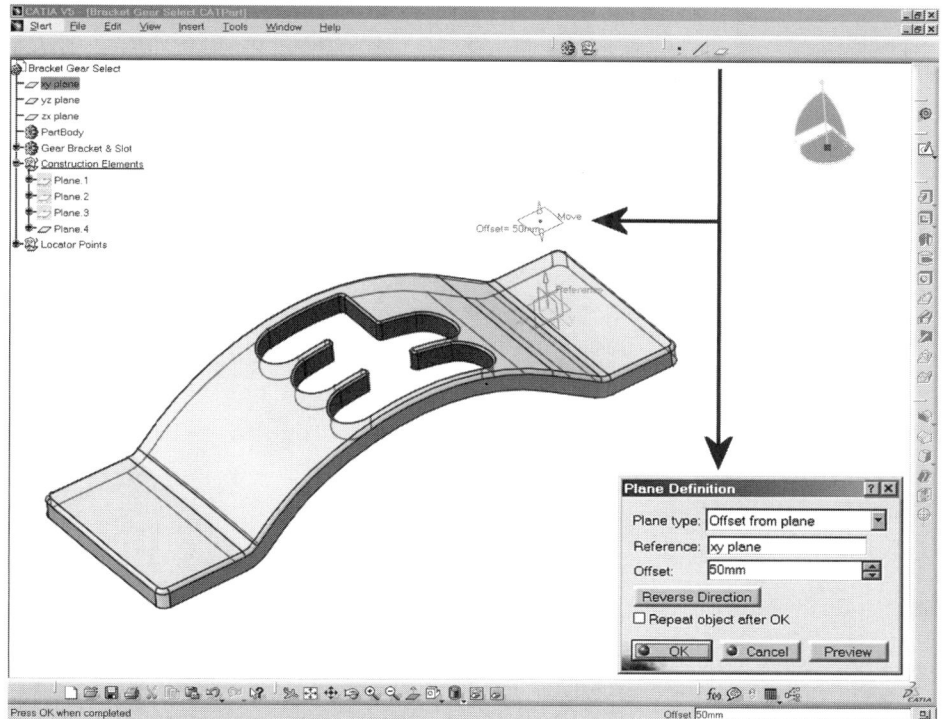

Fig. 5-92. Plane named Boss Construction Plane.

37 Insert a new part body and name it *Boss*. Make sure this body is active for the creation of new features.

38 Create a new sketched profile by selecting *Boss Construction Plane* and clicking on the Sketcher icon. On the Profile toolbar, select Circle. Create a circle and constrain it with a dimensional value of 10.65 mm. Constrain it coincident with one of the four centerlines (A through D), as shown in figure 5-93. Exit the Sketcher.

39 Click on the Pad icon, and select the previously sketched boss profile. Enter an offset value of *42* (mm). Select Reverse Direction and then click on OK. The resultant boss feature is shown in figure 5-94.

40 Click on the Edge Fillet icon, and then pick the top edge of the boss. Enter *1* (mm) for the fillet radius value. Click on OK.

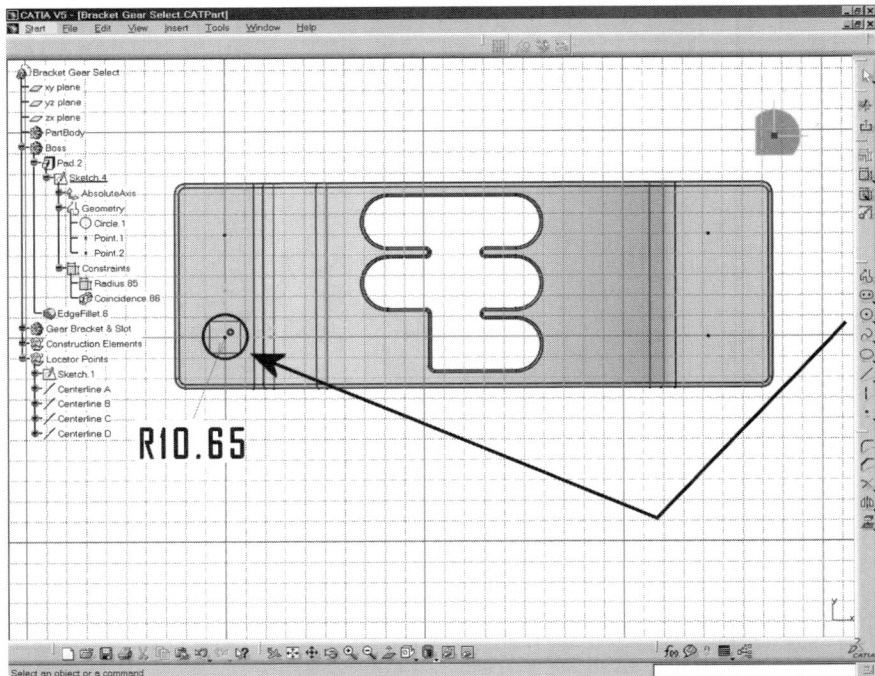

Fig. 5-93.
Sketcher
profile Boss.

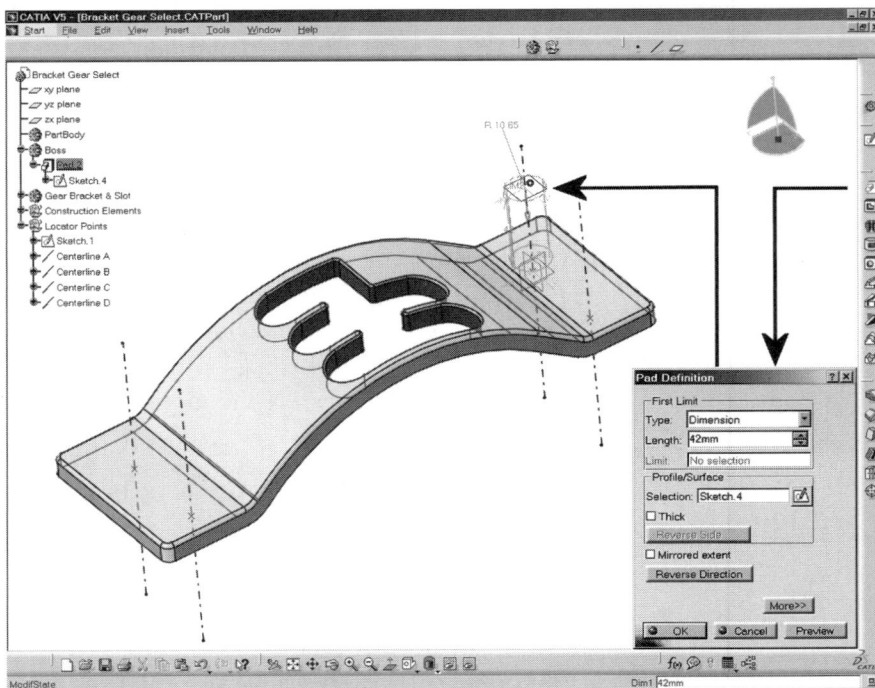

Fig. 5-94.
Boss feature.

41 On the Transformation toolbar, click on the User Defined Pattern icon, and then select *Sketch.1* as the position. Click on OK.

An instance of the boss and all related features will be copied to each of the points in the sketch, as shown in figure 5-95.

Fig. 5-95. Patterned features.

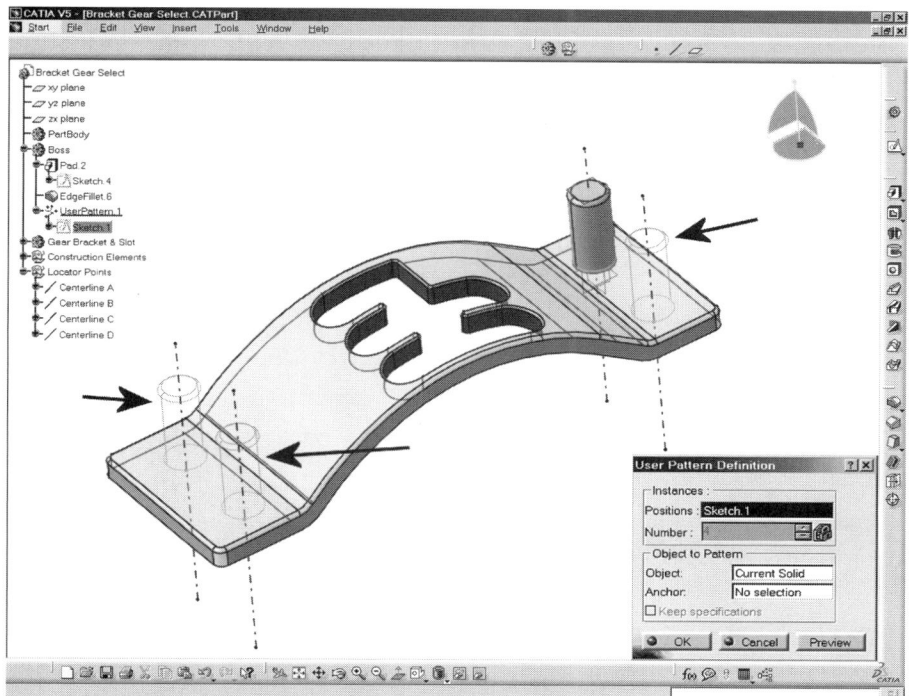

42 Insert a new part body and name it *Complete Cast Bracket*. Make sure this body is active for the creation of new features.

43 Using MB3, click on *Gear Bracket & Slot* and then select Object/Assemble in the pull-down window. Select *Complete Cast Bracket*, and then click on OK. This Boolean assembly is shown in figure 5-96.

44 Using MB3, click on *Boss* and then select Object/Assemble in the pull-down window. Select *Assemble.3*, and then click on OK to complete this Boolean assembly, shown in figure 5-97.

45 Click on the Edge Fillet icon, and then on the four lower edges between the boss and the main bracket. Enter *3* (mm)

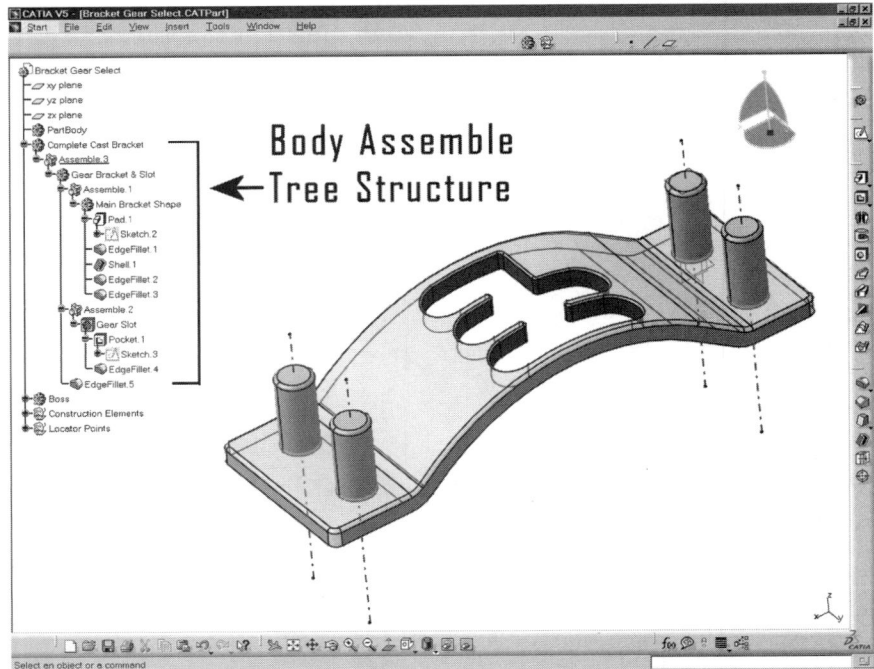

Fig. 5-96. Boolean assembly of Gear Bracket & Slot *and* Complete Cast Bracket.

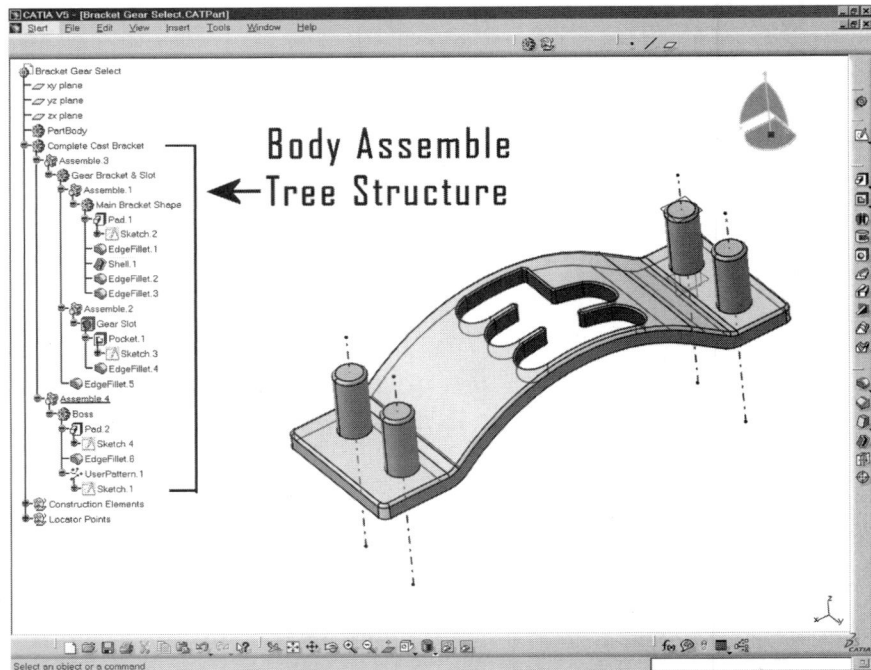

Fig. 5-97. Boolean assembly of Boss *and* Assemble 3.

for the fillet radius value. Click on OK. The edge fillets are shown in figure 5-98.

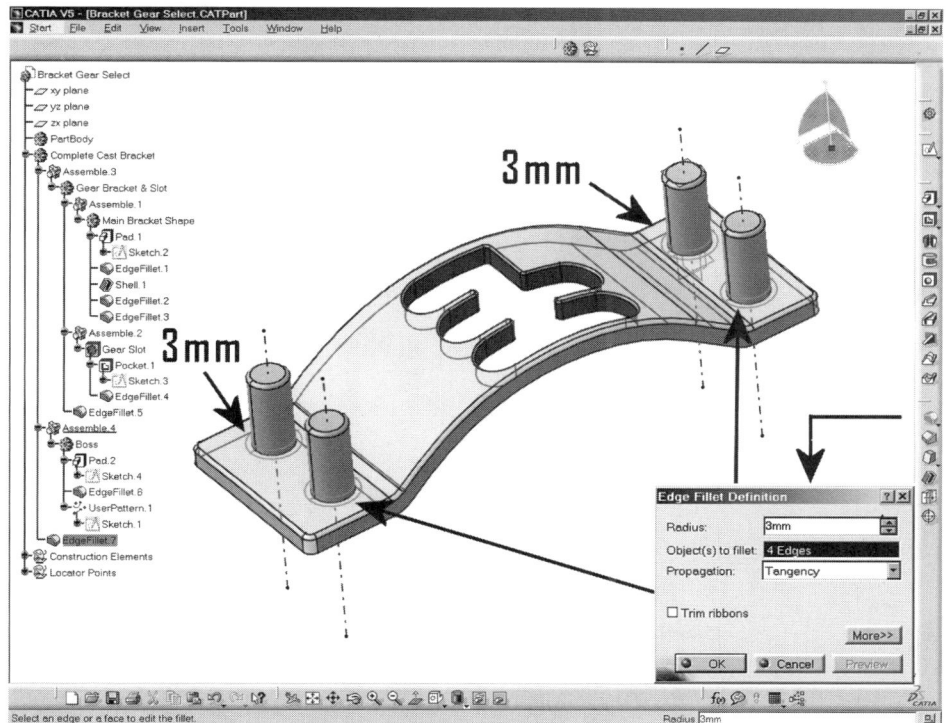

Fig. 5-98.
Edge fillets.

46 Insert a new part body and name it *Machined Holes*. Make sure this body is active for the creation of new features.

47 Click on the Hole icon, and then on *Boss Construction Plane*. Enter the values shown in the Hole Definition window. The resultant hole feature is shown in figure 5-99.

NOTE: *Wherever the mouse selection is made on the plane is where the hole sketch will be initially anchored.*

48 Click on the Sketcher icon in the Hole Definition window to position and constrain the center point of the hole to the centerline of the boss, as shown in figure 5-100.

49 Within the Sketcher window, click on the Constraint icon and pick the hole center point (yellow axis) and the centerline. Select the constraint option Coincident from the pop-up

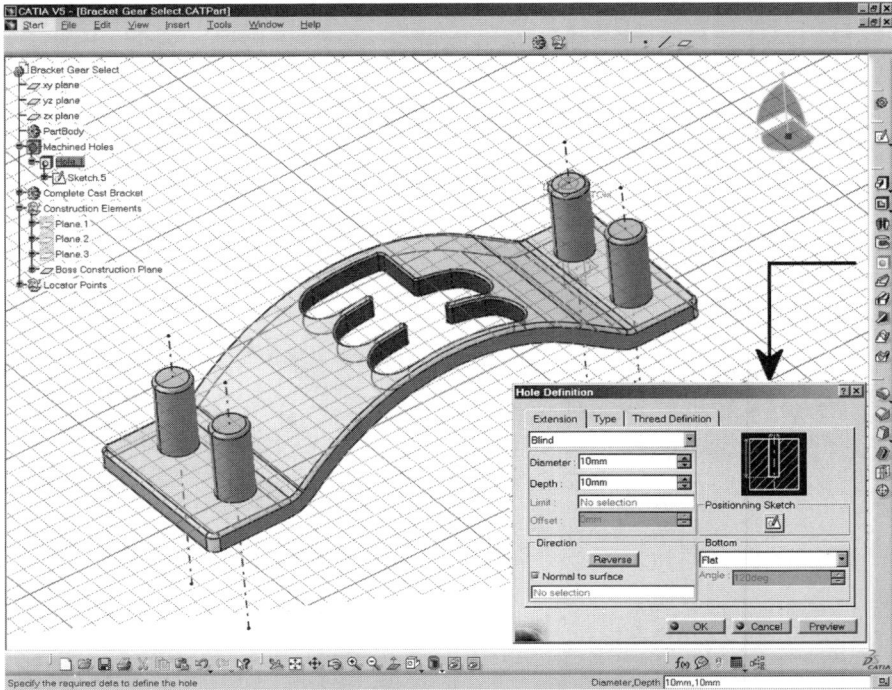

*Fig. 5-99.
Hole feature.*

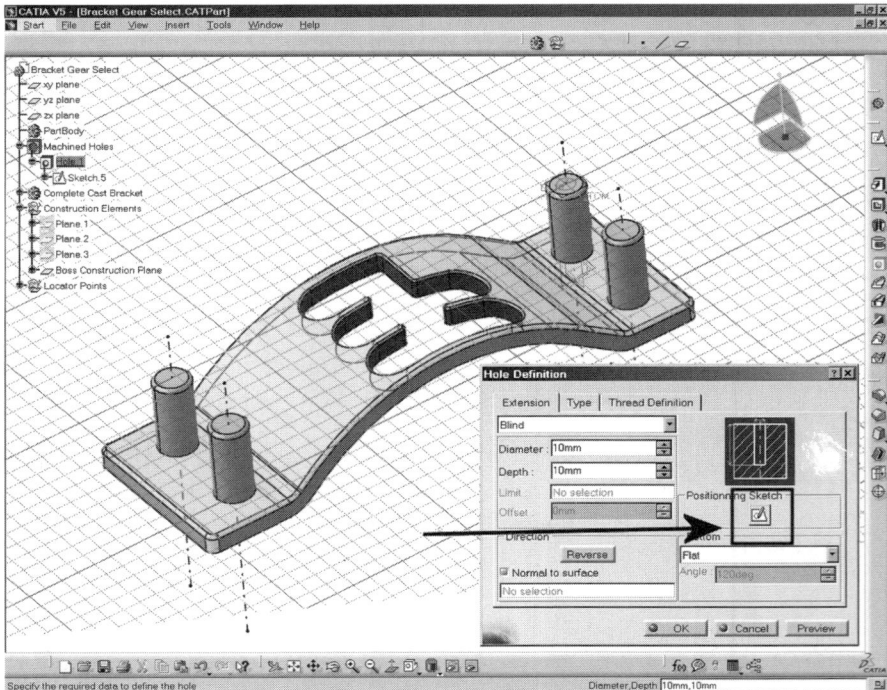

*Fig. 5-100.
Hole center
point
position.*

menu, as shown in figure 5-101. Click on OK. Exit the Sketcher.

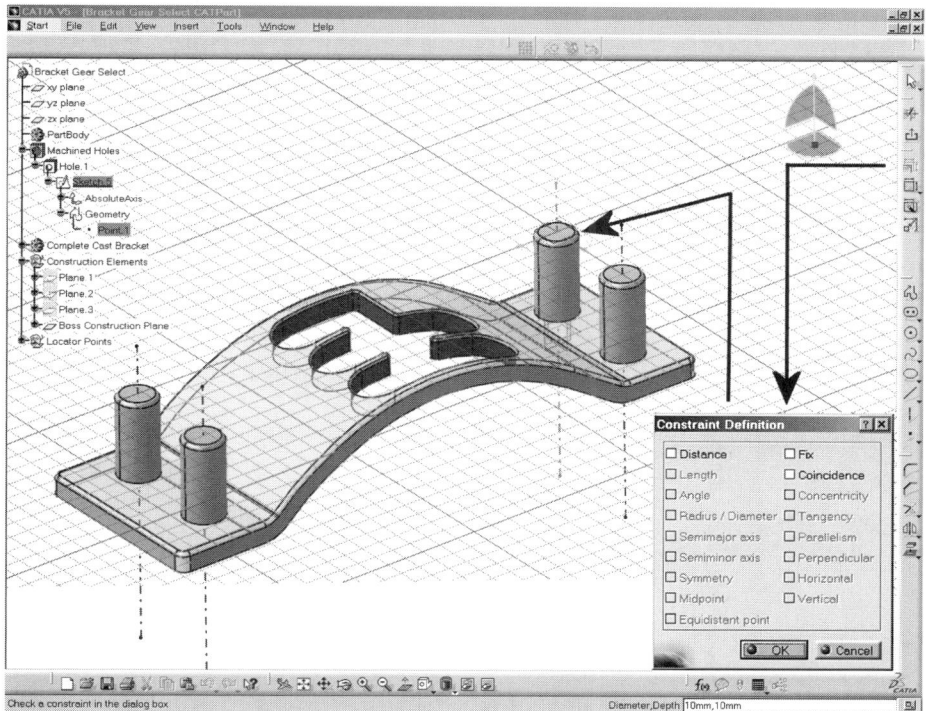

Fig. 5-101. Revised hole feature.

50 Select the User Defined Pattern icon to pattern the hole per the remaining bosses. Select *Sketch.1* as the position. *Sketch.1*, which resides in the open body *Locator Centerlines*, is a sketch of the centerline points. *Sketch.1* is referenced by the *Boss* and the *Machined Holes* part body.

NOTE: *A change in the location of any one of the centerline points will affect all three.*

51 Insert a new part body and name it *Complete Machined Part*. This body will contain the final Boolean operation defining the finished bracket.

52 Using MB3, click on *Complete Cast Bracket*, and then select Object/Assemble in the pull-down window. Select *Complete Machined Part*, and then click on OK.

53 Using MB3, click on *Machined Holes*, and then select Object/ Assemble in the pull-down window. Select *Assemble.5*, and then click on OK.

54 Create the finished part by assembling *Machined Part* and *Part Body*. The resultant bracket is shown in figure 5-102.

Fig. 5-102. Completed bracket.

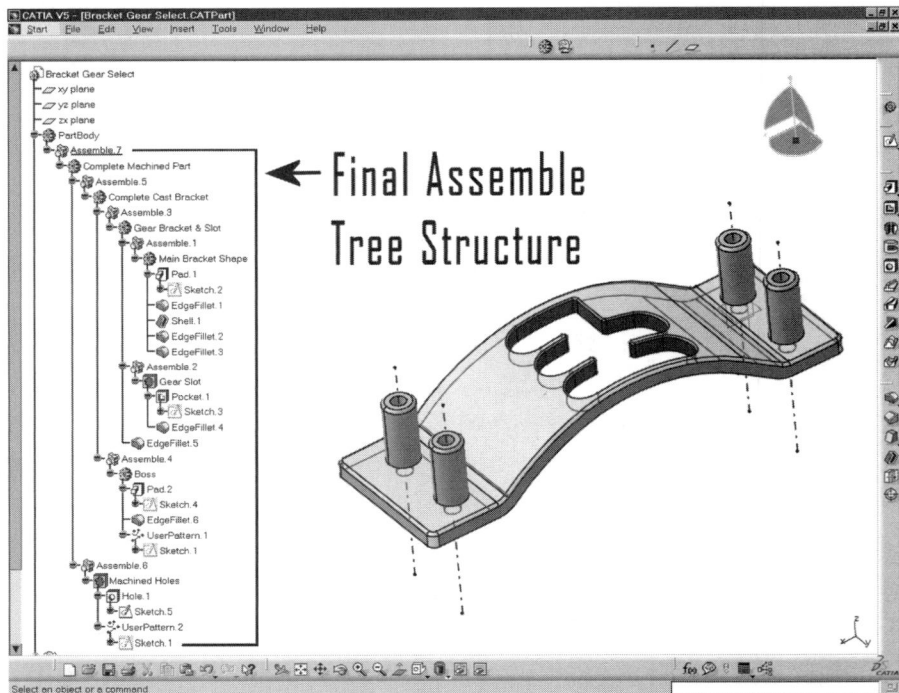

Example Exercises

After completion of the previous step-by-step tutorials, apply the learned techniques on the following examples. These models can be found on the companion CD-ROM.

Example 1: Shifter Knob

This part, shown in figure 5-103, relies on the creation of part bodies that hold unique geometric features. This part will involve using the following features.

- Creating points and lines using the Reference Elements toolbar

- Using Sketcher to define rotation axis and contour profiles

- Creating Shafts to be assembled and removed

- Renaming Bodies and created elements

- Adding fillets as high in the tree as possible

Fig. 5-103. Shifter knob.

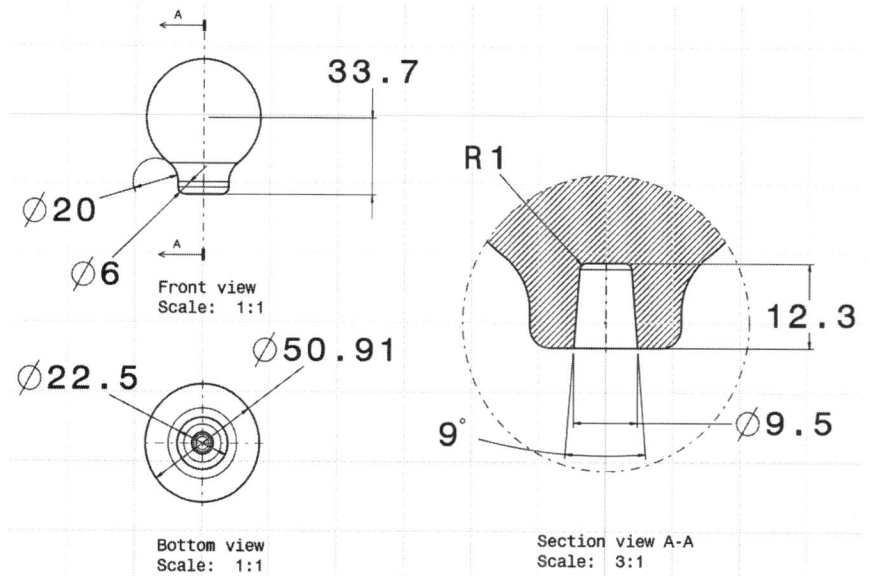

Example 2: Shifter Lever

This part, shown in figure 5-104, relies on the creation of part bodies that hold unique geometric features. This part will involve using the following features.

- Use of the Reference toolbar to create constructions elements

- Use of the Sketcher workbench to create profiles

- Creation of part bodies to hold exclusive features and related operations

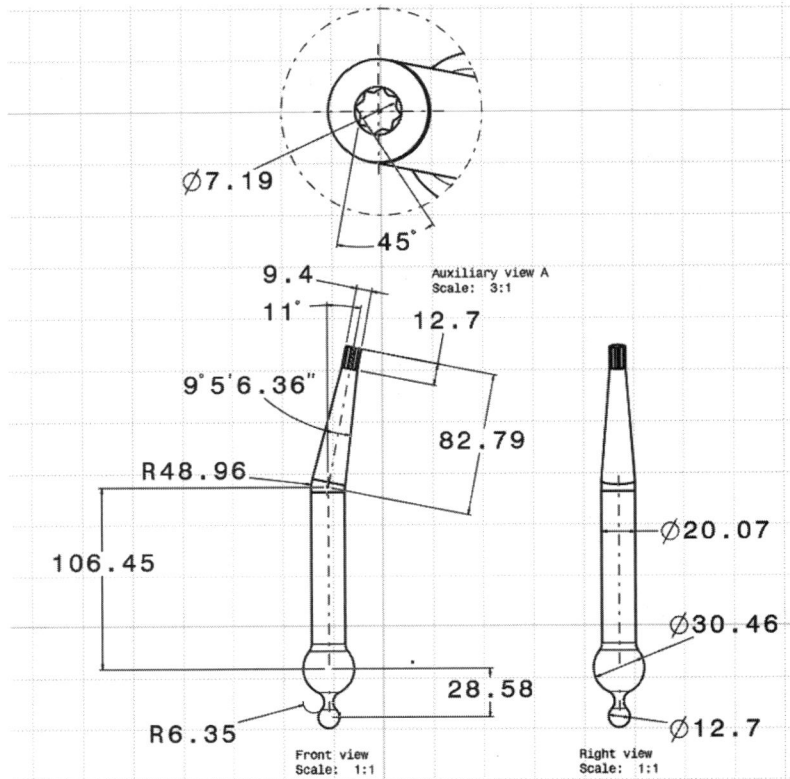

Fig. 5-104.
Shifter lever.

- The creation of pads and shafts as sketch-based features
- The use of the loft sketch-based feature to create a bridge between upper and lower areas

Summary

One of the most important concepts within CATIA V5 is the organization of data within the specification tree. The specification tree in CATIA V5 is the container for all of the information related to a model. It is also the diagram of how that model was constructed. Good model organization is extremely important for model flexibility and future manipulation of the design for downstream feature and assembly operations. Design standards, methods training, and attention to detail help promote good modeling practices. In the event a model becomes disorganized, CATIA V5 offers some flexibility in reorganizing the model by moving features to other bodies, reordering within a body, and refining elements and rerouting references. The key is developing consistent methods of modeling parts and refining these methods over time.

Review Questions

1   What is the specification tree within CATIA V5?

2   What are the two types of bodies that reside in the specification tree?

3   (True or false?): Dress-up features can reside in an open body.

4   Why is body organization essential in the design of a quality model?

5   List several benefits of body model organization.

6   What are some rules of thumb in organizing sketcher and dress-up features within a part design model?

7   What are the benefits of good feature organization?

CHAPTER 6

ROBUST MODELING

Introduction

CATIA V5 HAS BEEN DEVELOPED with variational modeling and feature-based design technology in mind. Variational modeling methodology requires more of a thought process and a greater range of modeling techniques to arrive at robust models. Producing robust, high-quality models requires planning for eventual change and incorporating detailed functional requirements as models are developed.

Building flexibility into a model is, for example, critical in facilitating downstream engineering changes. This chapter explores the development of good design habits and approaches, including modeling Do's and Don'ts. A practical approach to developing good modeling habits is essential for developing robust models flexible enough to accommodate change, including their use by others within the design community.

Objectives

This chapter explores the following.

- The design blueprint
- Root modeling changes
- Facilitating downstream changes

The Design Blueprint

This section explores the design blueprint, which is the plan developed for the design before (and possibly as) the design matures. Concepts involved here include breaking down the design into manageable sections, keeping the design simple at beginning stages, "flexing" the model, and root modeling changes. These issues are examined in the sections that follow.

Breaking Down the Design

The idea behind "breaking down the design" is to break the design model into smaller, more manageable functional groups. This involves consideration of all production, assembly, and manufacturing requirements for the design. The blueprint serves as the guideline for identifying the functional requirements to be incorporated in the design.

One of the first tasks is to determine how many component models might be required to represent the overall product or part. For example, this might be necessary if the product were to require two or more production, assembly, or manufacturing states (e.g., casting and machining). This task involves answering such questions as "Will a model exist that describes the part before its features are machined, or will the design be models as machines?" Multiple linked models might be required to fulfill manufacturing process requirements. Depending on the efficiency of methodologies applied, the number of required component models and the relationships between and among these models will vary.

Gross, Fine, Net

In dealing with the types of issues previously discussed, we recommend a "gross, fine, net" (GFN) approach to model construction, wherein no single component of the model represents an overly complicated or overwhelming challenge. The "gross" part of this approach refers to beginning with a high-altitude view of the basic construction of the model, and planning for the use of associativity and parameterized relationships in the design process.

The "fine" part of this approach refers to assessing the functional requirements and features of the model in the context of design intent, toward developing a blueprint that will satisfy such needs. The "net" part of the approach refers to assessing issues related to the nitty-gritty details of a final design that will produce a model that incorporates all features required to support functional requirements.

Process Development

It is recommended that you develop and establish a standard method of assessing how a model should be constructed and how it should be grouped (if necessary) according to functional and other requirements—whether it is GFN or some other way of approaching the problem. A high-level view as a component of the blueprint helps in setting up the proper infrastructure of the model. Another critical component of any blueprint is a plan for handling the nature and level of associativity required by the overall design, and required between and among its component models.

This can further be defined by how much geometric associativity versus how much formula, relationships, or parameter driven associativity. Unlike CATIA V4, CATIA V5 offers multiple means of creating both geometric associativity and parametrically driven relationship associativity. Design specifications can be linked via parameters and formulas to create associated geometry within a model or among multiple models. This type of functionality builds in control and flexibility and should be incorporated into the traditional geometric approach.

As the design model begins to mature, a more detailed (or granular) view of the model becomes necessary. An example of granularity would be the lessons learned from Chapter 5 regarding model organization. This forces the user to delve deeper into the thought process about feature and body organization within the specification tree. The goal is to develop a process and adjust accordingly to make the CATIA V5 experience more productive and enjoyable.

Keeping It Simple at the Start

Remember the old design acronym KISS. It stands for "Keep it simple, stupid." We would like to change that to "Keep it simple to start." The premise behind this is very similar. CATIA V5 is a variational type of modeler. What this means is that there is no need to fully constrain model geometry at the beginning or even at the end of design work. At the start of design work, geometry should be created if at all possible with critical constraints only, or at most with those simple few of initial concern.

Under CATIA V5, constraints can be added at any time during the product creation process. This provides a high degree of flexibility, in that you do not have to initially fully constrain the model. The message here is to not be too concerned about fully constraining the model until the functional requirements of the design are fully realized. Evaluate the constraints along the way, and mature the model in accordance with requirements.

In CATIA V5, the majority of base features are developed from sketched profiles. Because of this, it is wise to cultivate good working habits and construction methodologies. A great deal of complexity should not be built into sketched profiles. There are two key points related to the creation of sketched profiles: the use of constraints and the geometric makeup of such profiles.

Constraints

In the initial (or concept) stage of design, do not feel that there is an urgent need to know how a sketch should be fully constrained. At first, develop only those constraints of immediate necessity. Such constraints might be dimensional constraints or geometric constraints (e.g., parallel). The natural progression of the design will drive the need for additional constraints and control within the model.

Valuable time can be saved in the early stages of the design process if you do not try to guess what constraints will eventually be required. The initial stage of design should concentrate on the conceptual development of the design, and should be largely free of constraints. Remember, CATIA V5 offers the flexibility to go back and add constraints at any time during the design process.

Sketched Profile Geometric Makeup

Points, lines, and curves are used to create various 2D sketched profiles. The way in which these elements interact with one another is extremely important in determining how they will react to design change. A simple sketcher methodology should be employed consistently within the model. Use only those elements in a sketcher that support the creation of the base feature.

As a general rule of thumb, detailed model features (such as rounds, fillets, chamfers, and drafts) should be kept out of the sketched profile. These types of features should be added later, as dress-up features, and organized appropriately within the specification tree. Only in rare instances should these features reside in the base sketched profile or feature. This allows for a higher level of feature control when you want to activate or deactivate features.

Flexing the Model

Flexing the model refers to testing the model's integrity; that is, its sensitivity to parametric change. It is essential to gauge the flexibility of the model during the product creation process, rather than at the end of the product creation process. A good design process involves flexing the model at key milestones during the product creation process. This allows you to test the behavior of constraints to see if such behavior conforms to the requirements of the overall design intent.

For example, try propagating a minor, normally occurring change to a model and witness the results. First, was the change processed, and did the model completely update? Was the desired result realized? Was the update fast or slow? Did the model need to be repaired due to a feature failure? These are the types of questions that need to be answered, and events that need to be monitored, during the flexing process.

Continue to flex the model with a higher degree of difficulty to test the design for change propagation. This is a good way of discovering critical flaws and issues with geometric relationships or constraints. Wrong constraints, too many parameters, or too few parameters are a few problems that can require that a model be reorganized. Making such fixes as a matter of design methodology (i.e., as you design) will incorporate flexibility into the model, which will result in a higher level of control. Flexing a model at regular intervals should be a standard operating procedure.

Root Modeling Changes

Every design model will have to be altered or changed at some point during the product creation process. No model is 100-percent perfect. However, proper methodology and technique typically reduce the amount of rework necessary. Because change is such a normal part of the product creation process, certain design philosophies should be adopted with respect to change management and model maintenance.

When you are faced with facilitating design changes, ask yourself a few questions prior to addressing the changes. First, what is the real intent of the change? Second, does this change to the design involve changing existing features within the model? If the answer is yes, root features may need to be modified. Primary root modeling changes involve modifying sketched profiles that drive base features. Secondary root modeling changes involve changing the dress-up features associated with these primary features. If the bodies and features within the model are organized correctly, finding and modifying these features will be easier and more intuitive.

Avoiding the Introduction of Unnecessary Features

One of the worst things a user can do is to introduce unnecessary features into the model. This adds unnecessary complexity to the design model, which further compounds problems associated with downstream modifications. These unnecessary features, and their possible misplacement, can also have a negative impact on the performance and stability of the model.

Adding layers of geometry is too often used as a quick fix to various modeling problems, rather than taking the time to properly redefine or modify a model. This excuse is generally supported by the lack of willingness, training, and diligence of the design community. To the contrary, however, it is easier and faster to modify an existing feature within the design model than to add geometry. In addition, modification of existing geometry can be preformed with a much higher level of confidence that the change will propagate correctly throughout the model.

Modifying Existing Features

In other instances, the degree of change required cannot be accomplished by a modification to the existing model geometry. If this is the case, there is a genuine need to add or delete features within the design model. In this scenario, you need to think about the best way to introduce new features within the model. For instance, where in the specification tree does the feature need to reside? Does the new feature require its own body? Does the feature need to be organized near related features?

Such considerations go by the name of "feature logistics," an important concept in regard to the addition of new features, particularly those that require a higher level of interaction. Issues of feature logistics should be scrutinized with the same care as the creation of the initial feature geometry.

Facilitating Downstream Changes

The ability to anticipate the requirements of potential change is built on experience, training, and a natural intuition on how the design may have to be modified. Two primary areas of concern should be addressed when you are attempting to anticipate change: feature completeness and the level of feature intelligence. These two issues are explored in the sections that follow.

Feature Completeness

This area of concern addresses the completeness of features within a model. If a high frequency of change is anticipated, it is recommended that you not have all detailed features defined in the model. This type of strategy is particularly relevant at the beginning (or concept phase) of the product creation process. More importantly, it applies to "dress-up" features that will be shared by two or more design features.

Dress-up features should be added when their locations can better be defined, and their required interactive relationships (i.e., between and among multiple features) better discerned. If dress-up features are interacting with a single design feature only, it may well be appropriate to create them. The chances of feature failure, repair, and rerouting are greatly reduced by keeping such issues of feature completeness in mind.

Feature Intelligence Level

This second area of concern addresses the intelligence level of features required in a model. This focuses on the use of parameters and formulas in building intelligent relationships between and among features. Many types of changes can be introduced into the model by changing specific parameters. These types of changes are typically functional changes to design features.

Designing for Change Exercise

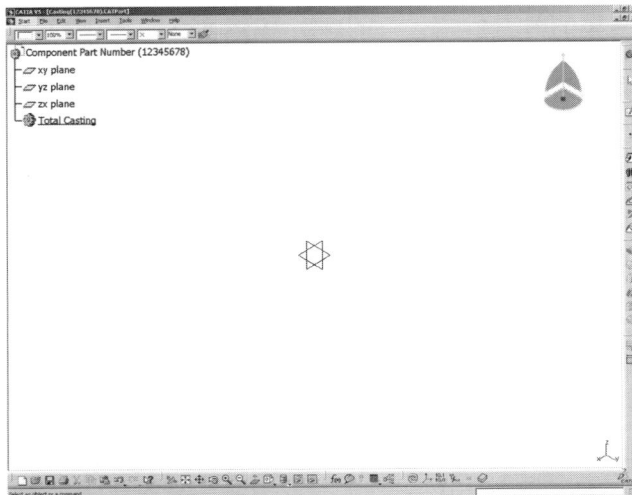

Fig. 6-1. New part.

This exercise explores both geometrically and parametrically driven relationships.

1   Start a new CATIA V5 session and create a new part design, as shown in figure 6-1. Name the default part number *12345678*, and name the default body *Total Casting*.

2   Access the Knowledge toolbar and select the f(x) icon.

This activates the Formula window, shown in figure 6-2.

3 In the box next to *New Parameter of type*, select the drop-down arrow button and then select Length, as shown in figure 6-3.

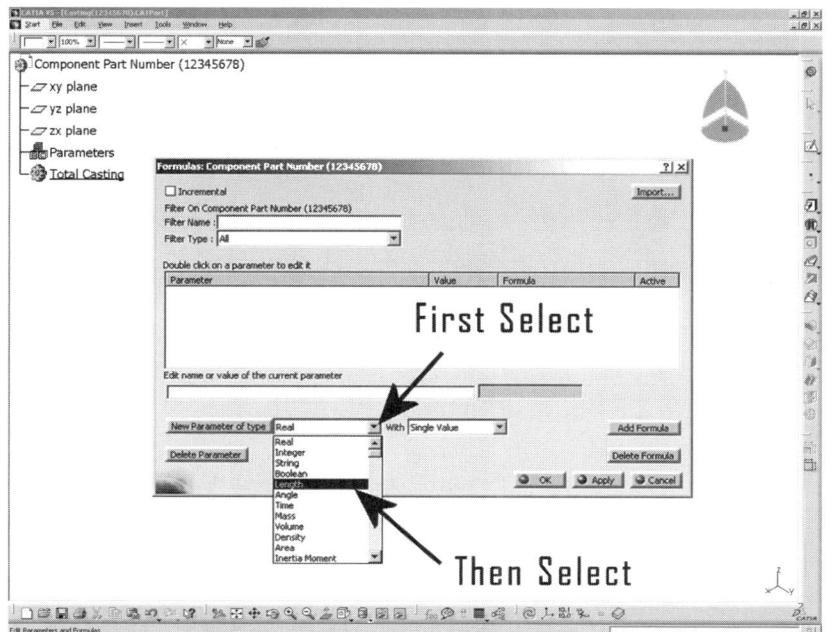

Fig. 6-2. Formula window.

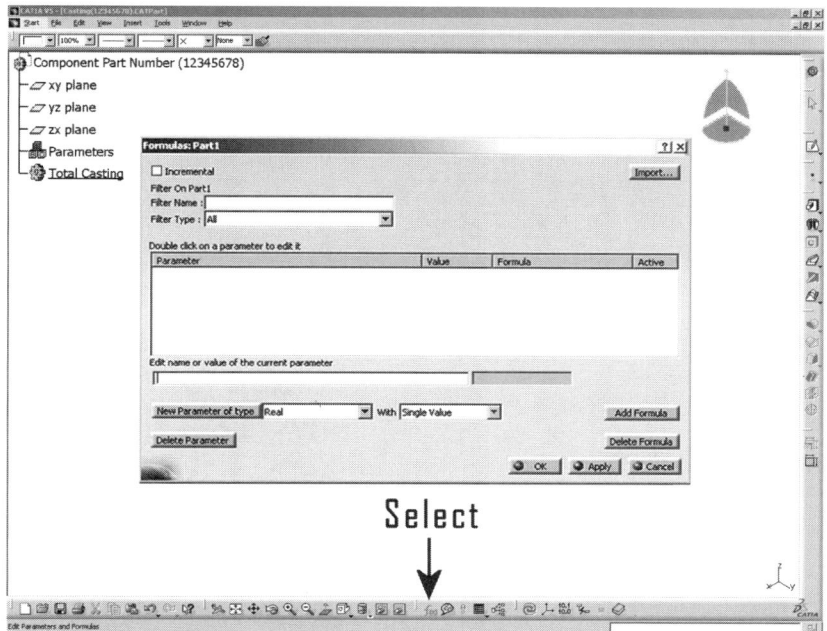

Fig. 6-3. Length parameter type.

4   Click on the *New Parameter of type* button to create the new parameter, located within the window. Within the parameter entry field, change the name of the new parameter to *Fastener Bolt Size*. Enter a value of *8* (mm) for the length, as shown in figure 6-4. Click on OK to create the parameter. The new parameter, named *Fastener Bolt Size = 8mm*, will be located with the specification tree.

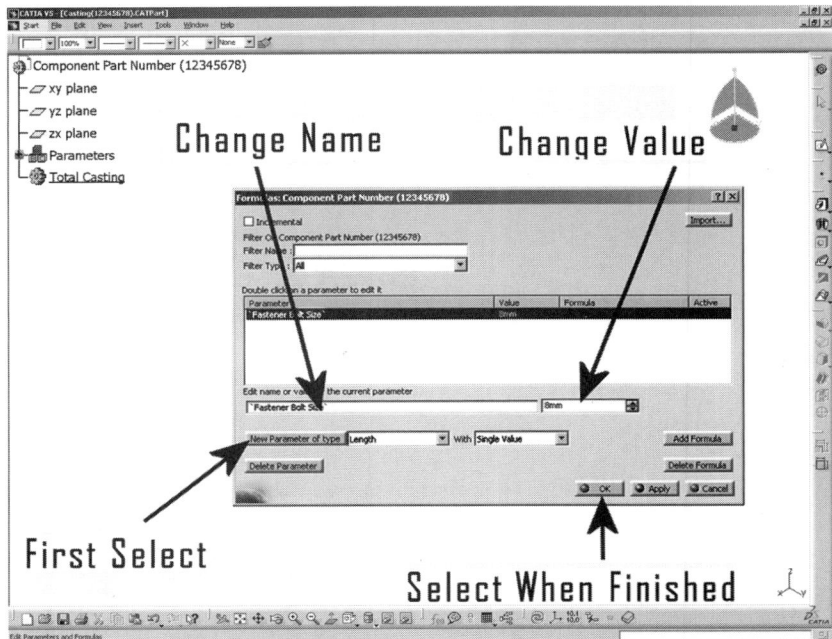

Fig. 6-4. Parameter Type box.

5   Create two more parameters, named *Wall Thickness = 4mm and Machine Stock = 3mm*. These are added to the specification tree, shown in figure 6-5. Insert an open body and name it *Construction Elements*. Create a new sketch profile referencing the XY plane. Sketch four points in a square pattern, with one of the points constrained to the main axis. Dimension the three points a distance of 100 x 100 units from the main axis.

6   Create four new line elements that are normal to the XY plane and placed through each point. Name the four new lines Centerline 1, Centerline 2, Centerline 3, and Centerline 4, as shown in figure 6-6. These centerlines will be used to locate boss features.

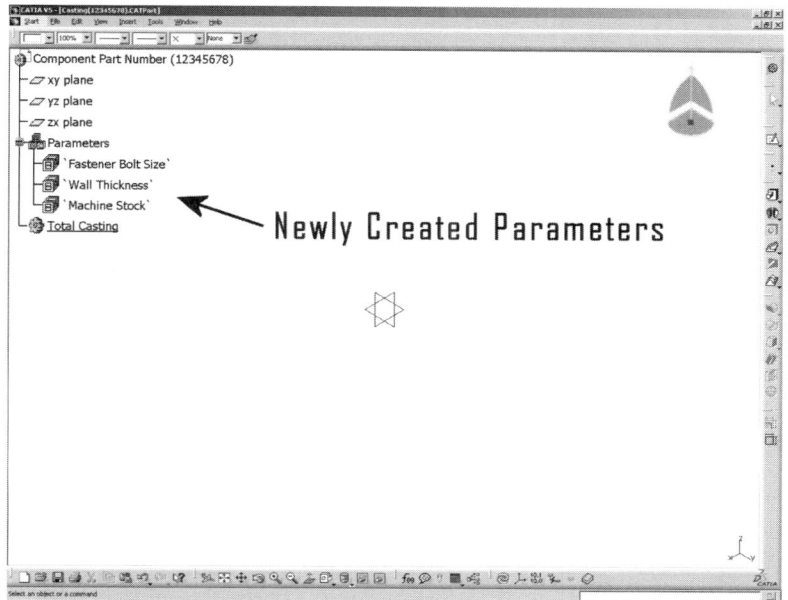

*Fig. 6-5.
Specification tree
showing parameters
established.*

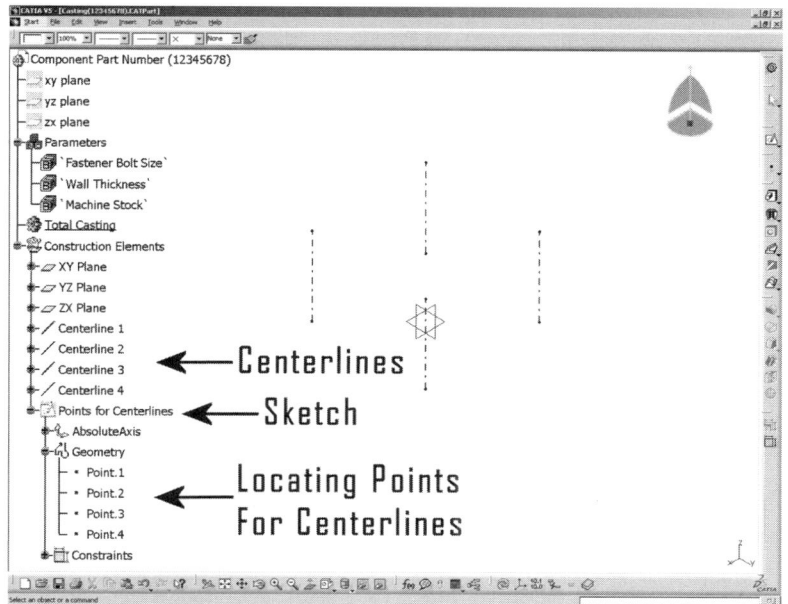

Fig. 6-6. Centerlines.

7 Insert seven (7) new part bodies, to be used to organize the features and elements within the specification tree Name these new part bodies, shown in figure 6-7, *Casting Features, Base, Mounting Bosses, Boss 1, Boss 2, Boss 3,* and *Boss 4.*

8   Using MB3, activate the Base body and then select the Define
    In Work Object option. Create a rectangular sketched profile
    that constrains each edge of the rectangle 20 mm from each
    centerline. Extrude a pad feature 20 mm from this sketch pro-
    file in the negative direction. Add a 5-degree draft feature and
    5-mm fillets to the model, as shown in figure 6-8.

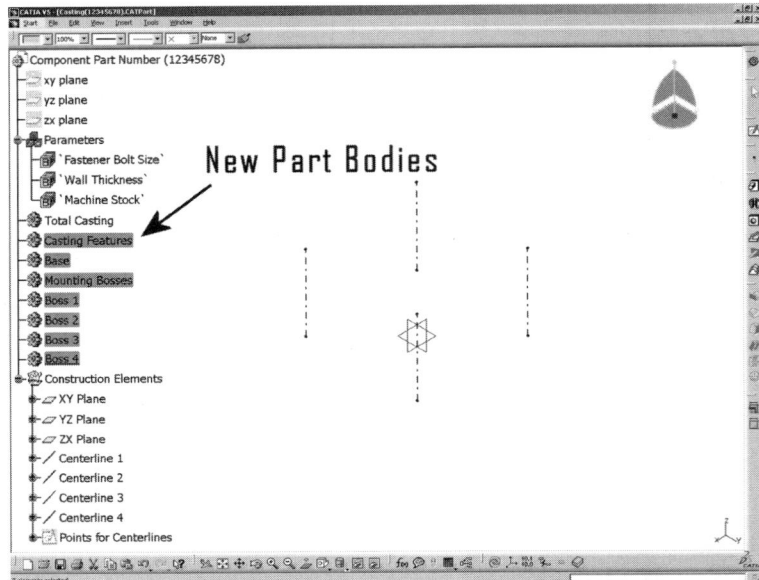

Fig. 6-7. Part bodies.

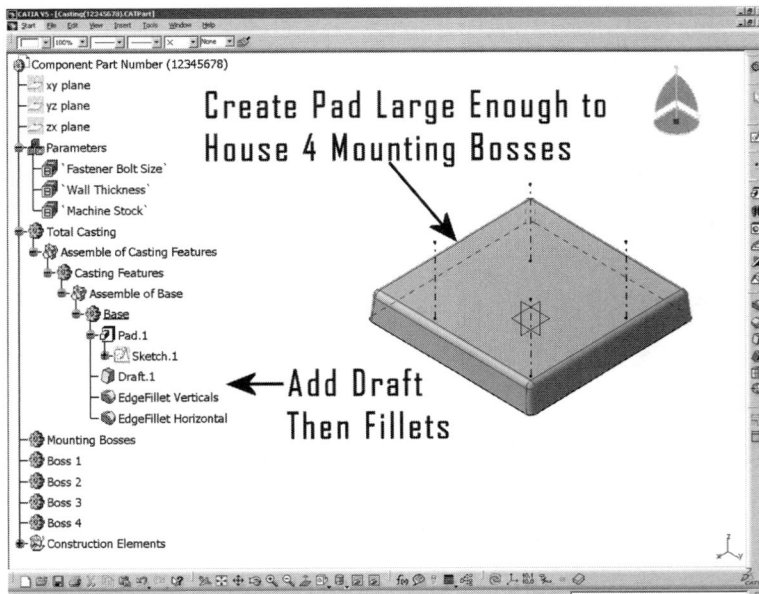

Fig. 6-8. Rectangular pad for mounting bosses.

NOTE: *Associate the outer pad profile sketch to the four centerlines in a logical manner.*

9 Assemble the Base body to the Casting body using a Boolean operation, as shown in figure 6-9.

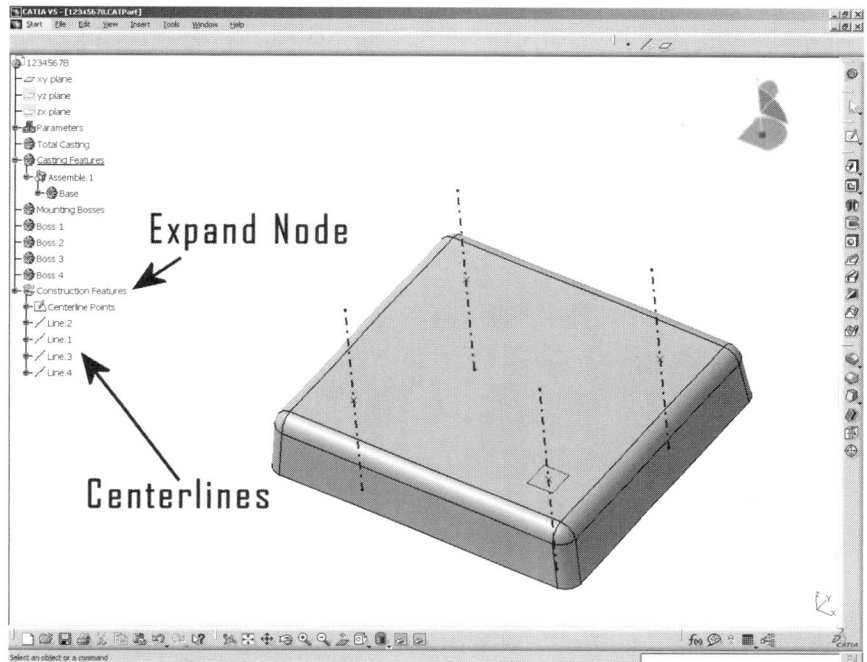

Fig. 6-9. Centerline construction elements.

10 Activate the Boss body and create a new profile using the XY plane as the reference plane for the sketch. Sketch a circle element near but not on top of one of the centerline features, as shown in figure 6-10.

11 Select the underlying centerline and then while holding down the Ctrl key select the center point of the circle profile. These elements should both be highlighted, as shown in figure 6-11.

12 Access the Constraints toolbar and select Constraints Defined in Dialog Box. The Constraint Definition dialog box should appear. The resultant menu displays the various constraint conditions available for selection. Only the conditions associated with those elements that are selected will be available. Check the Coincidence box and then click on OK. The result is shown in figure 6-12.

Fig. 6-10. Circle
profile.

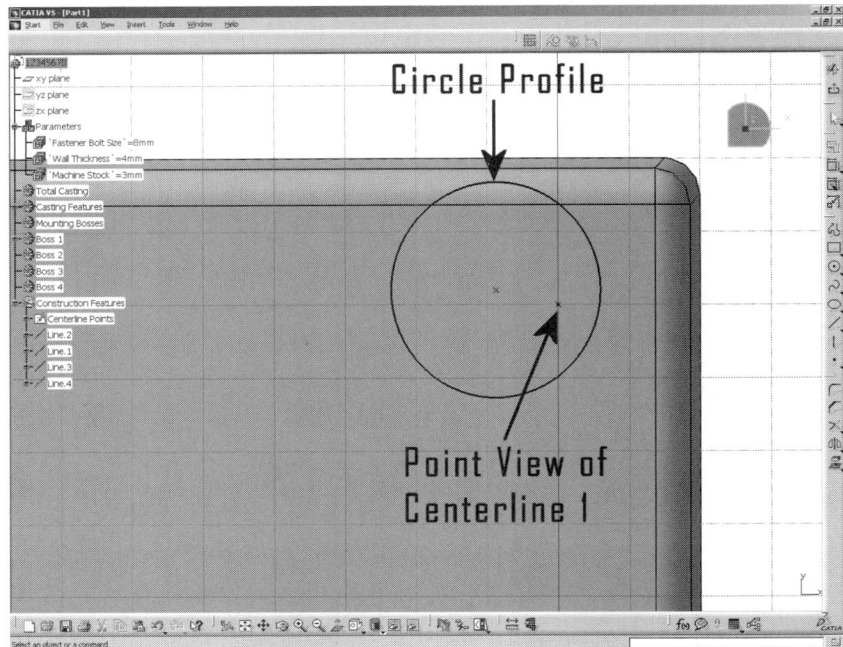

Fig. 6-11.
Highlighted
elements.

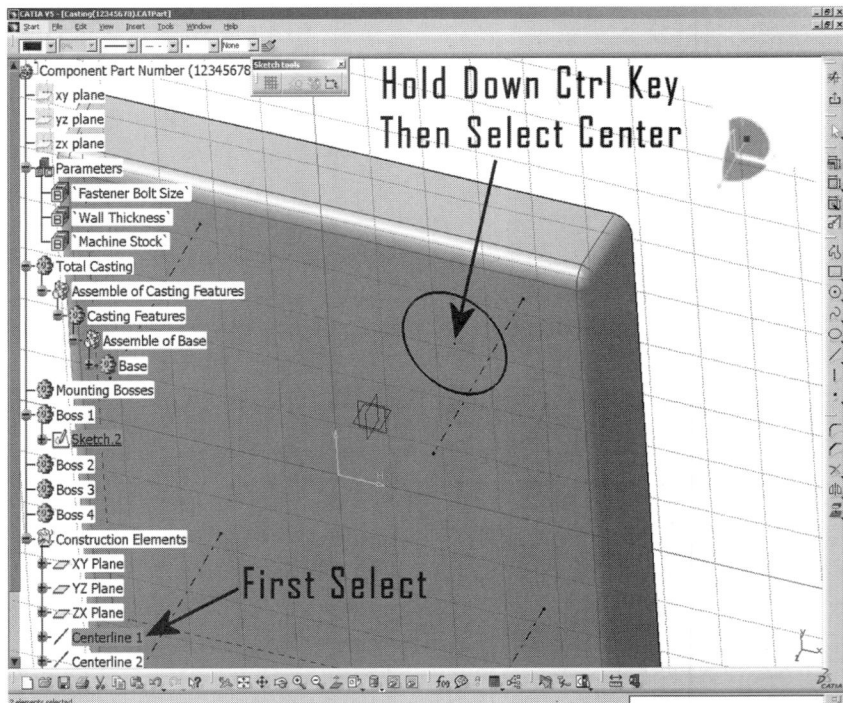

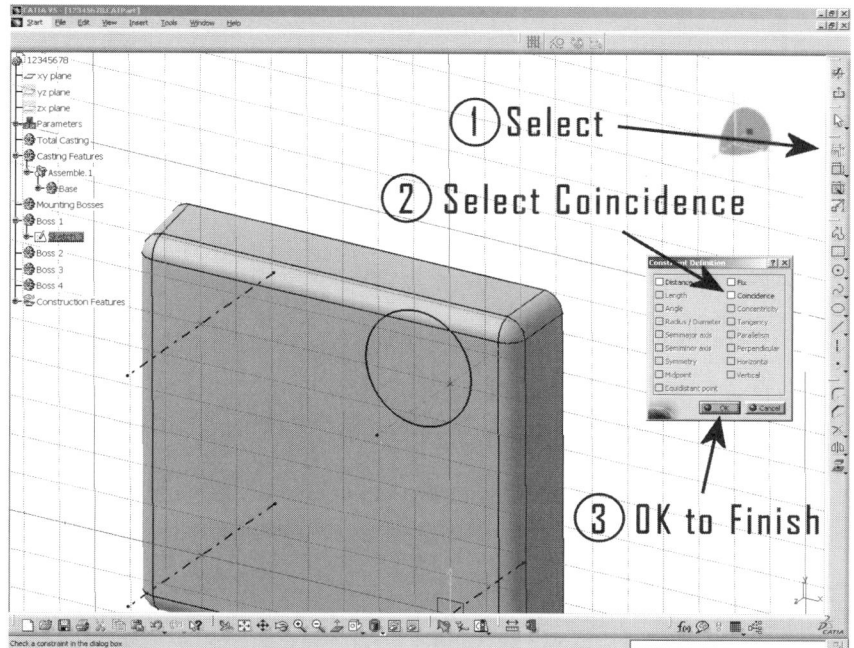

*Fig. 6-12.
Coincident
constraints
established.*

NOTE: *The sketched profile should have snapped the center point onto the underlying centerline axis. This means that the center point of the profile is now linked by a coincidence constraint to that centerline. The centerline is now the locating element for this profile. If this centerline ever moves, the profile will follow.*

13 In the Constraints toolbar, click on the Constraint icon. Select the circle profile. This will create a diameter dimensional constraint for the circle. Place the dimension using MB1 and selecting within the window, as shown in figure 6-13.

14 Place the cursor on the dimension and while holding down MB3, select the radius object and then select Edit Formula, as shown in figure 6-14. The Formula editor will appear. This will be used to impose a formula on this diameter dimension.

In the first box below the Incremental checkbox you will see the CATIA definition of the profile. Note that the definition is called out as a radius. At the end of this box is an equals sign, and below it an empty box. This empty box is where the formula, as shown in figure 6-15, will be defined.

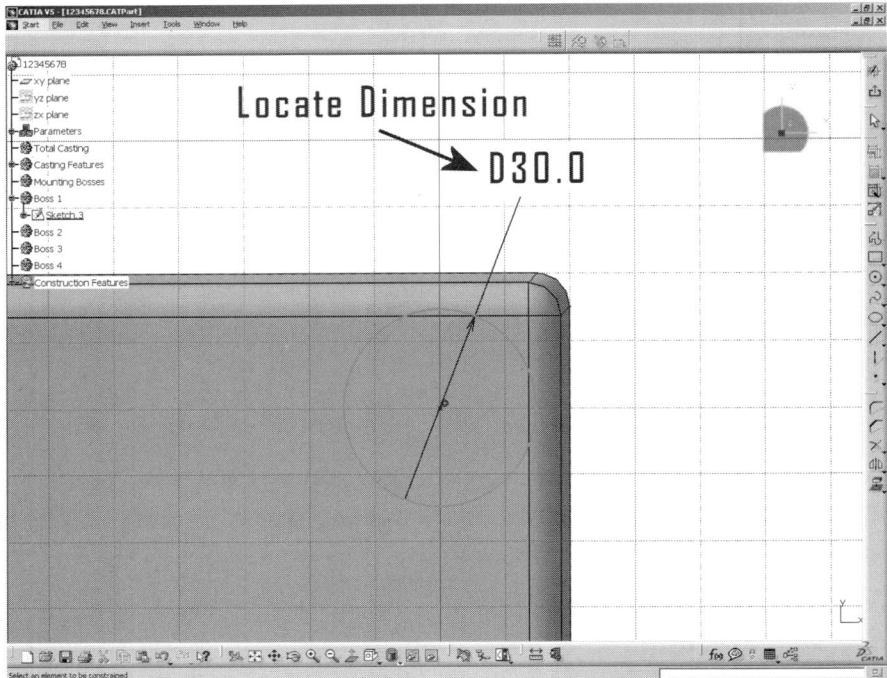

Fig. 6-13. Dimensional constraint established.

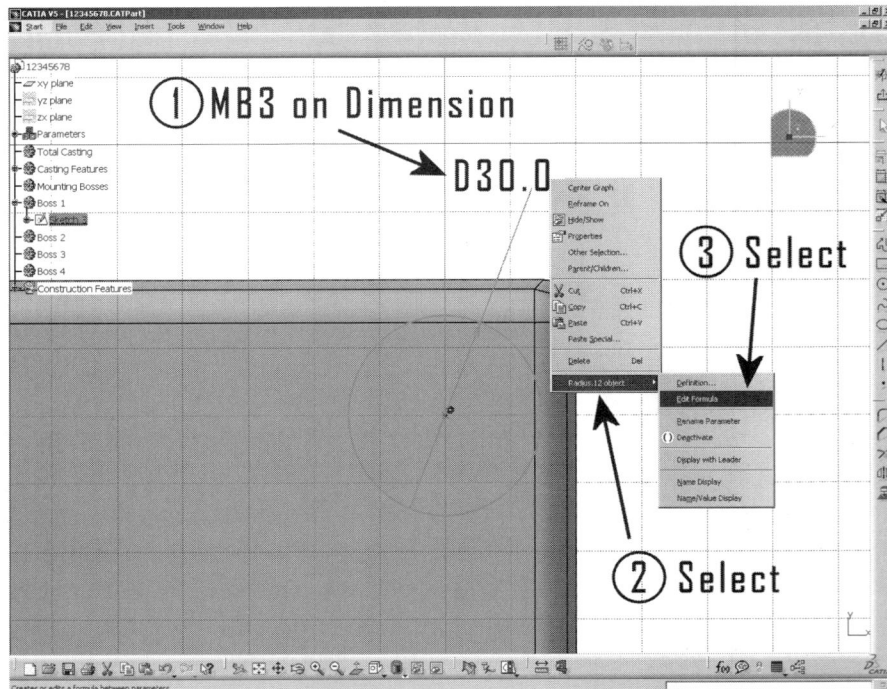

Fig. 6-14. Edit Formula option.

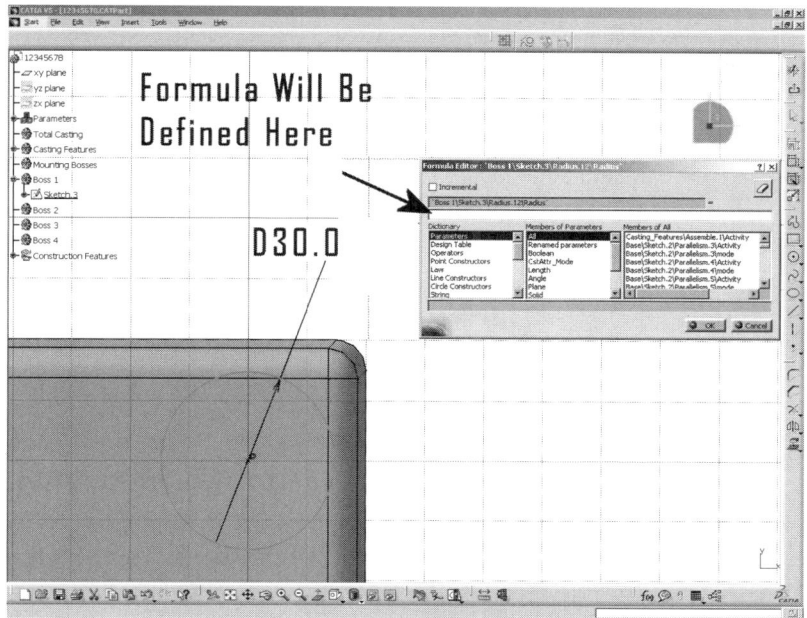

Fig. 6-15. Formula input box.

15 Access the V5 Specification tree and select the parameter *Fastener Bolt Size*. This parameter should translate into the box, as shown in figure 6-16.

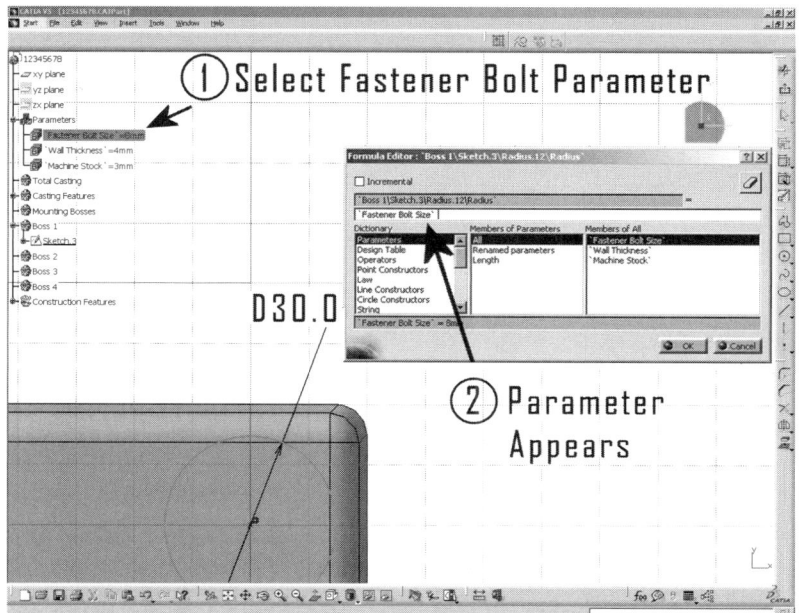

Fig. 6-16. Parameter selection.

16 In the box, enter */2+* immediately following the *Fastener Bolt Size* parameter selection name. Return to the V5 Specification tree and select *Wall Thickness*. This parameter should translate after the */2+*, as shown in figure 6-17.

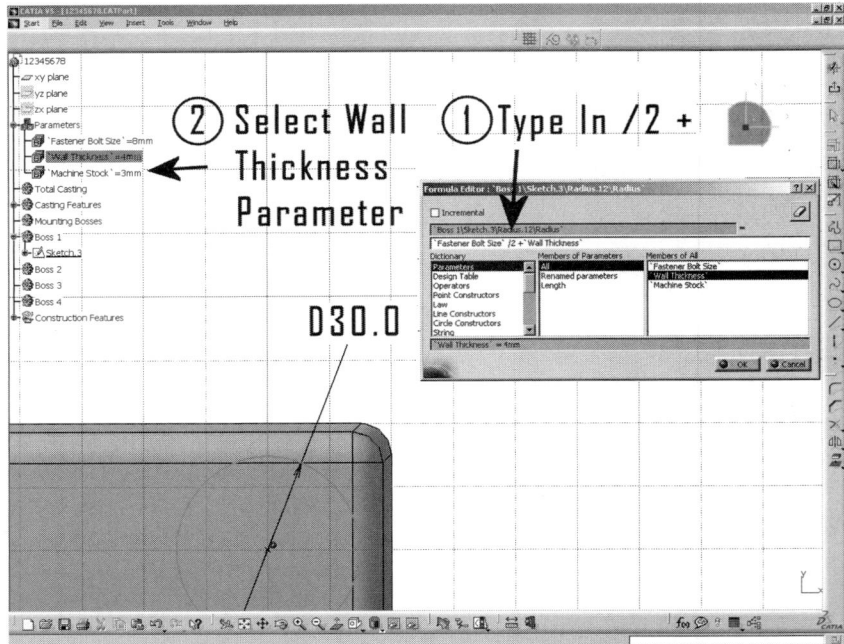

Fig. 6-17. Wall Thickness *parameter.*

17 Click on OK to complete the formula. Exit the Sketcher to complete the profile.

You should now have a dimension that reads *D16 f(x)*, which indicates a formula has been introduced to calculate this dimension, as shown in figure 6-18.

Remember that the profile is based on a radius. The parameter for the bolt size is based on a diameter. Thus, the formula we created is *R=10/2+4*, which equals a diameter of 18.

18 Create a solid feature by clicking on the Pad icon and selecting the previous sketch profile. A temporary feature will be created and the Pad Definition box will appear, offering various input options, as shown in figure 6-19.

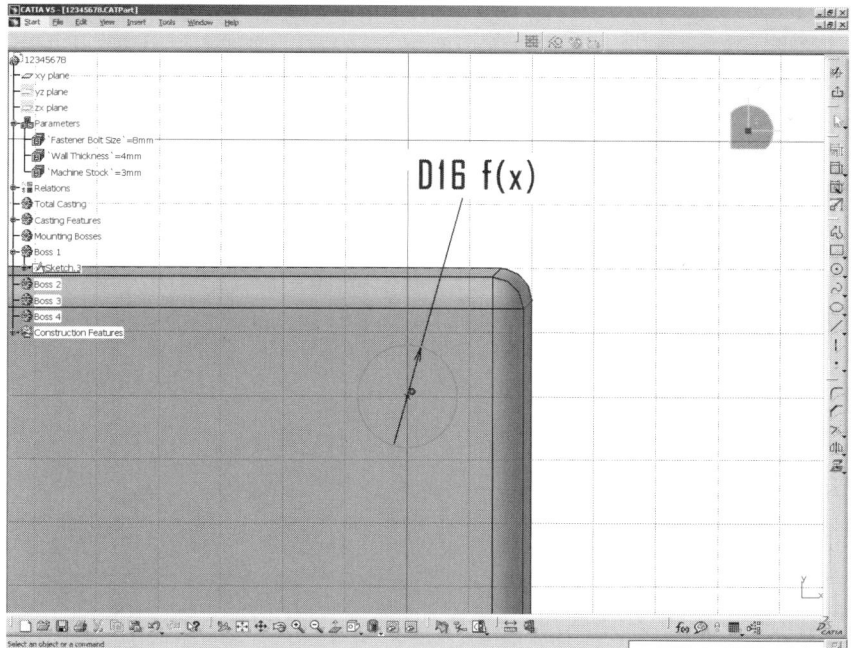

Fig. 6-18.
Dimension
formula.

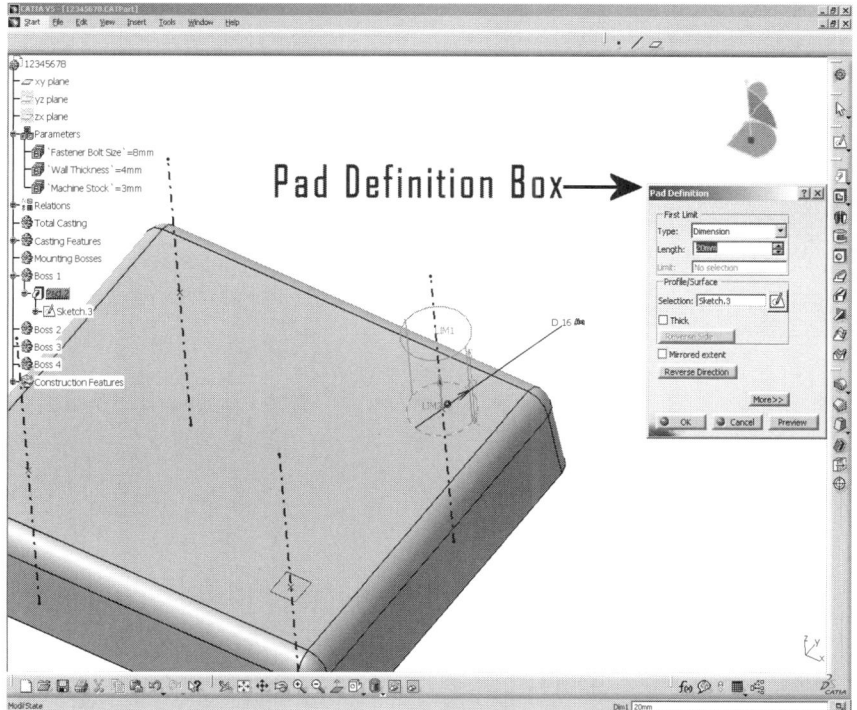

Fig. 6-19. Pad
Definition box.

19 Place the cursor over the length input section. Hold down MB3 and select Edit Formula from the pop-up box, shown in figure 6-20. (The Formula editor will appear to allow for the creation of a relationship.)

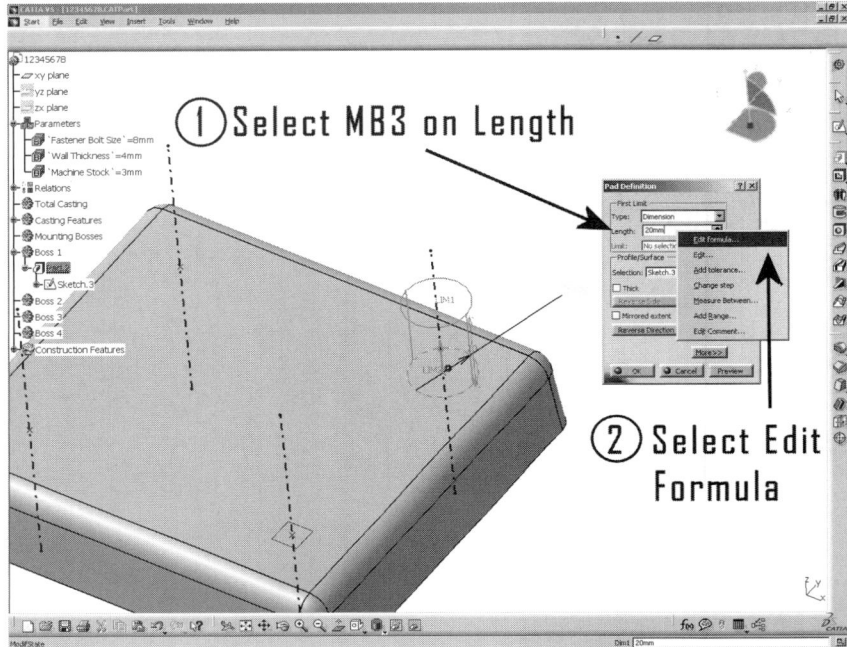

Fig. 6-20. Pad Formula menu.

20 In the first box below the Incremental checkbox you will see the CATIA definition of the pad. Access the V5 Specification tree and select the parameter *Fastener Bolt Size.*

This parameter should translate into the box, as shown in figure 6-21.

21 In the box, enter *1.5+ immediately following the parameter selection name. Return to the V5 Specification tree and select *Machine Stock.* Type a plus sign (+) again, and then return to the V5 Specification tree and select *Wall Thickness.* See figure 6-22.

22 Click on OK to complete the formula.

The Pad Definition box reappears. The length value in the Pad Definition box is now deselected (not highlighted) and cannot be

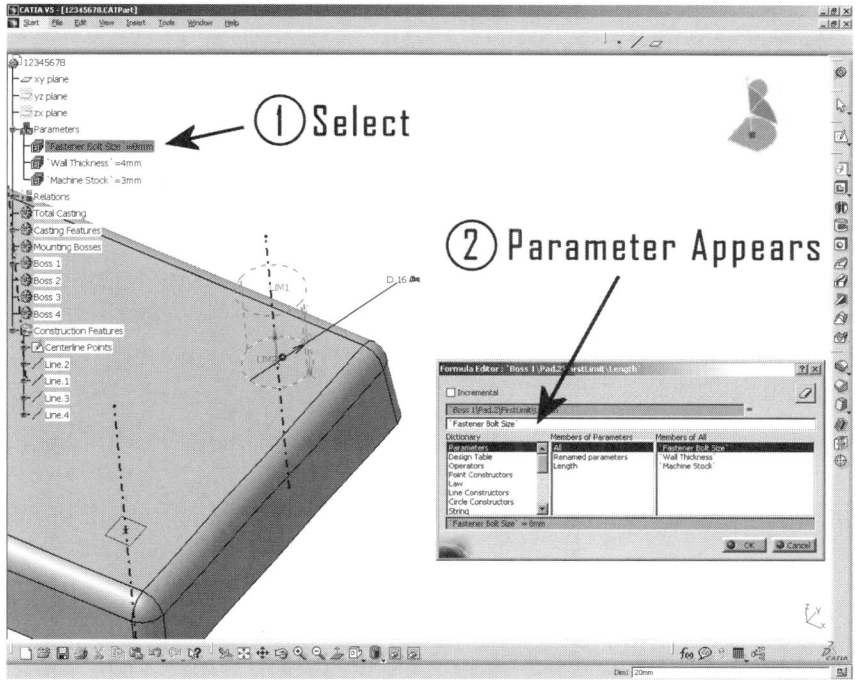

Fig. 6-21.
Fastener Bolt
Size *parameter*
selection.

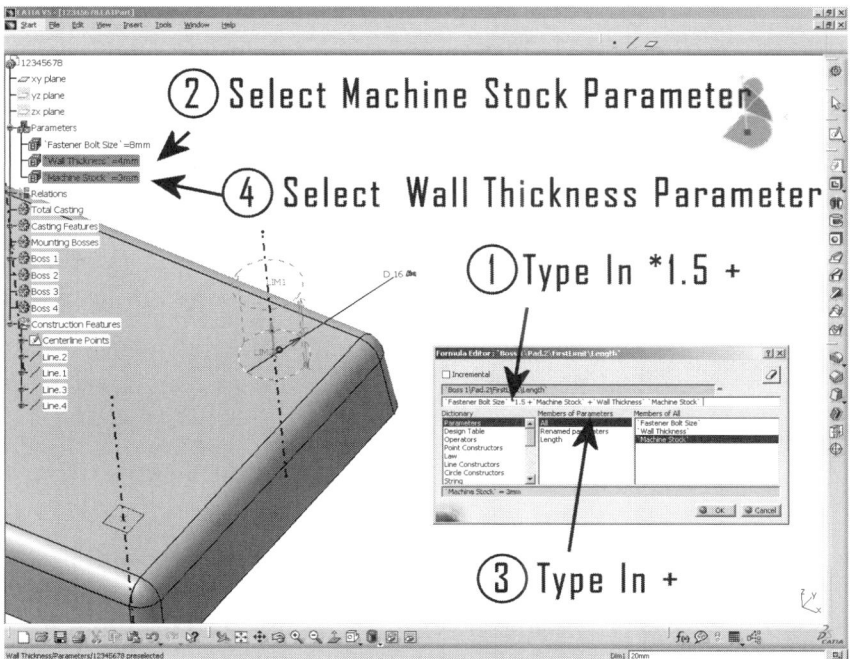

Fig. 6-22.
Machine Stock
parameter
selection.

selected. You will also note a new f(x) symbol next to the input box, as shown in figure 6-23.

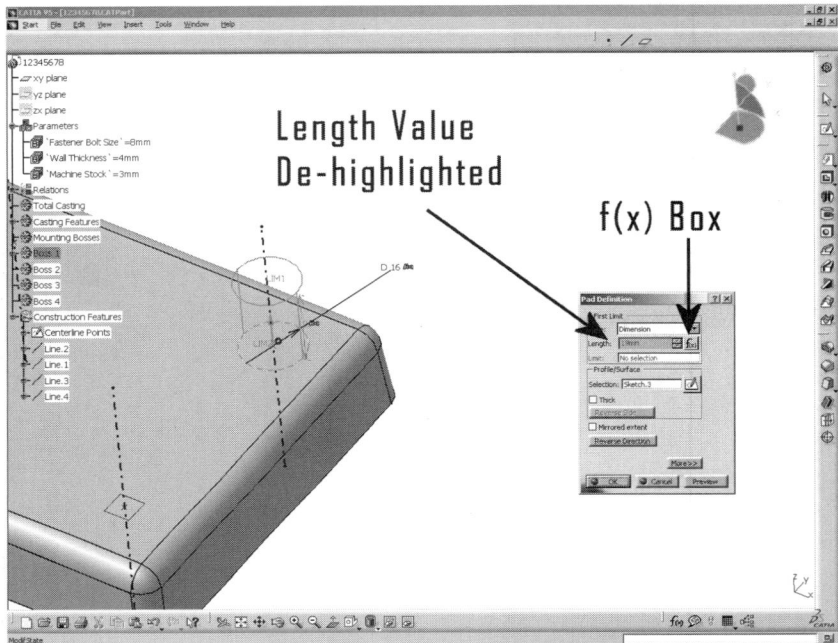

Fig. 6-23. Pad Definition box showing formula.

23 Selecting the f(x) box will put you back into the Formula Editor box for any modifications. Click on OK in the Pad Definition box to complete the feature.

24 Add a 5-degree draft to the sides and a 3-mm round feature to the top of each boss.

25 Create three additional corresponding boss features, draft, and fillets in their respective bodies (bosses 2, 3, and 4).

26 Using the Assemble Boolean operation, assemble each boss body (1 through 4) into the *Mounting Bosses* body. Assemble the *Mounting Bosses* body to the *Casting Features* body.

27 With the Casting Features body active, create a fillet at the intersection of each boss and the base plate.

The *Casting Features* tree structure should appear as shown in figure 6-24. The diameters and lengths of these bosses can now be modified by simply double clicking and modifying the parameters from the Specification tree.

28 As the final operation, assemble the *Casting Features* body to the *Total Casting* body as the last step in the creation of the cast model. The completed model should resemble that shown in figure 6-25.

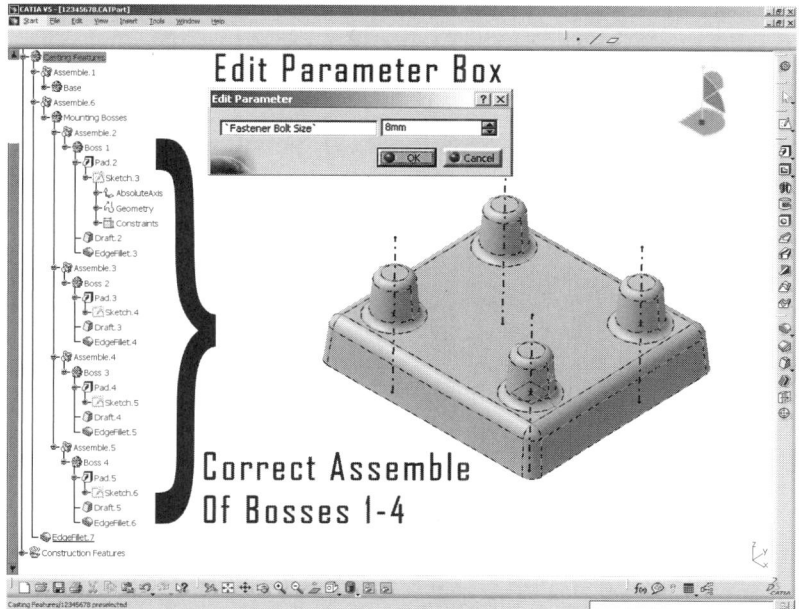

Fig. 6-24. Casting Features tree structure.

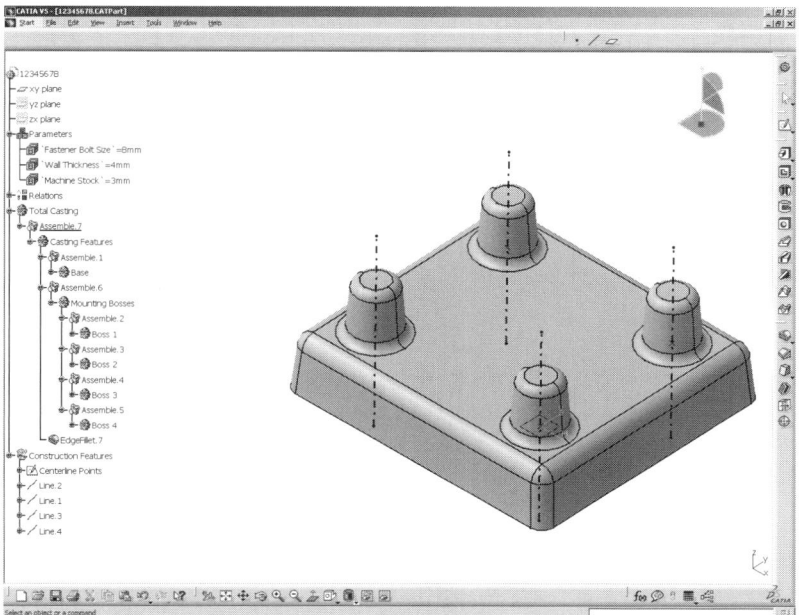

Fig. 6-25. Completed cast model.

If your model looks good at this point, we can take this example one step further and create an associated machined part from this cast part. Verify that the cast model has updated and has been saved to your hard drive, and then continue with the following steps.

29 Create a new part design model and name it *Machined Part.* Select Window | Tile Vertically.

This window arrangement, shown in figure 6-26, will make the copying from the cast part to the machined part easier.

Fig. 6-26. Revised window arrangement.

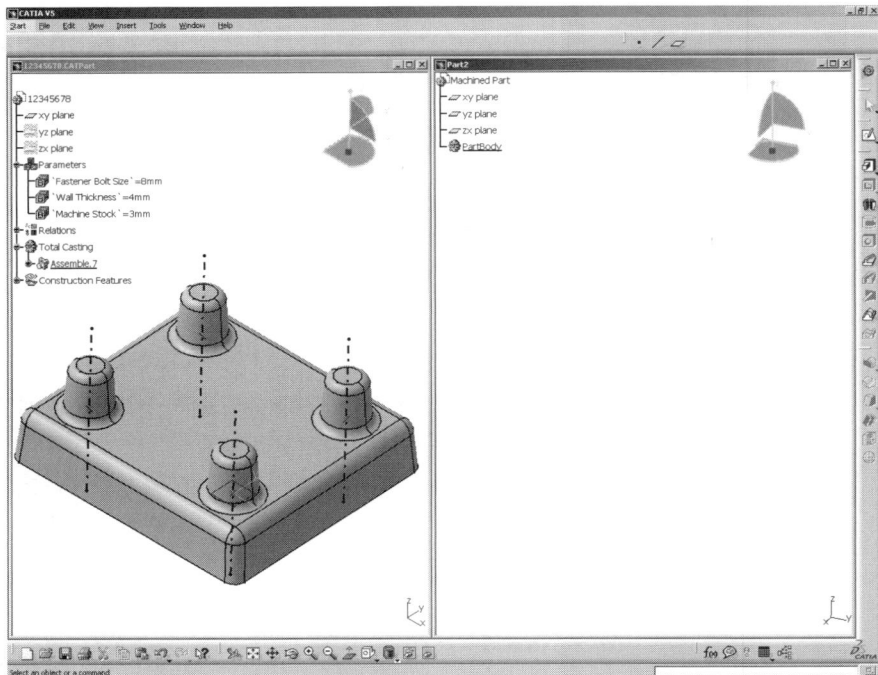

30 Enter the cast part model, and while holding down the Ctrl key use the cursor to highlight all three parameters. With the cursor on one of those parameters, hold down or click MB3 and then select Copy. See figure 6-27.

31 In the Machined Part window, position the cursor on top of the part name at the top of the Specification tree. Hold down or click MB3, and then select Paste Special. Select the As Result With Link option from the menu, shown in figure 6-28.

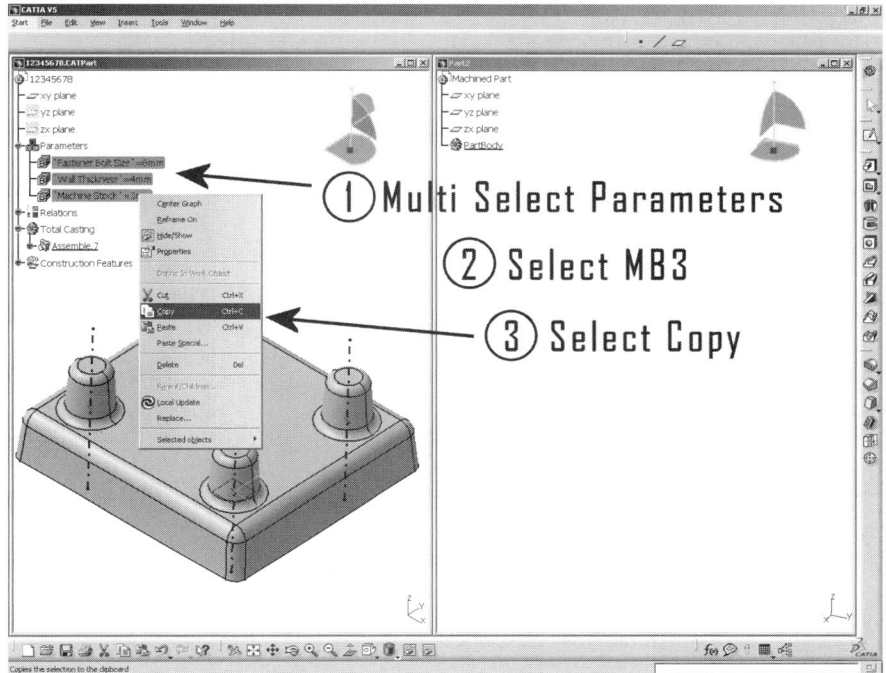

Fig. 6-27.
Copying
parameters.

① Multi Select Parameters
② Select MB3
③ Select Copy

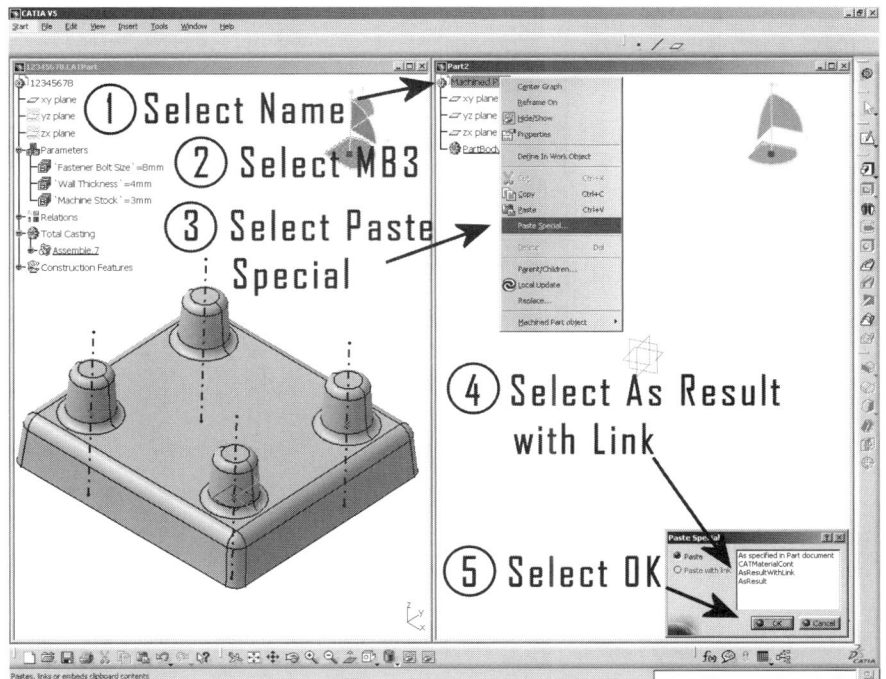

Fig. 6-28. Paste
Special options.

① Select Name
② Select MB3
③ Select Paste Special
④ Select As Result with Link
⑤ Select OK

32 A node will appear on the specification named *External Parameters*. Expand this node and you will see all of the copied/ linked parameters from the cast part, as shown in figure 6-29.

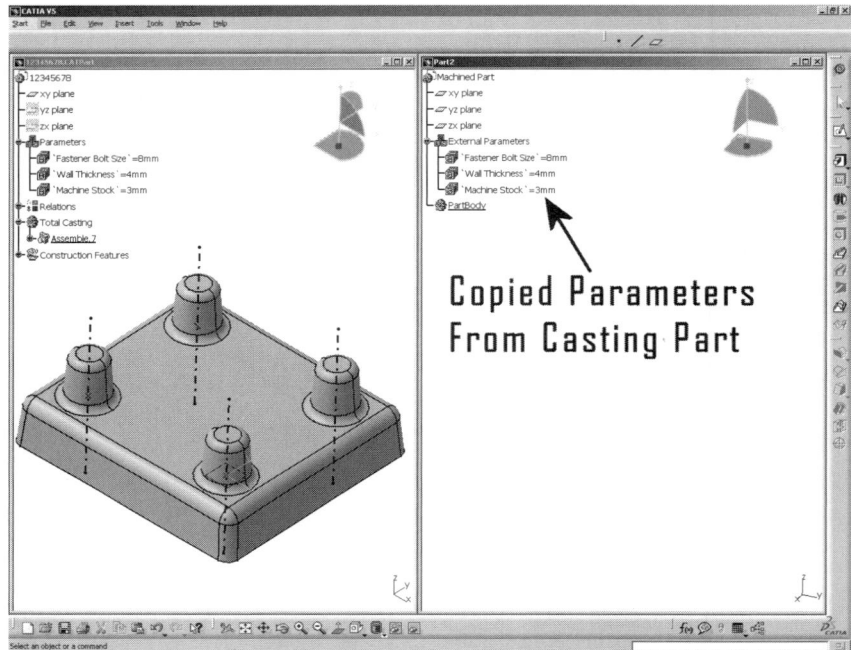

Fig. 6-29. Linked parameters from cast part.

33 Follow the same steps as the previous operation and copy the four centerlines from the cast part. In the Machined Part specification tree you will notice a new node named *External References*. If you expand this node, shown in figure 6-30, you will see the copied/linked centerlines from the cast part.

34 Follow the same steps and copy the *Total Casting* body from the cast part. In the Machined Part specification tree you will notice a new node named *Result of Total Casting*. If you expand this node, shown in figure 6-31, you will see *Solid.1*. Expand the *Machine* model to full screen.

Everything that has been copied from one model to the other is indicated by an icon in front of its name. The icons with green diamonds on them, as shown in figure 6-32, mean that these elements are linked and synchronized with the model they came from.

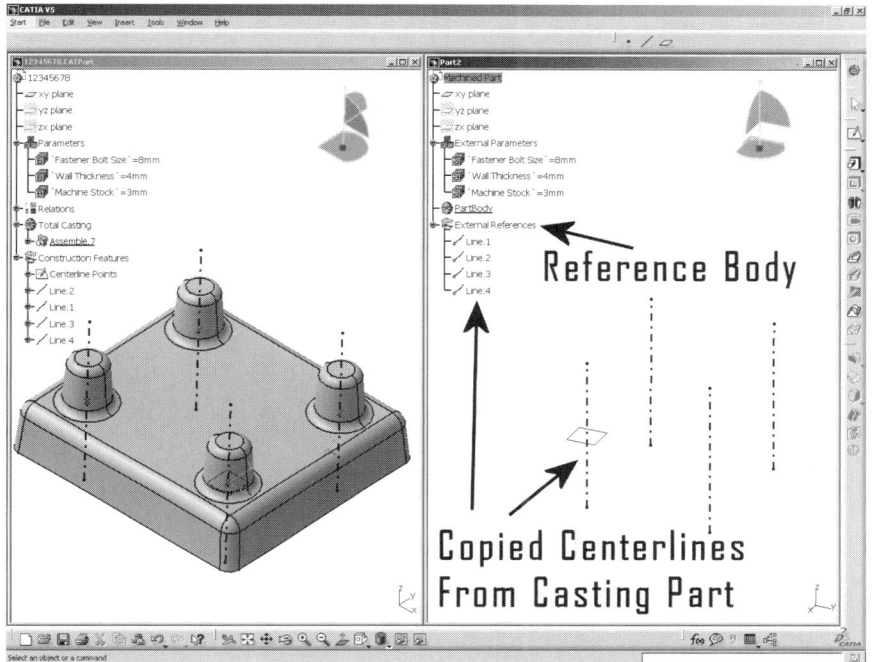

Fig. 6-30. External References *node.*

Reference Body

Copied Centerlines From Casting Part

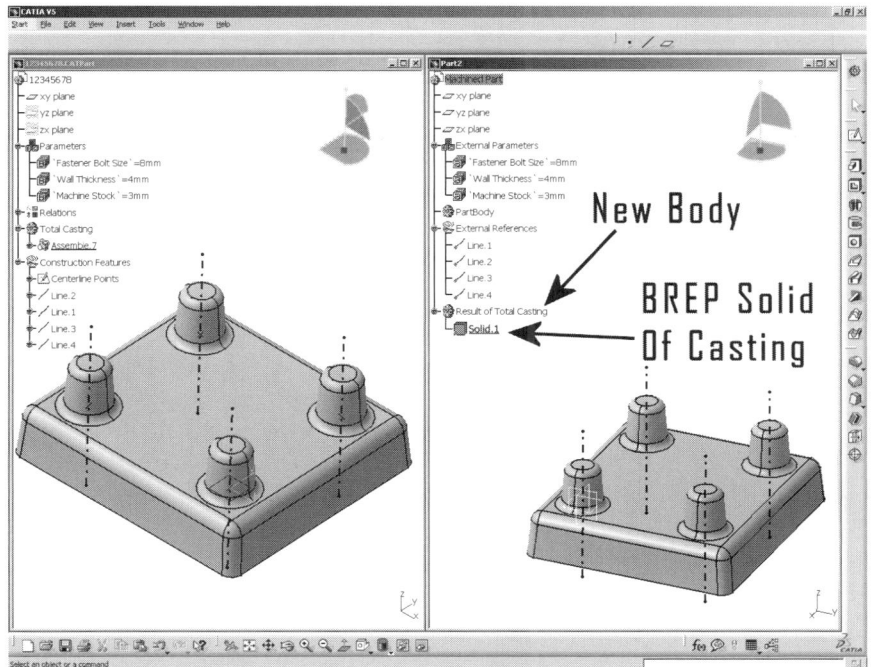

Fig. 6-31. Total Casting *copy.*

New Body

BREP Solid Of Casting

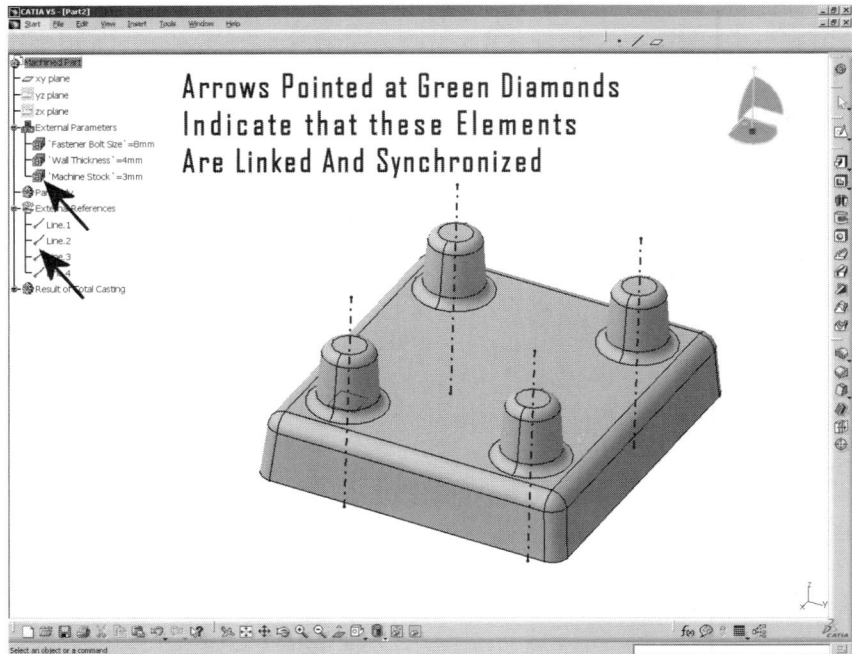

Fig. 6-32. Icons
indicating
linked
elements.

35 Insert two (2) new part bodies into the model and name them
Machine Stock Cut and *Machined Holes*. Make the *Machine Stock
Cut* body active. Select the top of one of the bosses and then
enter the Sketcher workbench. Create a rectangular profile
larger than the four mounting bosses. Constrain the sides of
the sketch so that the walls of the rectangle are constrained to
the centerlines of the cast model. Exit the Sketcher workbench to complete the profile. Click on the Pocket icon on
the Sketch-Based Features toolbar.

The Pocket Definition box appears, with the two new part bodies
indicated, as shown in figure 6-33.

36 Place the cursor in the length value input box. Holding down
MB3, select Edit Formula from the contextual pop-up menu,
as shown in figure 6-34.

The Formula editor will appear. This will be used to impose a formula relationship. Continue with the following steps.

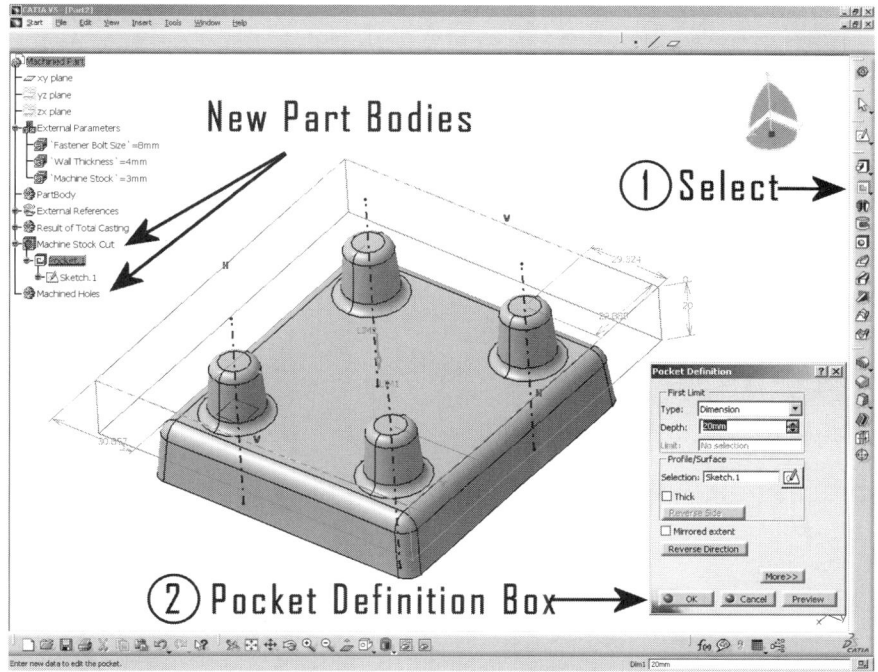

Fig. 6-33. New part bodies indicated in Pocket Definition box.

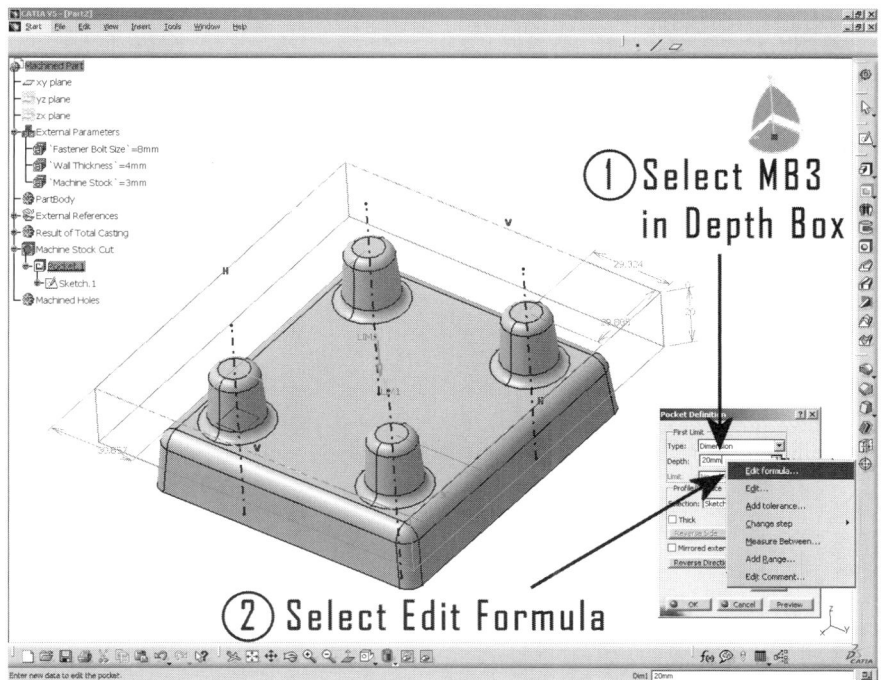

Fig. 6-34. Selecting Edit Formula for the length value.

37 In the first box below the Incremental checkbox you will see the CATIA definition of the pocket. Access the V5 Specification tree and select the external parameter *Machine Stock*. This parameter should translate into the box, as shown in figure 6-35. Click OK to complete.

Fig. 6-35. Machine Stock parameter selection.

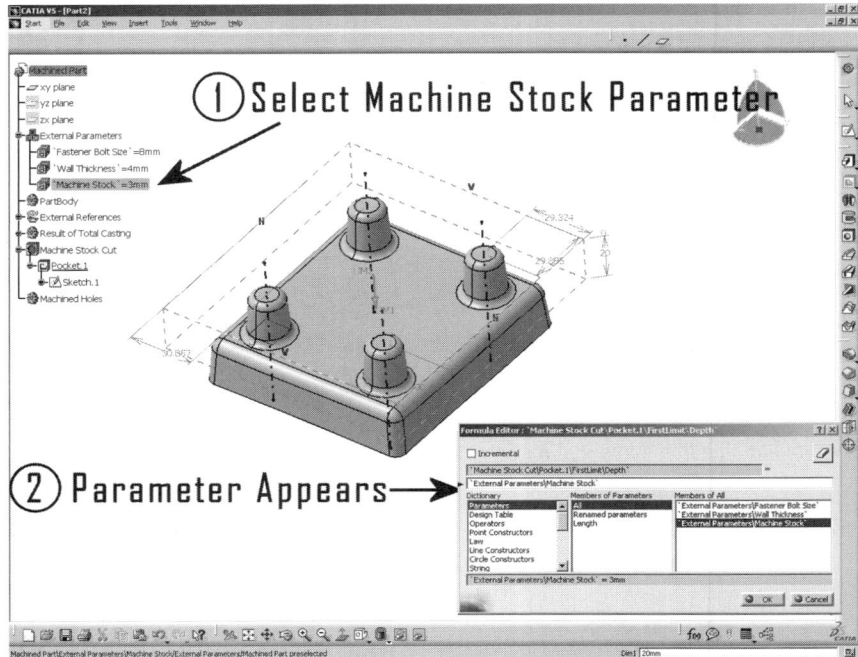

38 Click on OK to complete the formula.

The Pocket Definition box reappears. The depth value in the Pocket Definition box is now deselected (not highlighted) and cannot be selected. You will also notice a new f(x) box next to it, as shown in figure 6-36.

39 In the Pocket Definition menu, click on the More button. In the Depth box, shown in figure 6-37, enter a value of *10*. Click on OK. The new pocket feature is created in the *Machine Stock Cut* body, as shown in figure 6-38.

40 Assemble the *Machine Stock Cut* and *Result of Total Casting* bodies. Note the result, shown in figure 6-39.

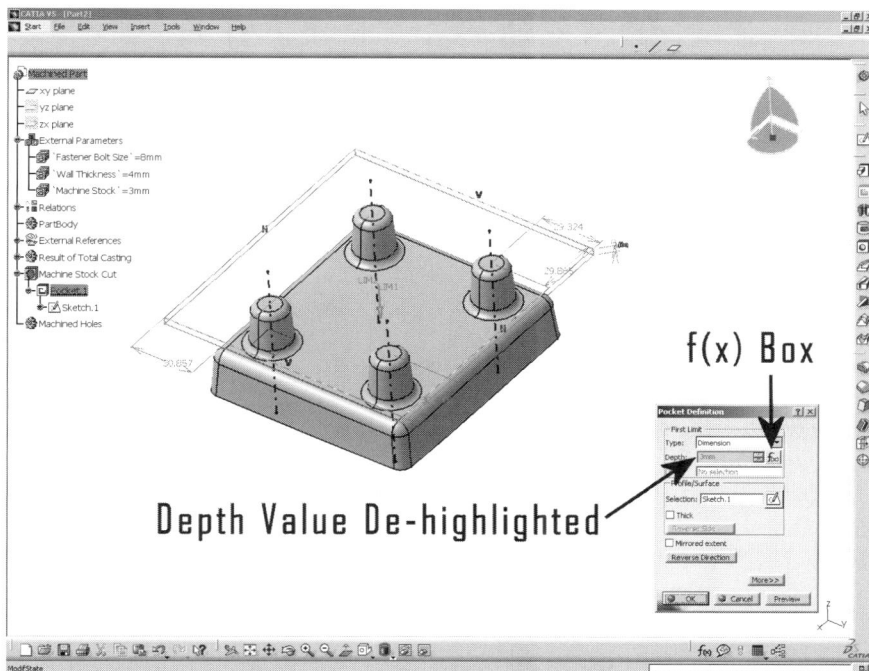

Fig. 6-36.
Depth value
deselected
and f(x) box
indicated in
Pocket
Definition box.

f(x) Box

Depth Value De-highlighted

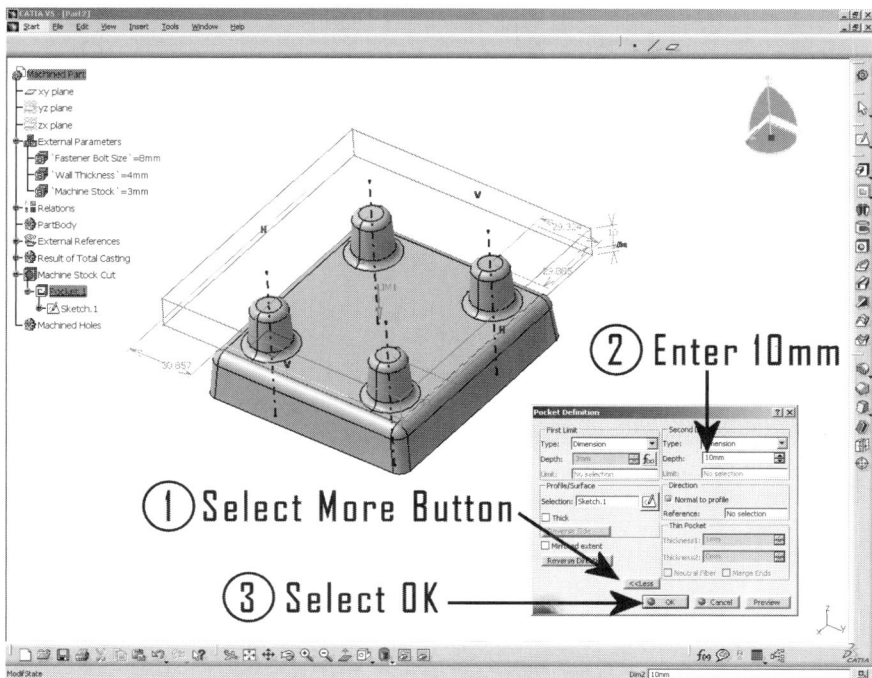

Fig. 6-37.
Second
direction
depth value.

② Enter 10mm

① Select More Button

③ Select OK

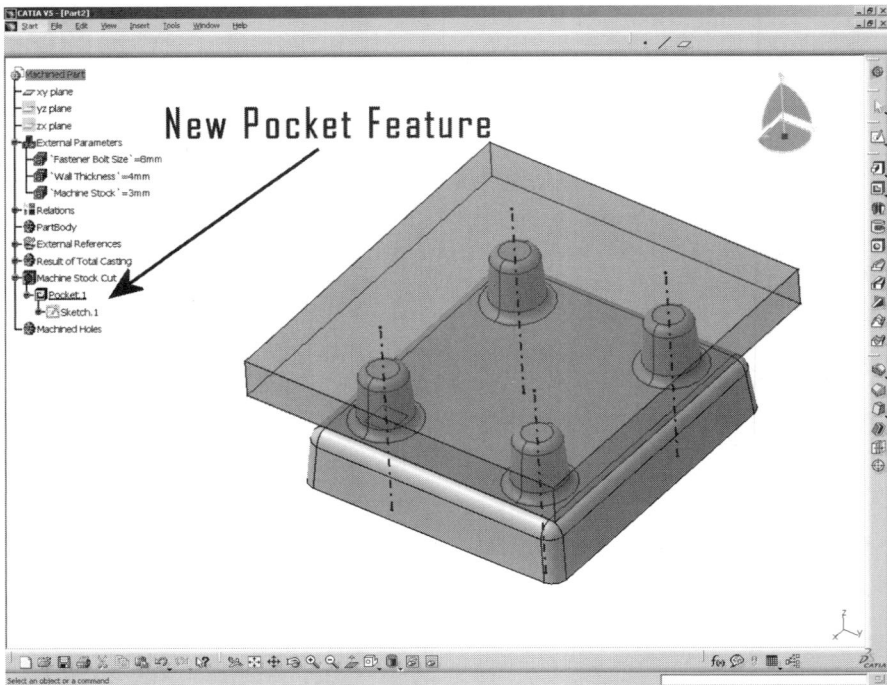

Fig. 6-38.
Pocket
feature
created in the
Machine
Stock Cut
body.

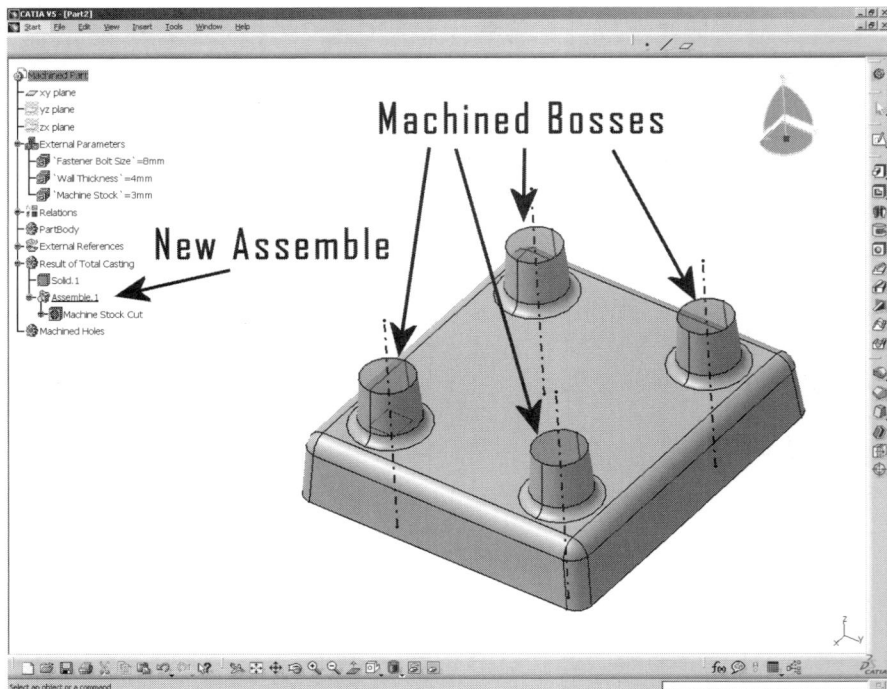

Fig. 6-39.
Result of
Machine Stock
Cut and Result
of Total Casting
assembly.

41 Next, place the machined holes in this part. Access Define In Work Object on the part body named *Machined Holes*. Select the top of what would be *Boss 1* if we were in the cast part. Click on the Hole icon on the Sketch-Based Features toolbar. The Hole Definition box appears, as show in figure 6-40.

Fig. 6-40. Machined Holes part body indicated in Hole Definition box.

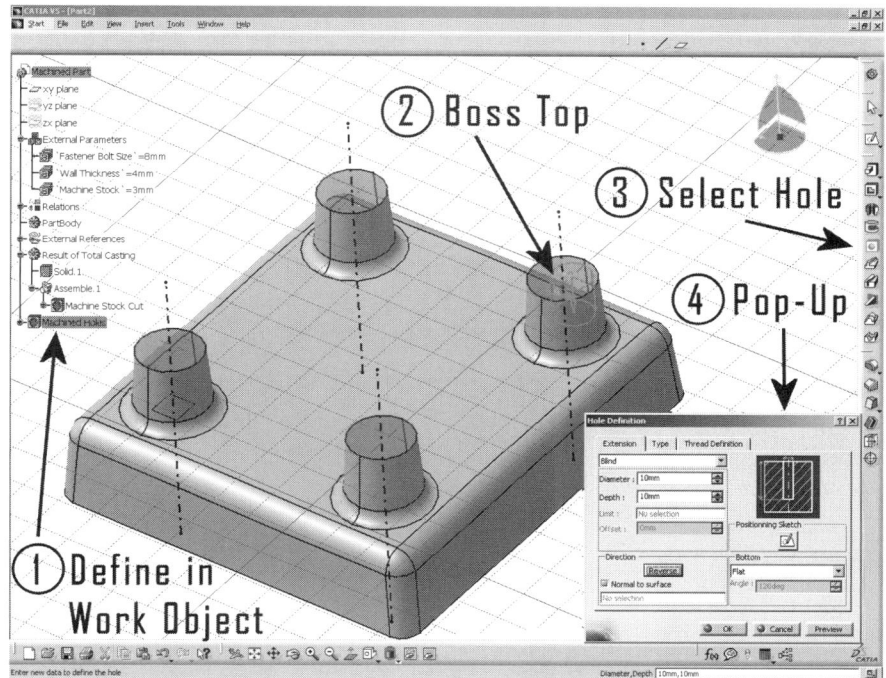

42 Place the cursor in the Diameter value box, and while holding down MB3 select Edit Formula, as shown in figure 6-41. The Formula editor will appear.

43 Within the Formula Editor menu, select the F*astener Bolt Size* parameter from the External Parameters section of the Specification tree, This parameter selection will appear in the empty box, as shown in figure 6-42.

44 Click on OK to complete the formula. The Hole Definition box reappears. The diameter value in the Hole Definition box is now deselected (not highlighted) and cannot be selected. You will also notice a new f(x) box next to it, as shown in figure 6-43.

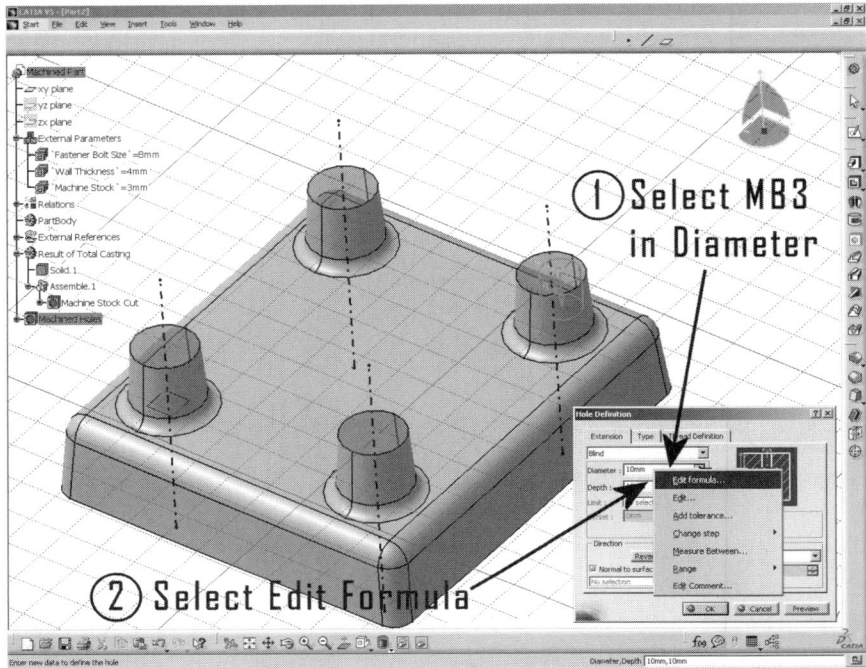

Fig. 6-41. Selecting Edit Formula for the diameter value.

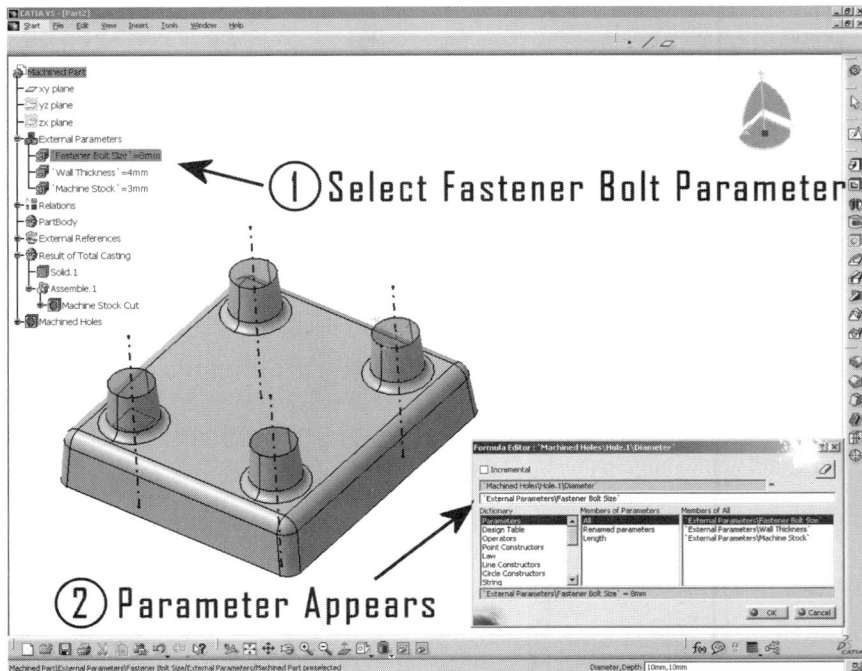

Fig. 6-42. Fastener Bolt Size parameter selection.

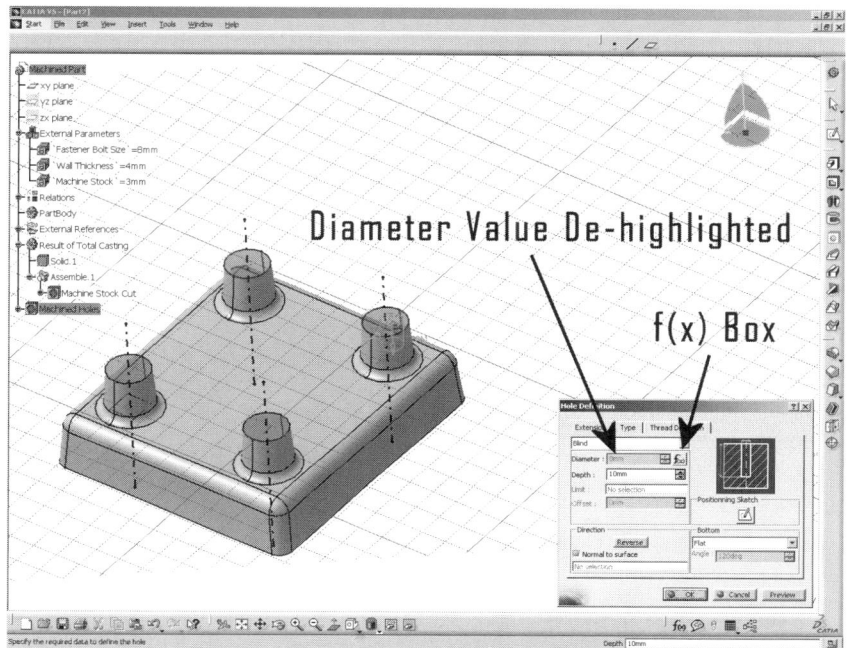

Fig. 6-43. Diameter value deselected and f(x) box indicated in Hole Definition box.

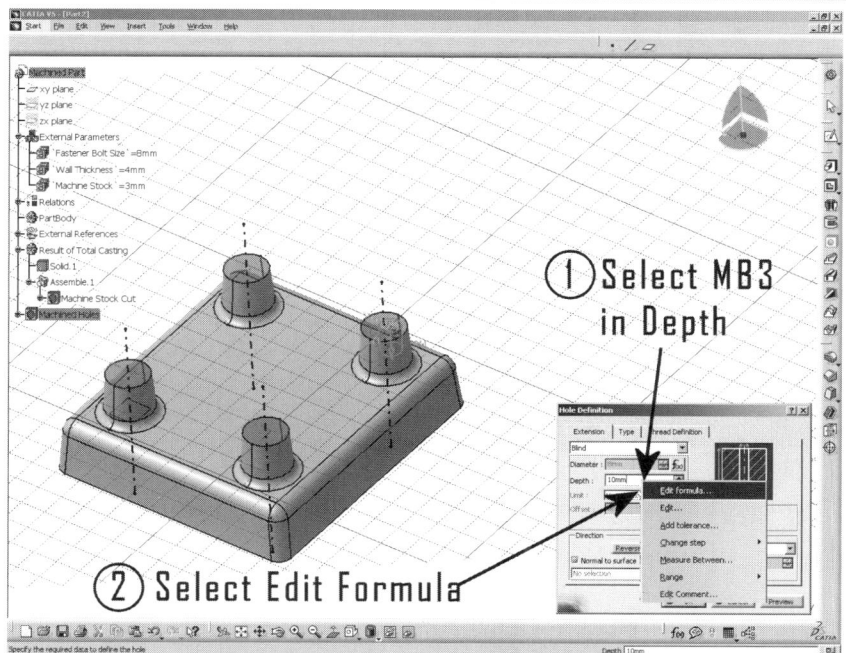

Fig. 6-44. Selecting Edit Formula for the depth value.

45 Place the cursor on the depth value. Holding down MB3, select Edit Formula, as shown in figure 6-44.

46 In the Formula Editor box, select the external parameter *Fastener Bolt Size* from the Specification tree and enter **1.5*. Click on OK. The defining hole is created, as shown in figure 6-45.

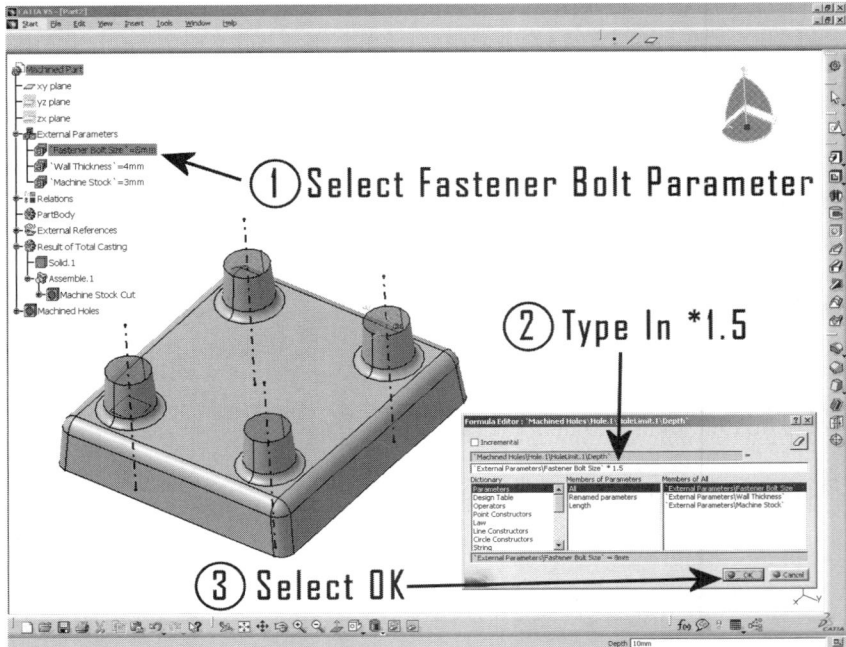

Fig. 6-45. The defining hole has been created.

We now need to Create an association that locates the hole to the cast boss feature. Continue with the following steps.

47 Click on the Positioning icon, located within the Hole Definition Window. (This automatically enters into the Sketcher workbench and the 2D sketched profile of the hole.) Locate the *Hole* feature within the Specification tree and the *Machined Hole* body. Expand the node to display the geometry branch that contains the Hole center point, as shown in figure 6-46.

48 Create a coincidence constraint between the *Hole* center point and the external reference centerline, as shown in figure 6-47.

49 Create three (3) additional holes with the *Machined Hole* body. Make sure to create the correct constraint to the appropriate external centerline to maintain the correct associative linkage, as shown in figure 6-48.

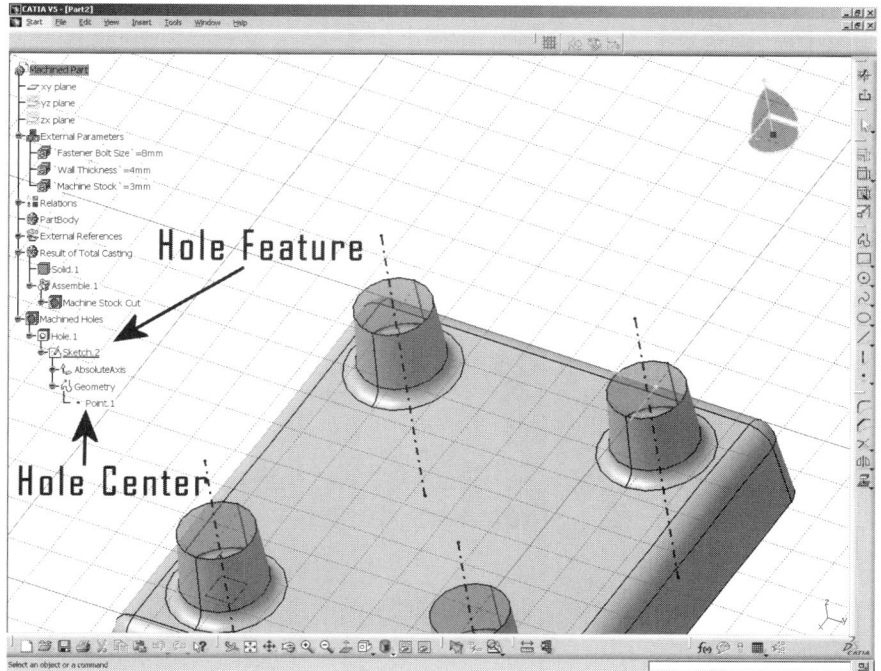

Fig. 6-46. Hole center Specification tree branch.

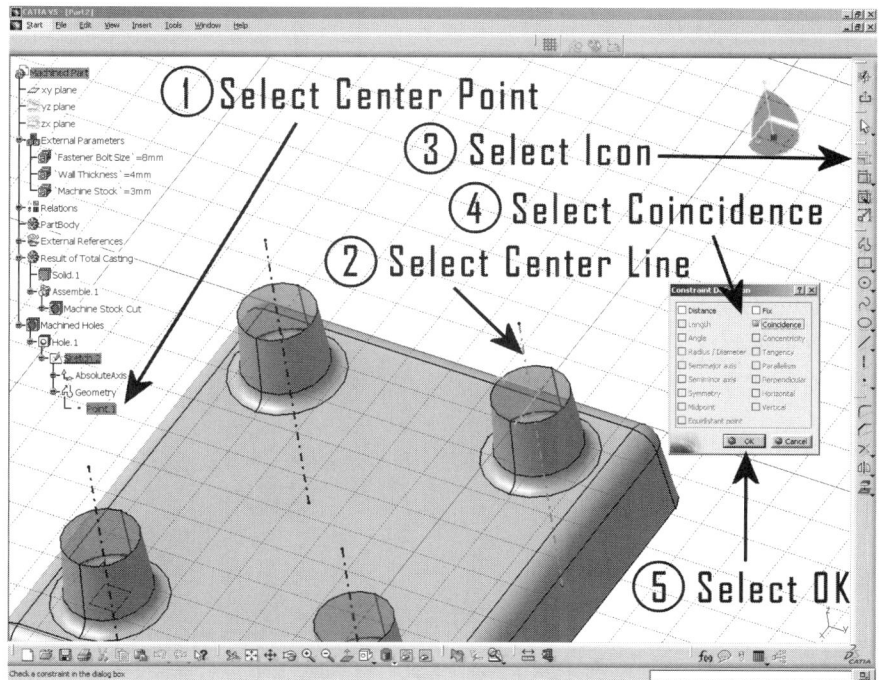

Fig. 6-47. Coincidence constraint.

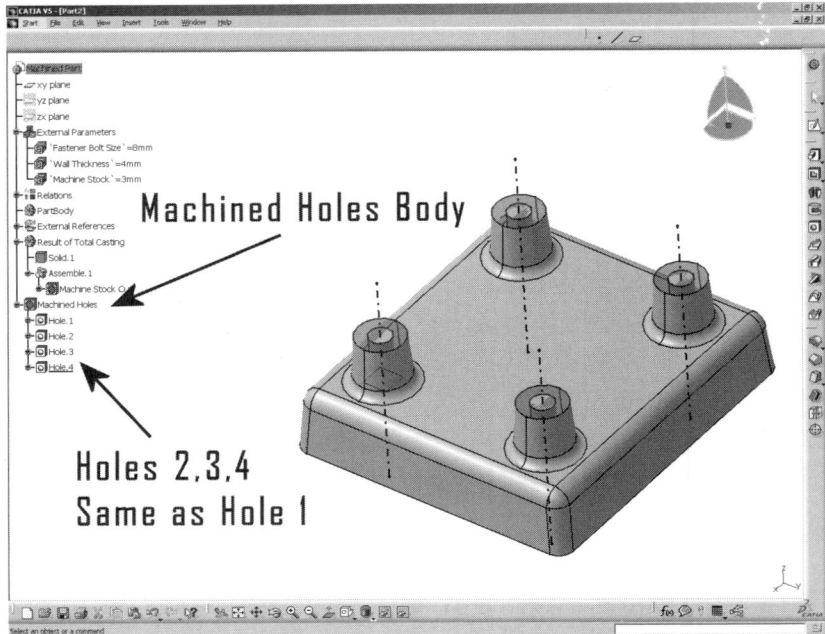

Fig. 6-48.
Additional hole
features.

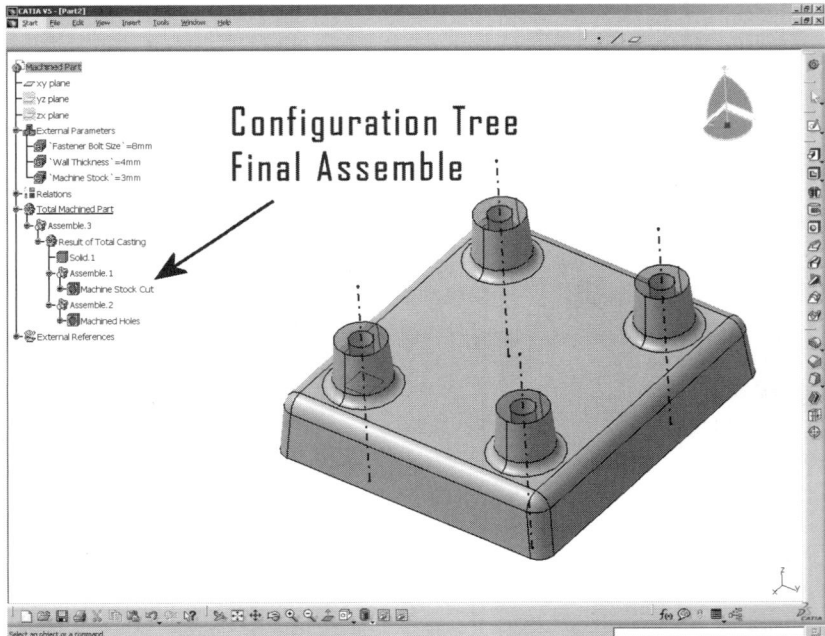

Fig. 6-49. Final
Specification
tree structures.

50 Assemble the *Machined Hole* body to the *Result of Total Casting* body, and as the final operation assemble the Total Casting body to the default part body. Name the default part body *Total Machined Part.* The final Specification tree structure should now resemble that shown in figure 6-49.

The holes are now constrained to the centerlines in the machined part. The centerlines are linked to the cast part centerlines. If the cast part centerlines ever move, the bosses in the cast part and the holes in the machined part will move with them. This is what is meant by intelligent associativity.

Summary

Building flexibility into a model is critical for effectively accommodating design changes. Many of the concepts explored in this chapter offer insights into developing a flexible modeling style. Model and modeling-process organization, as well as anticipation of potential changes, are issues that need to be addressed in regard to any project.

The issues of feature completeness and the use of constraints are particularly relevant at the beginning (concept phase) of a design project. An example of a strategy pertinent here is to stay away from too much feature definition and sharing of dress-up features until you can be reasonably certain of their final locations. This allows the model to be as flexible as possible in meeting the challenges of design change, which otherwise might require significant rebuild (or even redesign).

This type of built-in flexibility also allows you to more easily manage existing features in relationship to further features created at any given point in the process. Combining this strategy of flexibility with techniques such as the use of parameters, formulas, and degrees of associativity (and with a well-thought-out design plan) arrives at a model that is maximally robust. All of the factors explored in this chapter impact the quality of the model and how robustly it can accommodate and facilitate design change.

Review Questions

1 Why is breaking down a design so important in building a robust model?

2 What are some of the benefits of keeping models simple and maturing them over time?

3 (True or false?): All geometry must be fully constrained to create a robust model.

4 Why is it important to flex a model throughout the geometry creation process rather than at the end of the design process?

5 Is it better to make root modeling changes to a model or to simply add layers of geometry on top of existing features?

6 Why is it important to name features when creating links between models?

CHAPTER 7

THE WIREFRAME AND SURFACE WORKBENCH

Introduction

THIS CHAPTER INTRODUCES the Wireframe and Surface work-bench, which allows you to create wireframe and surface elements at any time to enhance the design of solid models. These elements enrich the capability of a feature-based design and complement the Part Design workbench from the standpoint of a hybrid surface-/solids-based modeling approach.

This hybrid approach offers a productive and intuitive design environment in which to capture and reuse design methodologies and specifications. Understanding the basics of combining wireframe, surface, and solid geometry is paramount is building robust models. This flexibility also offers the ability to incorporate post-design and local 3D parameterization in capturing part design intent.

Objectives

This chapter covers the following.

- Modeling methods
- CATIA V5 surface functionality
- Wireframe and surface features
- Wireframe and surface operations

Modeling Methods

The sections that follow explore modeling methods. These include solid modeling, surface modeling, and a hybrid of the two.

Solid Modeling

Solid modeling systems have become very popular over the last ten years. This approach to mechanical design is relatively easy, wherein a new user can begin to create simple to intermediate-level designs very quickly. The term *solid modeling* refers to a method whereby parts are designed by combining solid objects or features into a 3D mechanical part. Solid models can be created by adding or subtracting primitive shapes, such as cones, toruses, spheres, and cylinders. Solid models are also created using 2D sketched profiles as a basis for pads, pockets, shafts, ribs, and loft features. The two primary types of solid modeling systems are constructive solid geometry (CSG) and boundary representation (B-rep).

The power of solid modeling resides in how efficiently an object can be defined by combining various objects to create the desired shape. Because the final design is a 3D solid volume, features of the model can be associated (linked) with one another to represent an overall linked topology. Dress-up features such as thickness, draft, and fillets are added as detail to enhance a 3D solid part.

The parametric associative design capability makes for quick and easy updates of geometric constraints within the design model. However, there are some limitations in modeling certain complex shapes by only having basic pad, pocket, shaft, rib, and loft functionality. Solid modelers are not very satisfactory at handling complex geometries with free-form, sculptured surfaces. Such shapes are essential to any product for which aesthetics or ergonomics are an important consideration.

Surface Modeling

Surface modeling is for designs that require sculptured shapes, surfaces with a higher degree of curvature change, and requirements for a higher level of control over the outer surface of the design. It is virtually impossible to describe some shapes using only solid model features (e.g., pads and pockets). Wireframe and surface features basically work by building a series of "skin" features that break the design down into its fundamental root shapes. Surface features may be created in the same manner as solid features (e.g., extrusions, revolves, sweeps, fills, lofts, and blends) or from a subset of underlying curves.

Creating surfaces from a wireframe subset inherently offers an advantage in that just about any shape may be created. However, this process offers a disadvantage in that it typically takes more time to create these complex surface shapes. Surface modeling also requires a higher level of understanding of the geometrical hierarchy of control curves, control points, tangent and curvature blending, knots and weights, trim curves, and much more. Furthermore, extra operations are required to rejoin these elements into a usable design. Table 7-1 compares solid and surface modeling.

Table 7-1: Comparison of Solid and Surface Modeling

Solid Modeling	Surface Modeling
Ease of use and learning	Ability to define and create complex geometry
Parametric/associativity	Ability to break down the design to root level
Quick feature creation and updating of model history	Quicker creation and definition of complex designs
Good for substantial, well-defined models	Excellent for creating aesthetic or free-form models

Hybrid Modeling

Hybrid modeling refers to design applications that employ both surface and solid modeling operations. They combine the ease of use of solids with powerful surface creation and editing tools. This provides the best of both methods and offers you unlimited control over the design. Models can contain traditional solid features (such as pads, pockets, holes, fillets, and chamfers) with open sets or bodies of wireframe and surfaces. Surfaces can be added, removed, or replaced to offer a complementary control as they interact with solid model features.

These surface model features may be parametrically associative and rejoined with the solid to capture the overall design intent of the part. This technology offers a new approach to design with respect to complex free-form shapes that require attention to aesthetics and ergonomics. CATIA V5 offers the best of both worlds in that a model may contain both solid and surface features bound by a seamless common interface and architecture. This hybrid approach combines the speed and ease of use of a solid modeling system with the power and design flexibility of a surface modeler.

CATIA V5 Surface Functionality

CATIA V5 offers several dedicated workbenches focused on wireframe and surface modeling, depending on the level of model complexity and functionality requirements. The following are the four most notable surface workbenches.

- Wireframe and Surface

- Generative Shape Design

- FreeStyle Shaper

- Automotive Class 'A'

NOTE: *Although, CATIA V5 offers a full suite of complex surface modeling functionalities, the intent of this chapter is to capture only basic wireframe and surface functionalities to enhance the solid modeling experience. Advanced surface techniques are beyond the intent of this chapter.*

Wireframe and Surface Workbench

Fig. 7-1. Wireframe and Surface Workbench icon.

The Wireframe and Surface workbench, activated via the Wireframe and Surface Workbench icon (shown in figure 7-1), offers wireframe and basic-level surface geometry creation functionality. These basic surfaces include extrusions, revolves, lofts, fills, and sweeps. These features are intended to complement and interact with solid model features to provide enhanced modeling capability. The functions of this workbench overlap seamlessly with the Part Design workbench.

Generative Shape Design Workbench

Fig. 7-2. Generative Shape Design Workbench icon.

The Generative Shape Design workbench, activated via the Generative Shape Design Workbench icon (show in figure 7-2), includes all functions and commands from the Wireframe and Surface workbench. In addition, this workbench provides an additional set of tools for creating and modifying surfaces used in the design of more complex shapes. This workbench helps you design advanced shapes based on a combination of wireframe and surface creation-and-editing tools.

Freestyle Shaper Workbench

Fig. 7-3. Freestyle Shaper Workbench icon.

The Freestyle Shaper workbench, activated via the Freestyle Shaper Workbench icon (shown in figure 7-3), provides a full suite of free-form surface creation and editing tools for creating aesthetic shapes. These tools can create surfaces from a set of scan points or interactive 3D free-form design. Real-time diagnostic and analysis tools are available for checking the quality of surfaces and how they interact with one another.

Automotive Class 'A' Workbench

Fig. 7-4. Automotive Class 'A' Workbench icon.

The Automotive Class 'A' workbench, activated via the Automotive Class 'A' icon (shown in figure 7-4), is a dedicated module for the creation of automotive exterior and interior surfaces. This workbench combines generative shape modeling with free-form modeling to provide a world-class set of tools for generating and analyzing automotive quality surfaces.

Wireframe and Surface Features

This section explores functionality as it relates to wireframe and surface features. Reference elements, curve projection, curve intersection, and various types of curve elements are discussed under the topic of wireframe features. The various types of surface features are then explained under that topic.

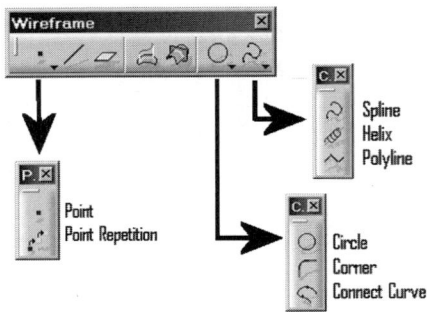

Fig. 7-5. Wireframe toolbar.

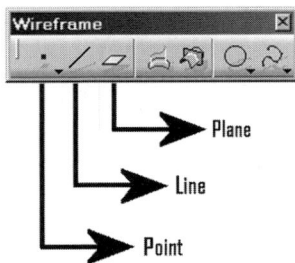

Fig. 7-6. Reference elements.

Wireframe Features

The Wireframe toolbar, shown in figure 7-5, allows for the creation of both construction and curve elements within a part model. These elements are non-sketched features and are used in creating a reference set of entities used to assist in the creation of more complex surface features.

NOTE: *The first time a wireframe or surface element is initiated, CATIA V5 automatically creates an open body within the specification tree that contains these elements.*

Reference Elements

Points, lines, and planes are all reference elements located on the Wireframe toolbar, as shown in figure 7-6. These elements are used to help set up the creation and construction of solid or surface features by further capturing the design intent. These elements typically reside in the open body portion of the history tree. See Chapter 4 for more information on point, line, and plane reference elements.

Curve Projection

The Curve Projection command, shown in figure 7-7, projects one or more geometric elements (points or curves) onto a surface. Projection types may be normal to the surface or along a selected direction. The projected element may be supported by another element to assist in the projection. If an exact projection cannot be achieved, you may have to select the option that creates the projection that is the nearest approximation.

The Curve Projection dialog box provides various curve-smoothing options as a curve is projected onto multiple joined surfaces. These options dictate the curve's blending at the point of intersection between two patch surfaces. The deviation value of the blending curves may also be adjusted to provide a higher level of manipulation. This is particularly important in creating class A automotive surfaces.

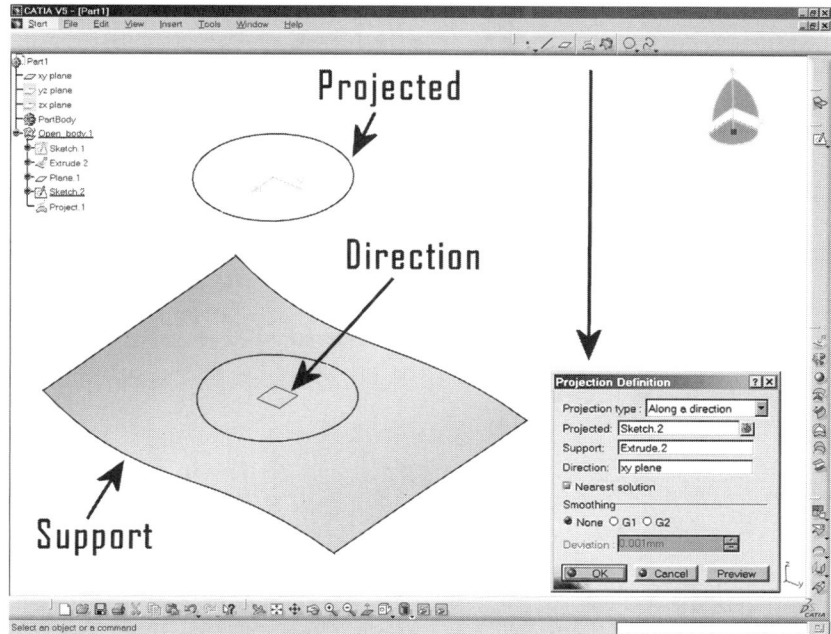

Fig. 7-7.
Curve
Projection
command.

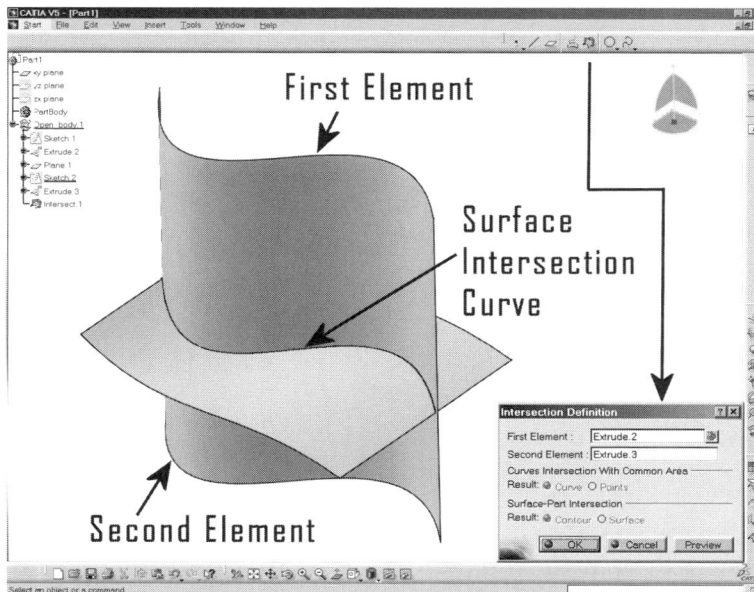

Fig. 7-8. Curve Intersection command.

Curve Intersection

The Curve Intersection command, shown in figure 7-8, creates a resultant curve from the intersection of two surfaces. A curve intersection can result in either a curve or a point. A surface-to-part intersection can result in either in a curve or a surface.

NOTE: *Surfaces must totally intersect for the Curve Intersection operation to execute.*

Circle Curve

The Circle Curve command, shown in figure 7-9, creates a 2D curve outside the Sketcher environment. Various circle creation types include Center and Radius, Center and Point, Two Points and Radius, Three Points, Bitangent and Radius, Bitangent and Point, and Tritangent. Circle-limiting options and start- and end-degree manipulation provides full control of the circle geometry.

Fig. 7-9. Circle Curve command.

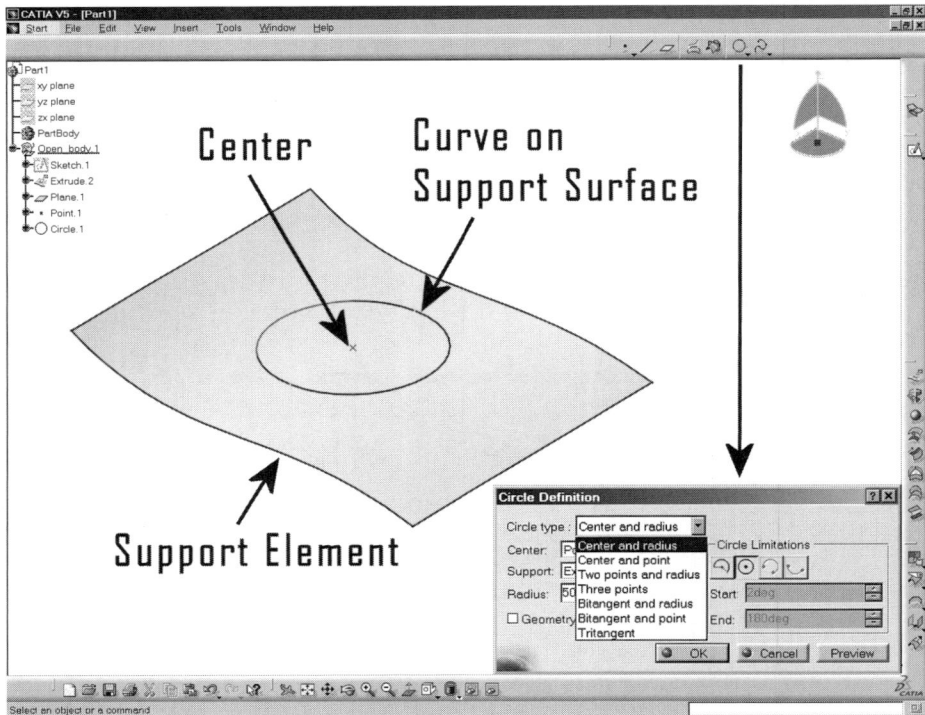

Corner Curve

The Corner Curve command, shown in figure 7-10, creates a radial fillet corner curve between two elements. Multiple solutions are presented, with the ability to toggle through them to locate the desired geometry. The *Trim elements* options provide the ability to trim the supporting curves upon creation of the corner element.

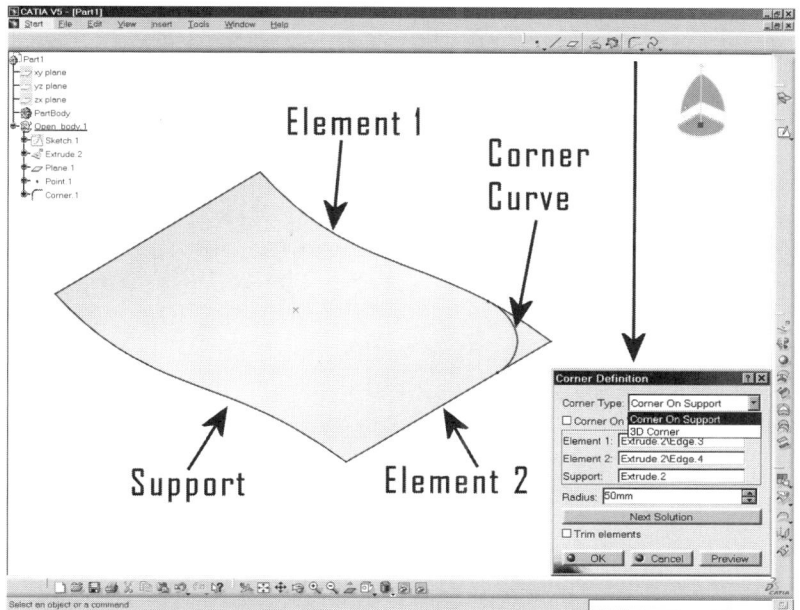

Fig. 7-10. Corner Curve command.

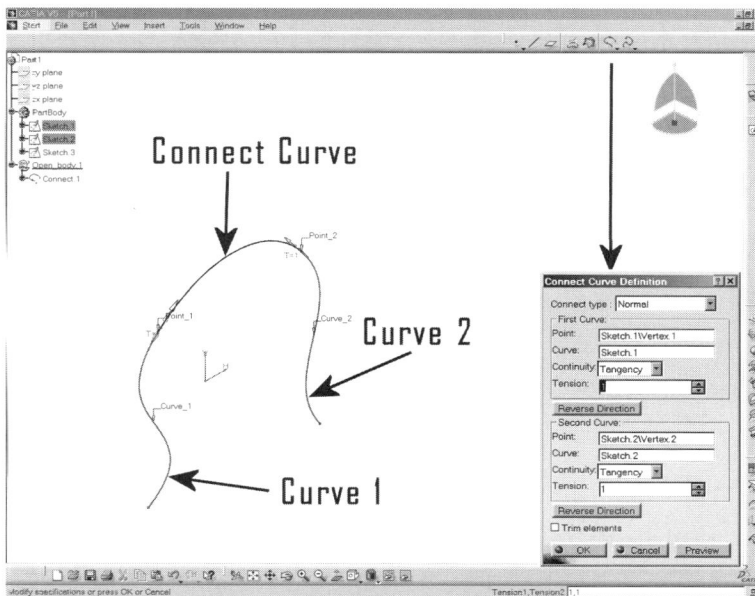

Fig. 7-11. Connect Curve command.

Connect Curve

The Connect Curve command, shown in figure 7-11, creates a connection curve between two points. The created curve can be supported with tangency or curvature blending at either the beginning or ending points. Blending tension may also be adjusted to control the magnitude of blending from start and end points. Trim options and reversing the direction of the blend are additional control features.

NOTE: *The Connect Curve command is used with no more than two points. If more points are required, use the Spline Curve command.*

Spline Curve

The Spline Curve command, shown in figure 7-12, creates a continuous spline through a subset of two or more points. Depending on the position of the points, the curve may be 2D or 3D. Points may be added or deleted within the Spline Curve dialog window. Tangency or curvature blending with tension control is also available for each point. The curve may be automatically closed by clicking on the Close Spline button.

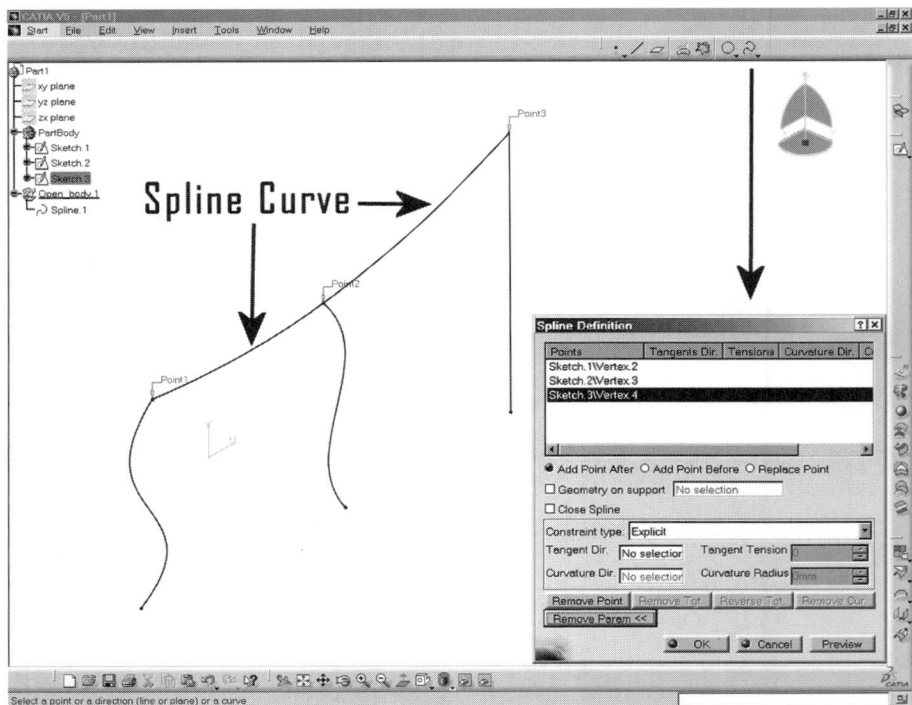

Fig. 7-12. Spline Curve command.

Helix Curve

The Helix Curve command, shown in figure 7-13, automatically creates a helix curve from a controlling set of parameters. Pitch, revolution, height, radius, orientation, starting angle, and stopping angle are a few of the controllable parameters. A law may also be defined to drive a more nonlinear shape into the helix curve.

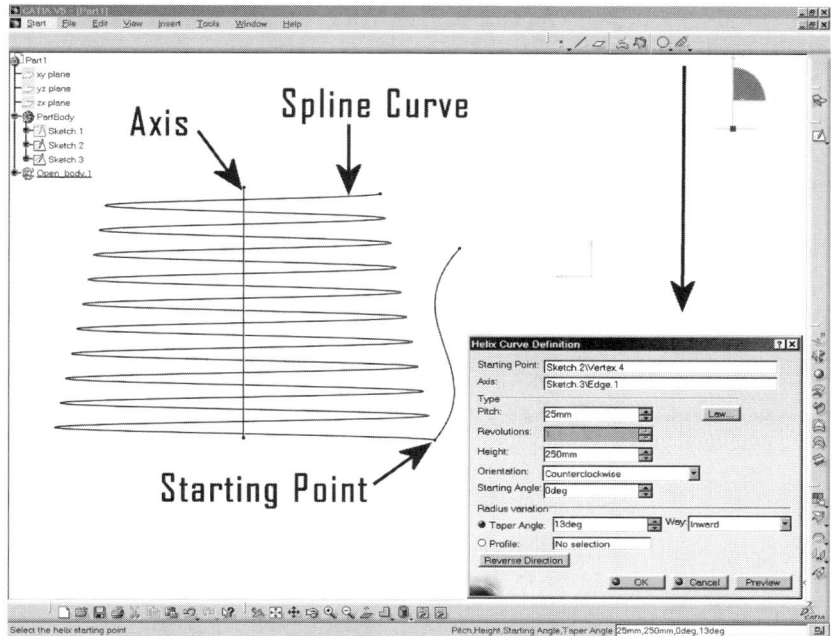

Fig. 7-13.
Helix Curve
command.

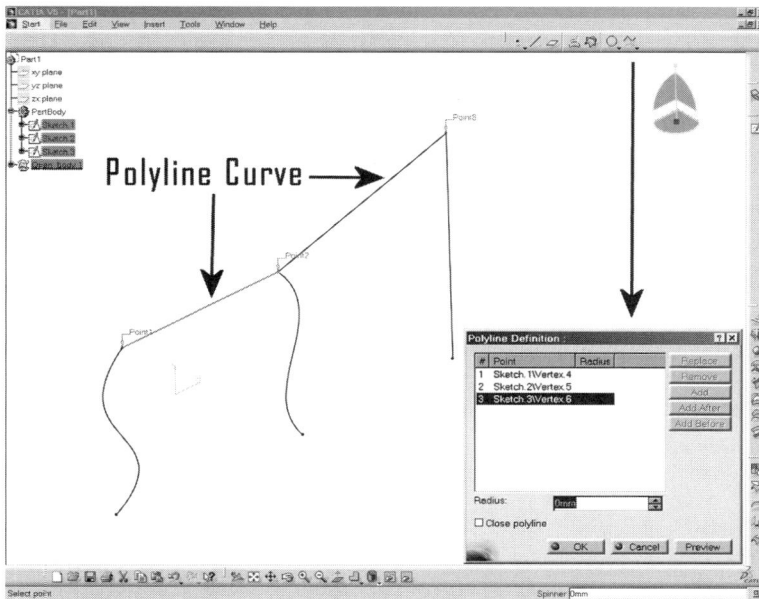

Fig. 7-14. Polyline Curve command.

Polyline Curve

The Polyline Curve command, shown in figure 7-14, creates a continuous linear connection curve through a subset of two or more points. Depending on the position of the points, the curve may be 2D or 3D. Points may be added or deleted within this feature. A radius element may also be defined at each polyline node. The curve may also be automatically closed by clicking on the Close Polyline button.

Surface Features

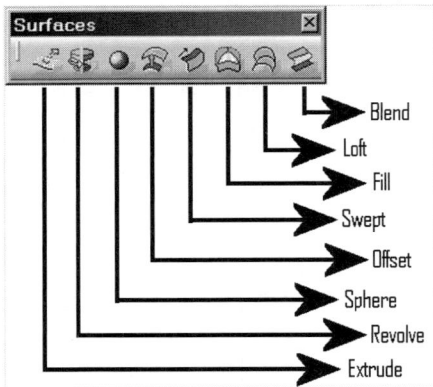

Fig. 7-15. Surface workbench.

The Surface workbench, shown in figure 7-15, allows for the creation of surface features within a part model. The majority of these surface elements require an existing sketched profile to drive the creation of the surface feature.

Extrude Surface

The Extrude Surface command creates a linear extruded surface from an existing sketched cross-sectional profile. This section may be an open or closed profile. The default direction is normal to the sketched element profile but may be altered to another direction by selecting a supporting directional element (curve or plane). The direction may also be reversed and set with limits in either direction. Figure 7-16 shows an example of an extruded surface.

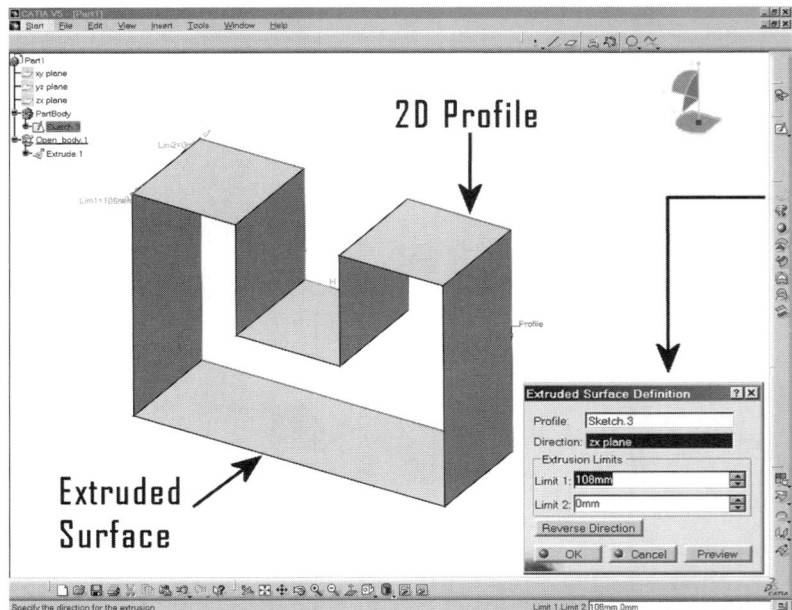

Fig. 7-16. Extruded surface.

Revolve Surface

The Revolve Surface command creates a surface revolution from an existing sketched profile section and a selected axis of rotation. The section may be open or closed, and cannot intersect the axis.

of rotation. Angular limits may be controlled in either direction from the profile plane. Figure 7-17 shows an example of a revolved surface.

Fig. 7-17.
Revolved surface.

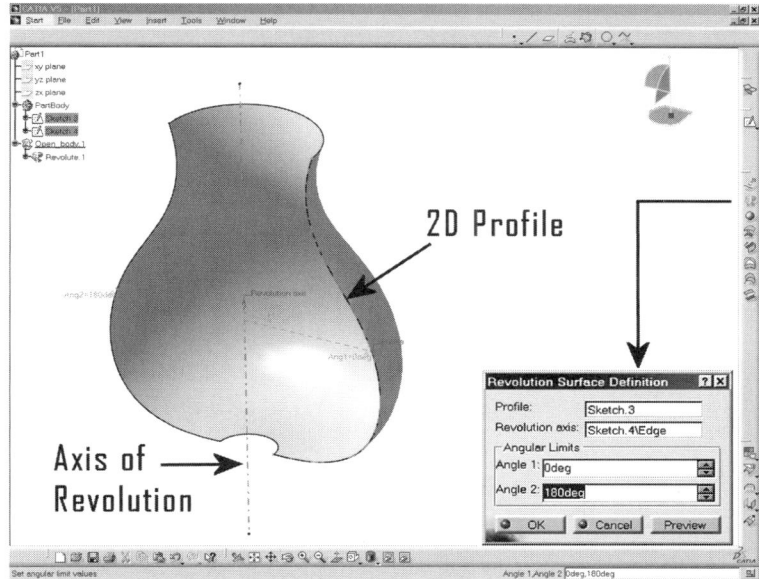

Sphere Surface

The Sphere Surface command creates a spherical surface at a given selected point. An axis of rotation is required to define this feature. The spherical radius must be defined to create the geometry. Angular control in all four quadrants is available to provide modification in each direction. A complete sphere may be quickly created by clicking on the Sphere icon in the Sphere Surface dialog box. Figure 7-18 shows an example of a spherical surface.

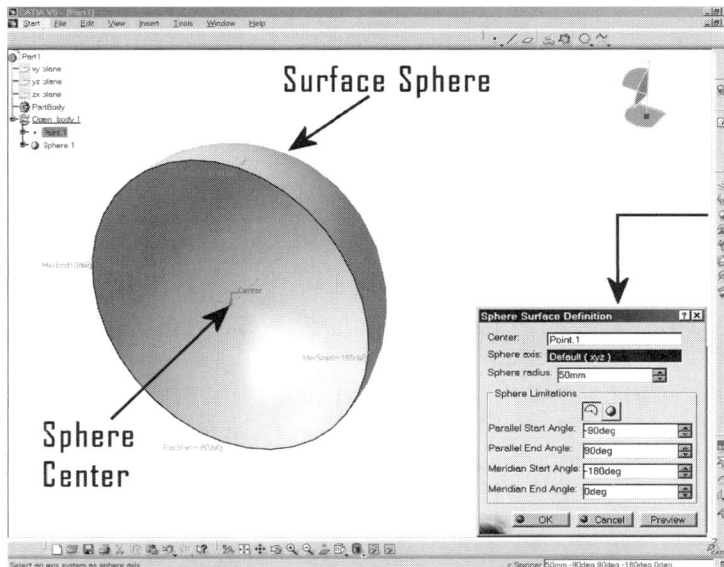

Fig. 7-18. Spherical surface.

Offset Surface

The Offset Surface command creates an offset surface from a singular patch or joined group of surfaces. The surface is offset parallel to the selected surface, and is modifiable within the Offset Surface dialog box. Subelements may be added or removed from a group of joined surfaces when an offset solution cannot be obtained for the entire group. Multiple offset surfaces may be created at one time by selecting the *Repeat Button after OK* option. Figure 7-19 shows an example of an offset surface.

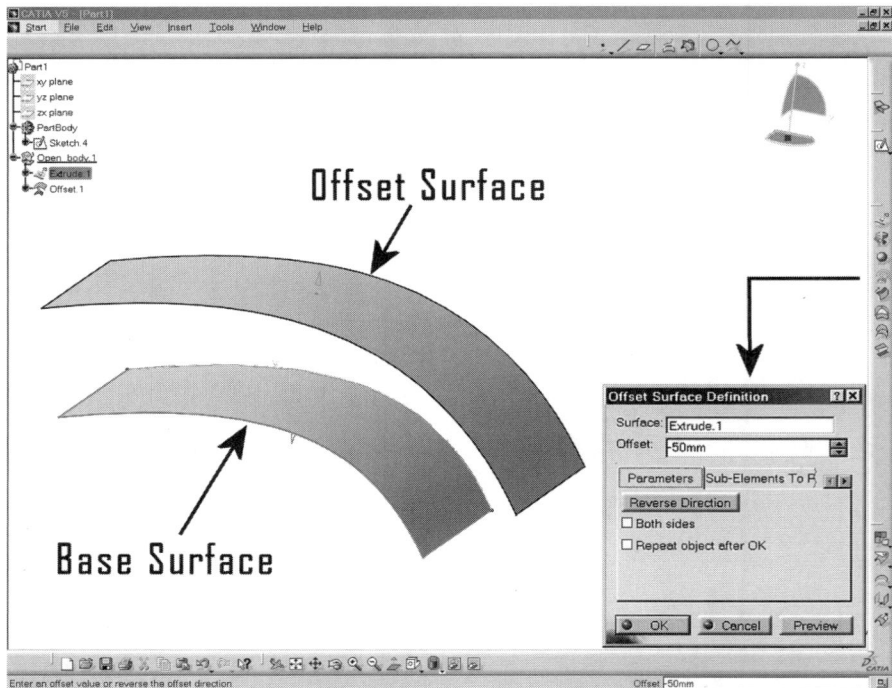

*Fig. 7-19.
Offset surface.*

Swept Surface

The Swept Surface command creates a profile swept surface along a selected guide curve. This option allows for the selection of an existing sketched profile. Further control of the sweep may be obtained by selecting an optional second guide curve, spine curve, or controlling surface. The selected profile's position may also be more tightly controlled with profile position parameters. Figure 7-20 shows an example of a swept surface.

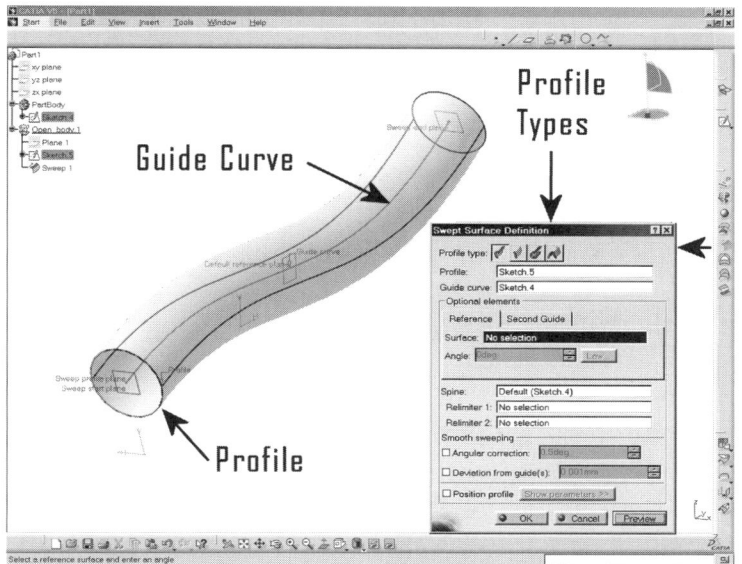

Fig. 7-20. Swept surface.

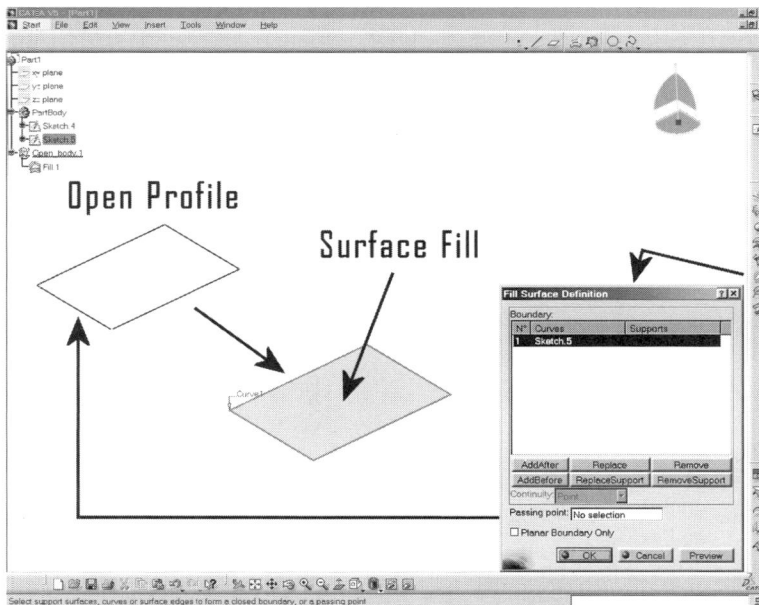

Fig. 7-21. Fill surface.

Fill Surface

The Fill Surface command creates a fill surface within a closed set of boundary segment curves. A support surface may be selected for each curve for the purpose of controlling point, tangency, or curvature continuity blending. Replace, Remove, and Add options allow for the management of all selected boundary element curves. Figure 7-21 shows an example of a fill surface.

Loft Surface

The Loft Surface command creates a lofted surface by sweeping two or more planar section curves along a computed or user-defined spine. Optional guide curves offer more control over the blending of the lofted surface. Tangency support, section coupling, and re-limiting options offer additional blend control. Figure 7-22 shows an example of a lofted surface.

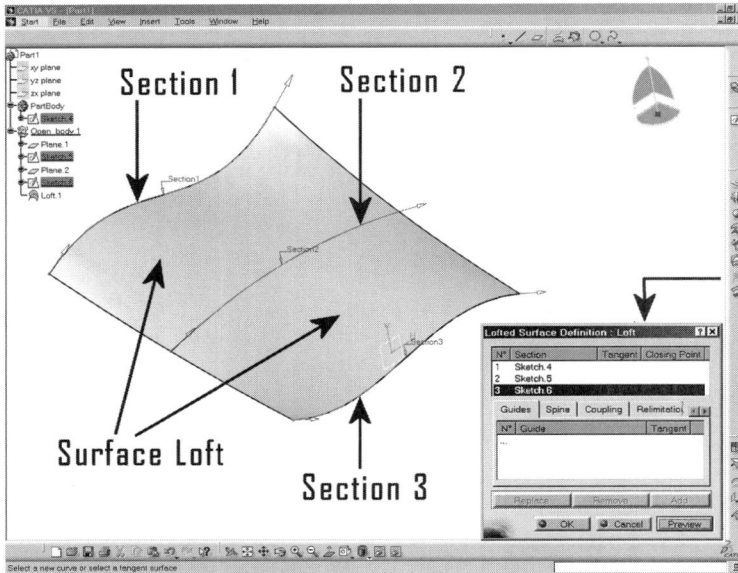

Fig. 7-22. Lofted surface.

NOTE: *The closing points of each section are linked to each other and changed by selecting any point on the section curve.*

Blend

The Blend command creates a blended surface between two selected curve elements. Only one edge may be selected at a time from each element. Support elements for each curve may be selected to provide point, tangency, and curvature blending control. Blending tension, coupling, and closing-point parameters are also available for achieving a higher level of blending control. Figure 7-23 shows an example of a blended surface.

Wireframe and Surface Operations

Operational features provide a means of manipulating wireframe and surface features after

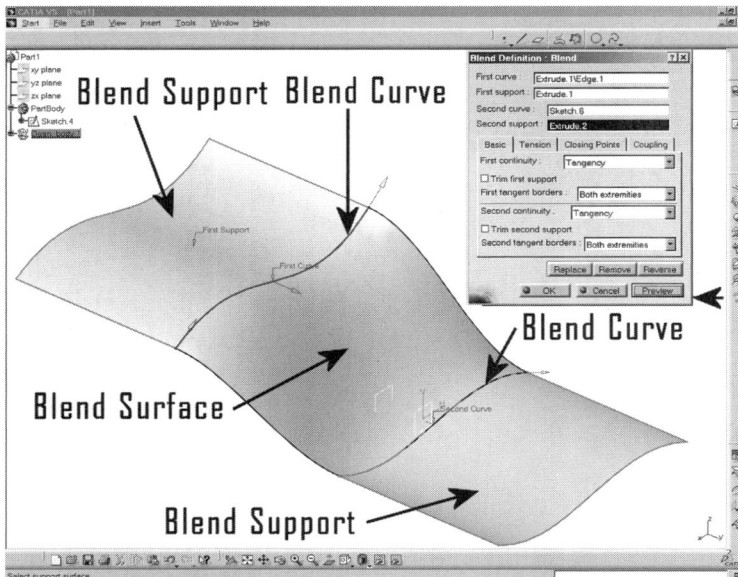

Fig. 7-23. Blended surface.

their initial creation. These operations, accessed from the Operations toolbar (shown in figure 7-24), are logged in the specification tree and are modifiable at any time. The sections that follow explore various categories of these operational features.

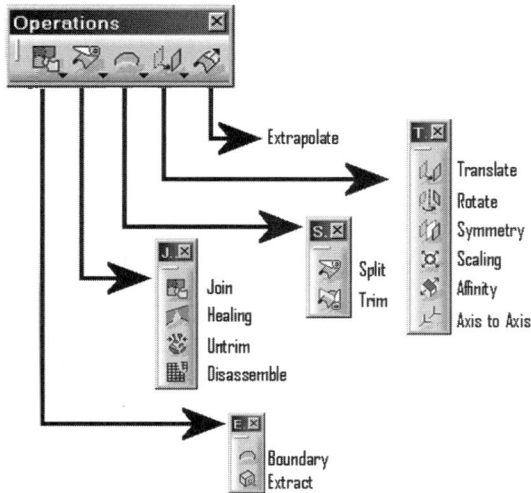

Fig. 7-24. Operations toolbar.

Joins and Healing

The sections that follow examine join and healing operations. Such operations include joins, surface heals, untrims, and disassembles.

Join

A join operation allows multiple independent elements (curve or surface) to be joined and treated as one element. Elements may be added or removed with great control within the Join dialog window. Subelements may also be added or removed based on selection. Parameter checks such as tangency, connexity, and merging distance are a few of the options that assist in controlling the quality of joined elements. Figure 7-25 shows an example of a join operation.

Surface Healing

A surface-healing operation attempts to join two surfaces at their common intersection boundary. In addition, this operation attempts to heal or repair any gaps or inconsistencies between the two selected surfaces. Figure 7-26 shows an example of a surface-healing operation.

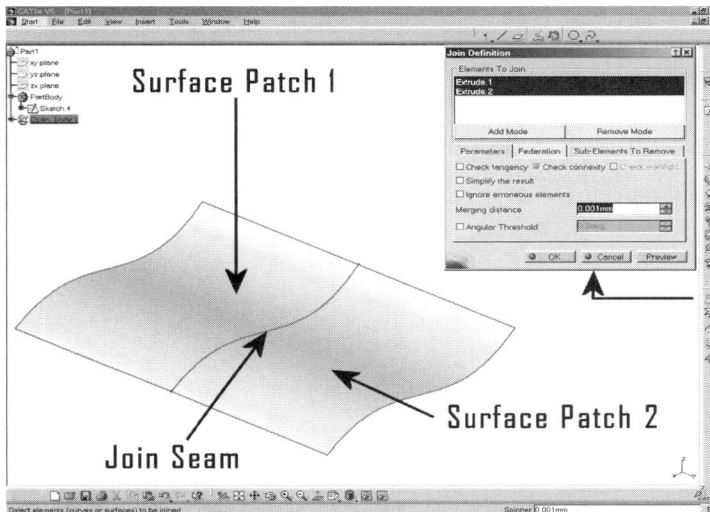

Fig. 7-25. Join operation.

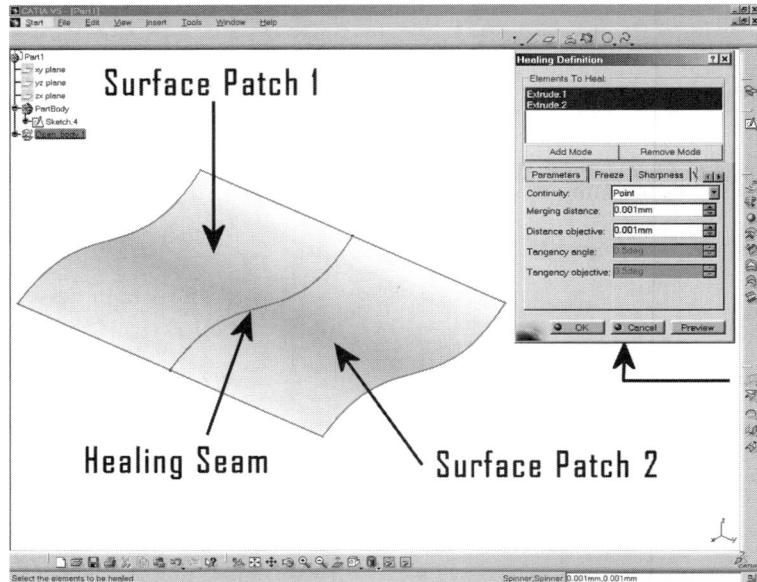

*Fig. 7-26.
Surface-healing
operation.*

Untrim

An untrim operation restores a trimmed element (curve or surface) to its original untrimmed state. Figure 7-27 shows an example of an untrim operation.

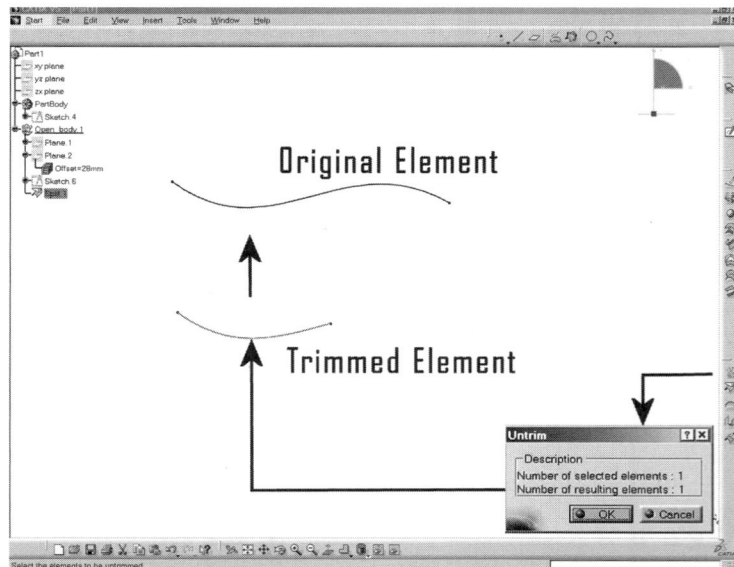

*Fig. 7-27. Untrim
operation.*

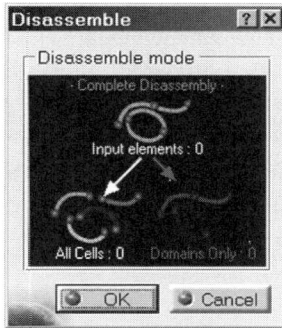

Fig. 7-28. Disassemble operation.

Disassemble Operation

A disassemble operation disassembles multi-element curves and surfaces into individual curves or surfaces. This is useful when working with geometry imported from another system. Figure 7-28 shows an example of a disassemble operation.

Split and Trim

The sections that follow examine split and trim operations. Splits and trims are performed on curves and surfaces.

Split

A split operation splits elements (curve and surface) with a cutting element. The *Elements to remove* and *Elements to keep* options allow for one or both sides of the element to be retained during the spit. The split operation splits (using another element) one element at a time. Figure 7-29 shows an example of a split operation.

TIP: *Before performing a split operation, first select the side of the element to be retained.*

Fig. 7-29. Split operation.

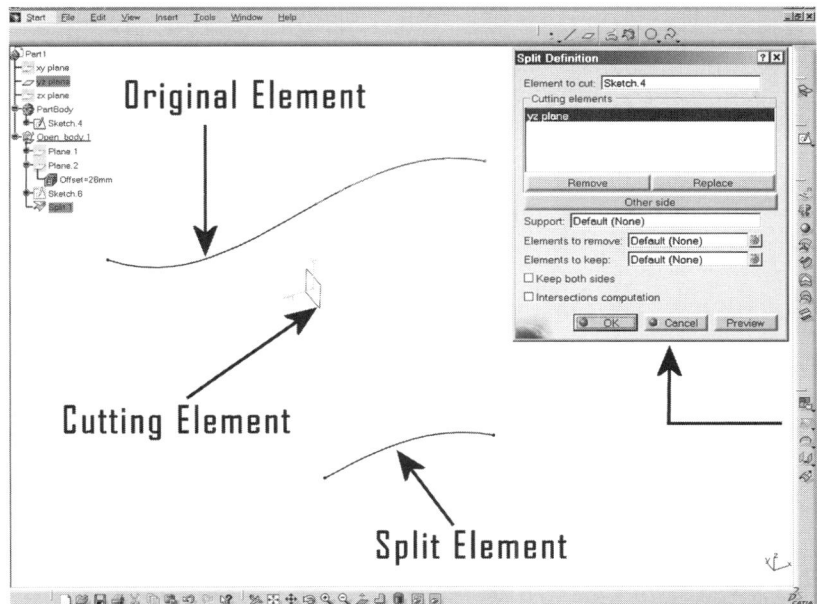

Trim

The trim operation trims elements (curve and surface) at their common area of intersection. The *Elements to remove* and *Elements to keep* options allow for one or both sides of the elements to be retained during the trim operation. The *Result simplification* button allows CATIA V5 to automatically reduce the number of faces in a resulting surface trim if a solution cannot be determined. Figure 7-30 shows an example of a trim operation.

Fig. 7-30.
Trim
operation.

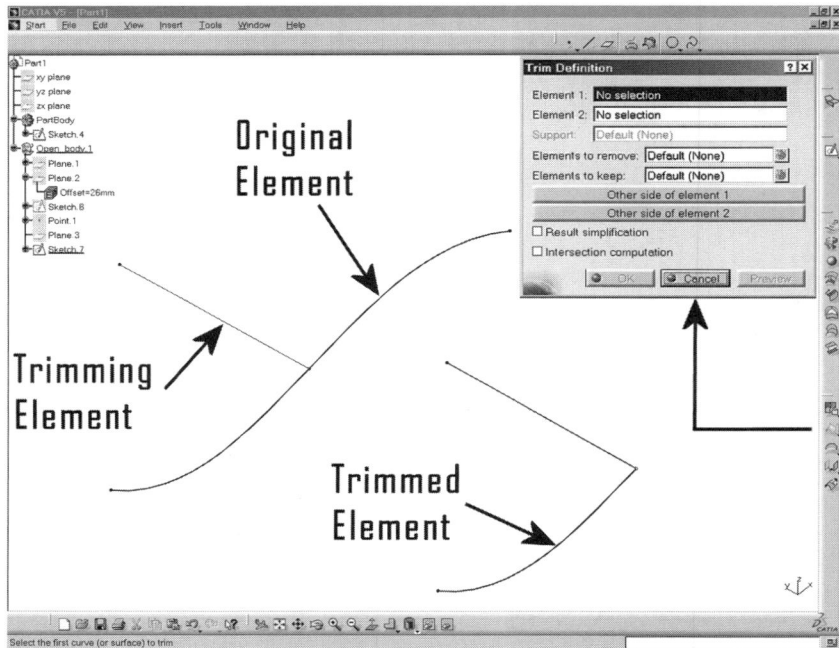

Extraction

The sections that follow examine extraction operations. Such operations include boundary extractions, face extractions, and surface filleting.

Boundary Extraction

Boundary extraction extracts a curve boundary from an existing edge or face of a surface. Various edge propagation options are available for limiting the boundary to one edge or all boundary edges. Figure 7-31 shows an example of a boundary extraction.

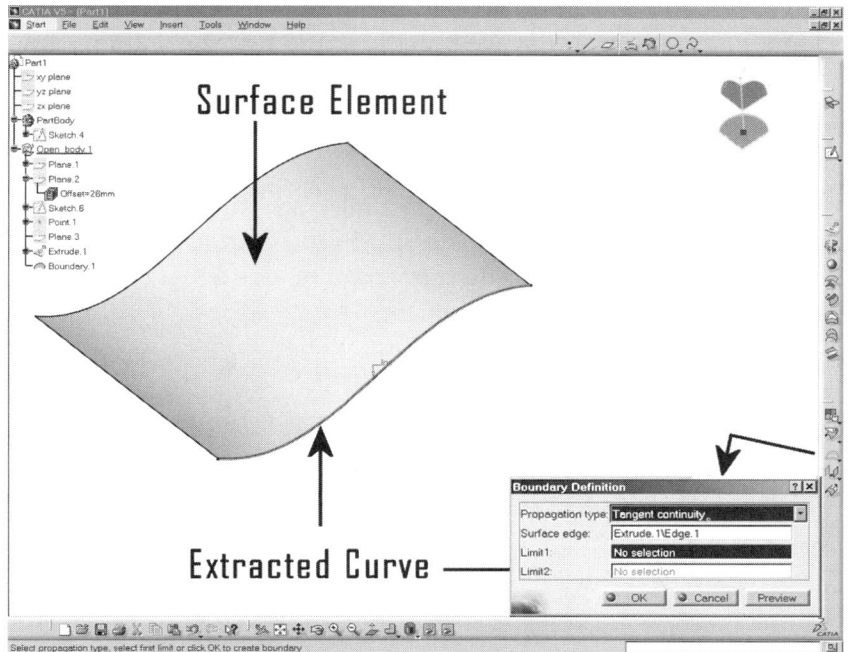

Fig. 7-31. Boundary extraction.

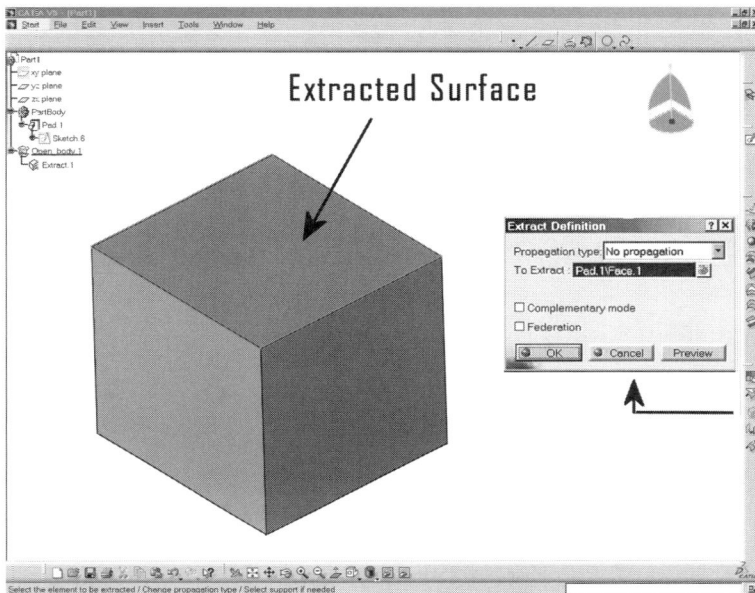

Fig. 7-32. Face extraction.

Face Extraction

A face extraction extracts a surface face from an existing surface or solid face. Various face propagation options are available for limiting the boundary to a single face or the complete element. Figure 7-32 shows an example of a face extraction.

Transformations

All transformation operations create a copy of the original element. The original element may be hidden or shown upon execution of the transformation operation. The sections that follow explore the transformation options Translate, Rotate, Symmetry, Scaling, Affinity, Axis to Axis, and Extrapolate.

Translate

The translate command, shown in figure 7-33, transforms geometry from one position to another along a given axis system. The vector direction can be defined along an edge, between points, or by coordinates. Use the *Repeat Object after OK* checkbox to create several translated elements.

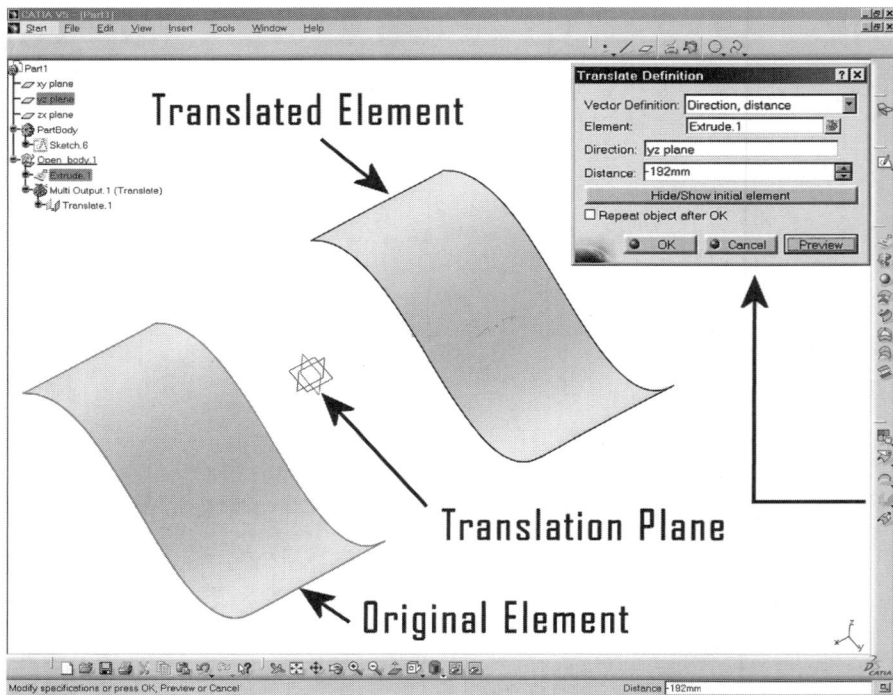

Fig. 7-33. Translate command.

Rotate

The Rotate command, shown in figure 7-34, rotates selected geometry around a selected axis at some determined angular dimension. Use the *Repeat Object after OK* checkbox to create several rotated elements at one time.

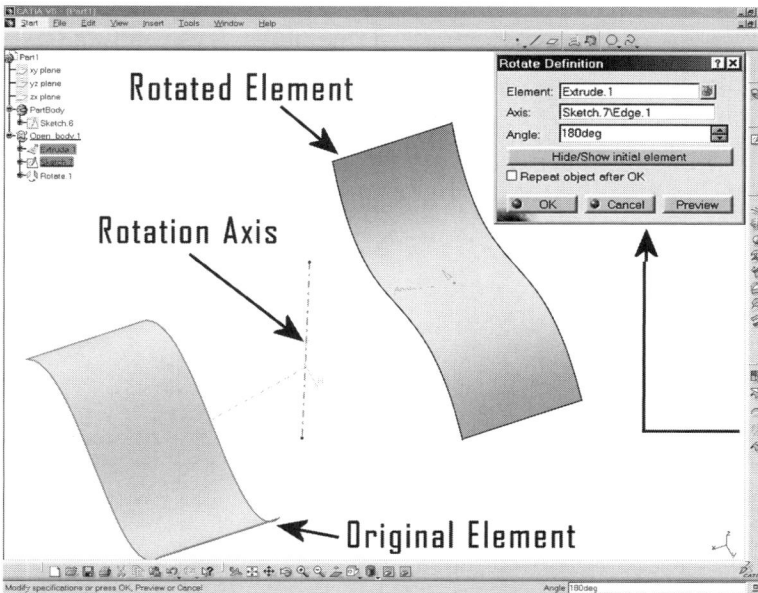

Fig. 7-34. Rotate command.

Symmetry

The Symmetry command, shown in figure 7-35, mirrors selected geometry about a reference plane.

Scaling

The Scaling command, shown in figure 7-36, scales selected geometry about a selected reference element. The scaling ratio is proportional in all directions if the reference element is a point. A plane or line may be used to influence the scaling ratio in a certain direction.

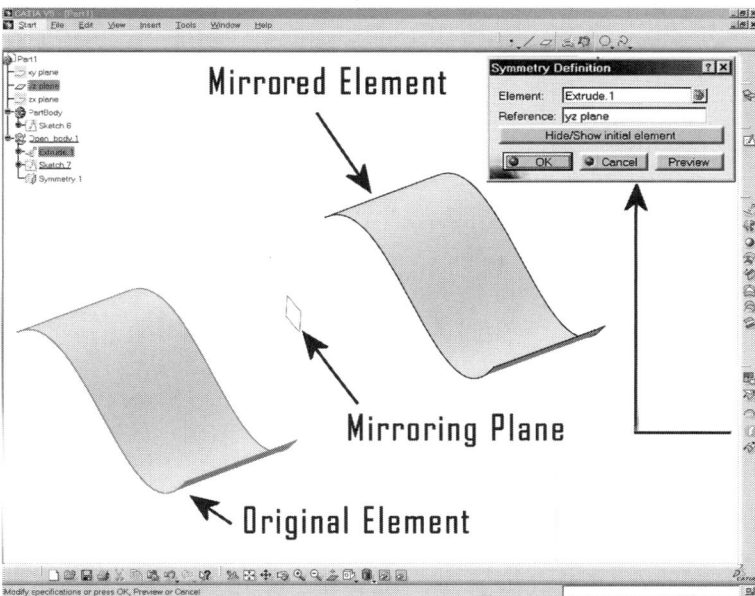

Fig. 7-35. Symmetry command.

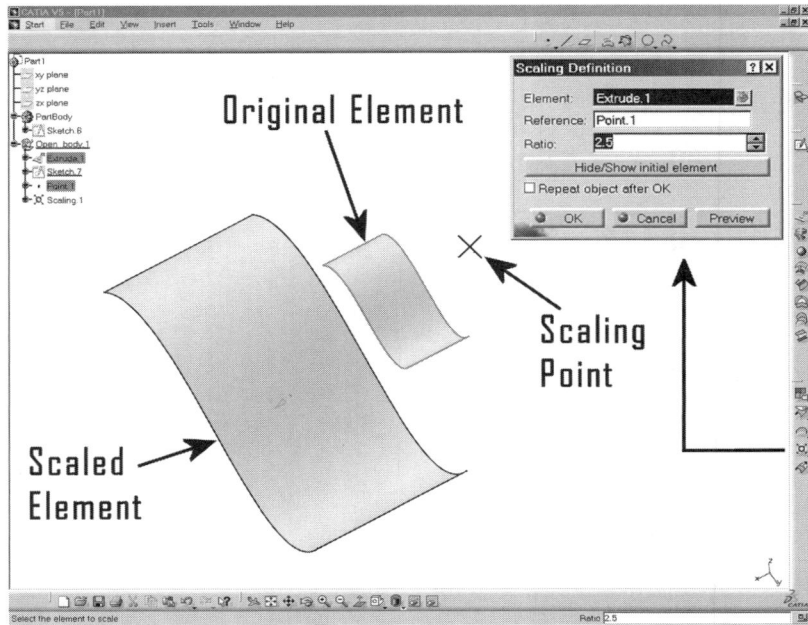

Fig. 7-36.
Scaling
command.

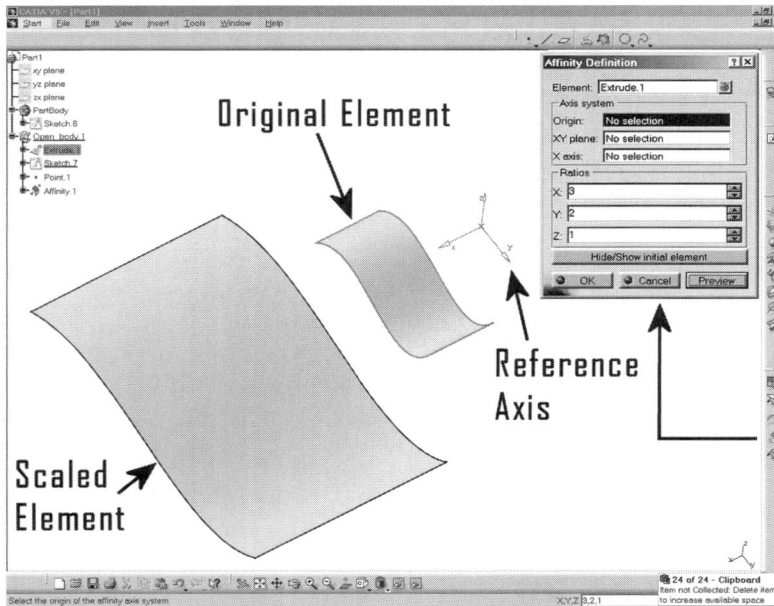

Fig. 7-37. Affinity command.

Affinity

The Affinity command, shown in figure 7-37, scales selected geometry about a selected reference element and with greater individual scaling control in X, Y, and/or Z directions.

Axis to Axis

The Axis to Axis command, shown in figure 7-38, provides for both translation and rotation of selected elements about a given axis system.

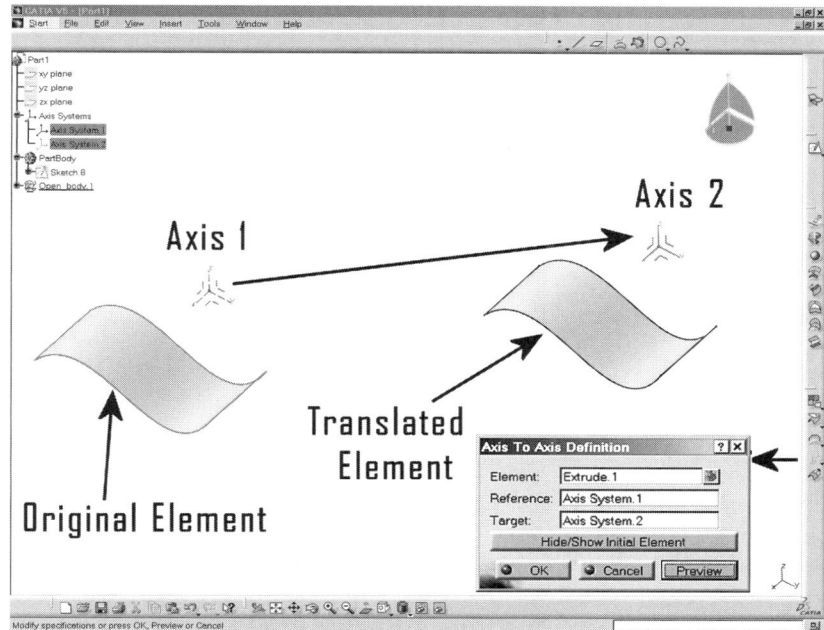

Fig. 7-38. Axis to Axis command.

Extrapolate

The Extrapolate command, shown in figure 7-39, allows for the extrapolation of a surface or of a boundary edge. Various options are available for controlling the type of continuity and transition of the extrapolate surface.

TIP: *It is generally not recommended that you extrapolate more than 25 percent of the original element. This is dependent on the curvature values at the edge of the surface to be extrapolated.*

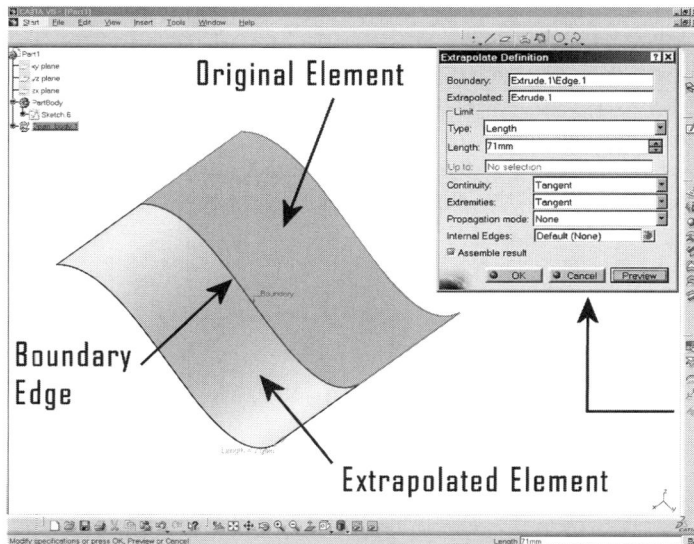

Fig. 7-39. Extrapolate command.

Wireframe and Surface Exercises

The sections that follow present wireframe and surface exercises. The first exercise involves the design of a plastic bottle. The second exercise involves the design of a fan blade.

Wireframe and Surface Exercise 1: Plastic Bottle

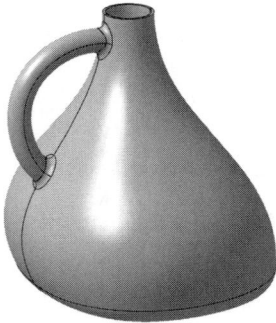

Fig. 7-40. Plastic bottle.

This exercise is intended to introduce basic wireframe and surface operations involved in the design of a plastic bottle, shown in figure 7-40. To create the plastic bottle, perform the following steps.

1 Create a new CATPart and rename the default part name to *Plastic Bottle*.

2 Activate the Wireframe and Surface workbench.

3 Rename the default open body name *Open Body* to *Construction Elements* by clicking on the Open Body icon using MB3 and modifying the properties.

4 Click on the Plane icon and select Offset from Plane. Select the XY plane and enter an offset value of *300* (mm). Rename the new plane *Bottle Top*. The offset plane is shown in figure 7-41.

Fig. 7-41. Offset plane.

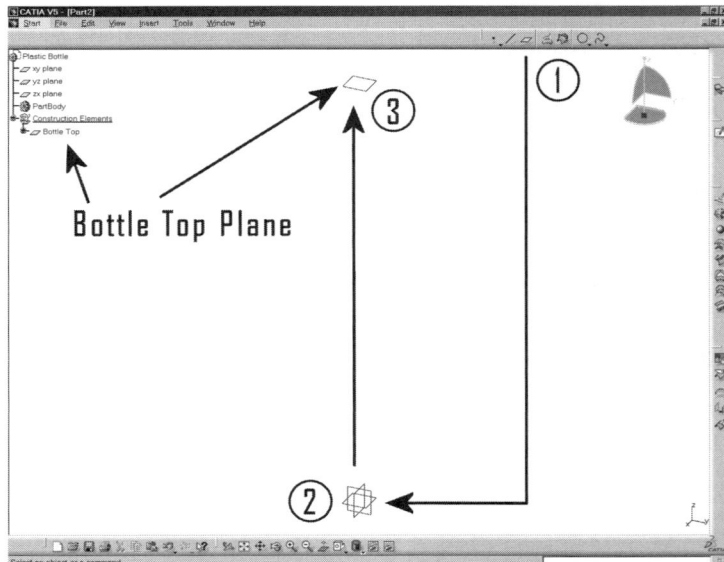

Bottle Top Plane

5 Create a new sketched profile by selecting the *Bottle Top* plane and then click on the Sketcher icon. Create an arc profile with the top and bottom segment about the YZ plane. Constrain the center of the arc on the axis system and create a 50-mm-diameter dimensional constraint. Sketch a horizontal construction line through the center of the circle and constrain it to the ZX plane. Sketch a point and constrain it to the intersection of the circle segment and the ZX plane. Exit the Sketcher when complete. The circle profile is shown in figure 7-42.

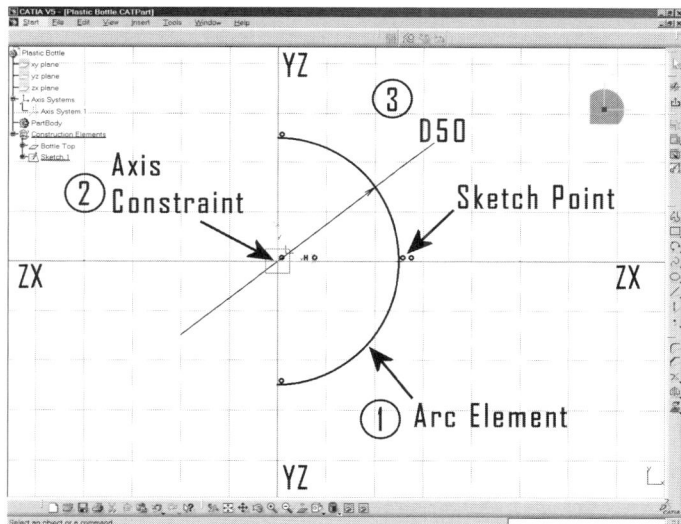

Fig. 7-42.
Circle profile.

6 Insert a new body and name it *Bottle Curves.*

7 Create a new sketched profile by selecting the YZ plane and then clicking on the Sketcher icon. Sketch a spline curve and create a coincidence constraint between the top point of the spline and the end point of the half-circle curve. Activate the Spline Modification menu by double clicking on the spline element. Impose tangency on the top point to display the tangent vector. Apply a vertical constraint to the tangent vector. Create a parallel constraint between the top of the spline curve and the vertical construction curve. This will ensure that the spline curve is normal to the top plane. Create another coincidence constraint between the bottom point of the spline curve and the XY plane. Create a dimension constraint of 100 mm from the bottom of the spline curve to the ZX plane. Exit the Sketcher when complete. This curve is indicated as curve 1 in figure 7-43.

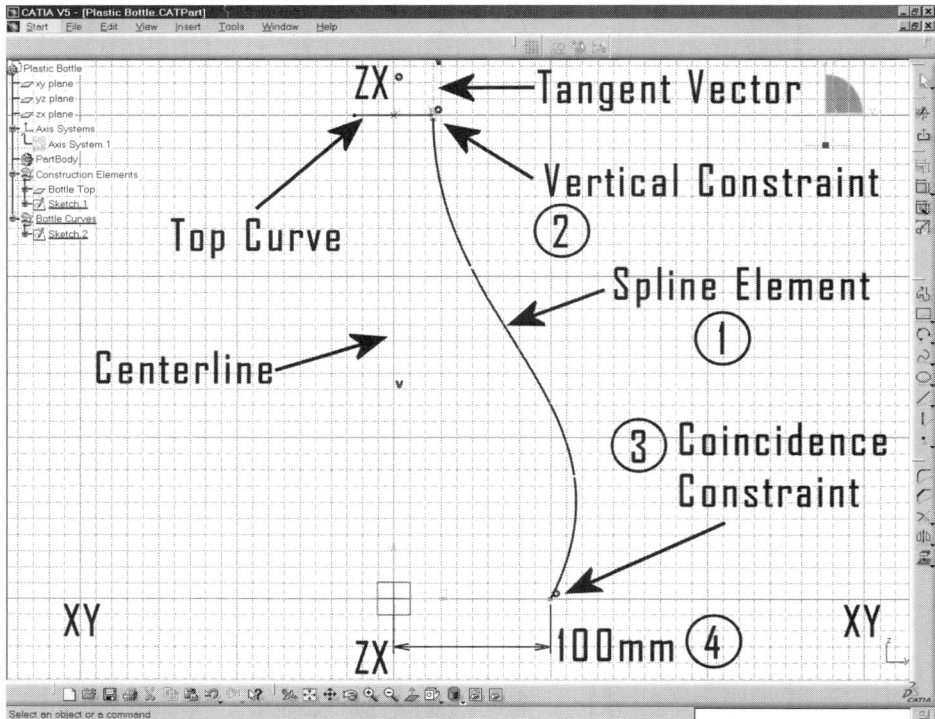

Fig. 7-43. Curve 1.

8   Create a second sketched profile by selecting the ZX plane and then clicking on the Sketcher icon. Sketch a spline curve and create a coincidence constraint between the top point of the spline and the midpoint, located halfway along the top sketched profile half-circle curve. Activate the Spline Modification menu by double clicking on the spline element. Impose tangency on the top point to display the tangency vector. Apply a vertical constraint to the tangency vector. This will ensure that the spline curve is normal to the top plane. Create a parallel constraint between the top of the spline curve and the vertical construction curve. This will ensure that the spline curve is normal to the top plane. Create another coincidence constraint between the bottom point of the spleen curve and the XY plane. Create a dimension constraint of 150 mm from the bottom of the spline curve to the YZ plane. Exit the Sketcher when complete. This curve is indicated as curve 2 in figure 7-44.

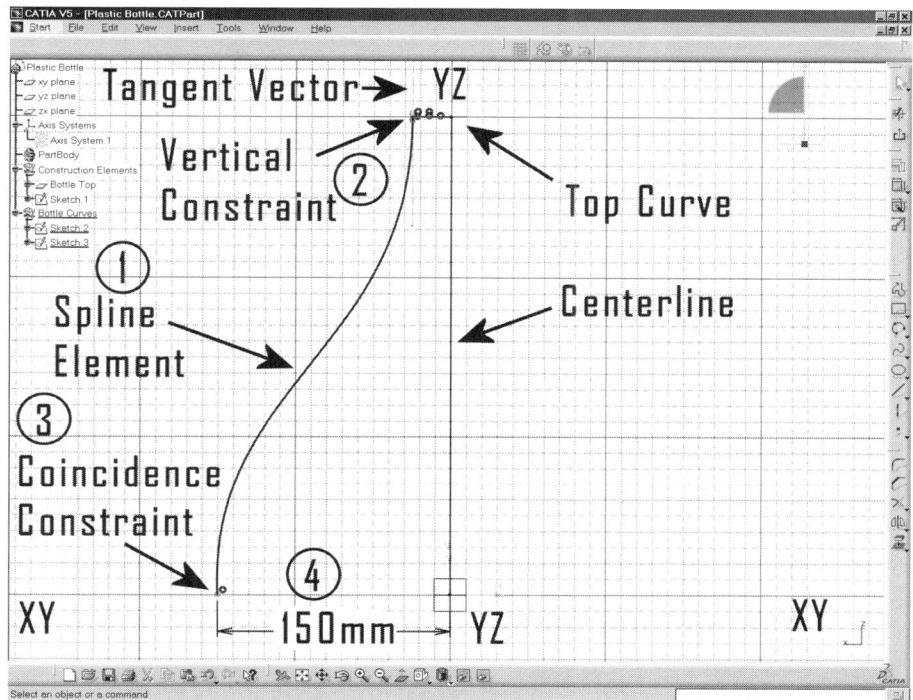

Fig. 7-44. Curve 2.

9 Create a third sketched profile by selecting the YZ plane and then clicking on the Sketcher icon. Sketch a spline curve and create a coincidence constraint between the top point of the spline and the other end point of the half-circle curve. Activate the Spline Modification menu by double clicking on the spline element. Impose tangency on the top point to display the tangency vector. Apply a vertical constraint to the tangency vector. This will ensure that the spline curve is normal to the top plane. Create a parallel constraint between the top of the spline curve and the vertical construction curve. This will ensure that the spline curve is normal to the top plane. Create another coincidence constraint between the bottom point of the spline curve and the XY plane. Create a dimension constraint of 150 mm from the bottom of the spline curve to the ZX plane. Exit the Sketcher when complete. This curve is indicated as curve 3 in figure 7-45.

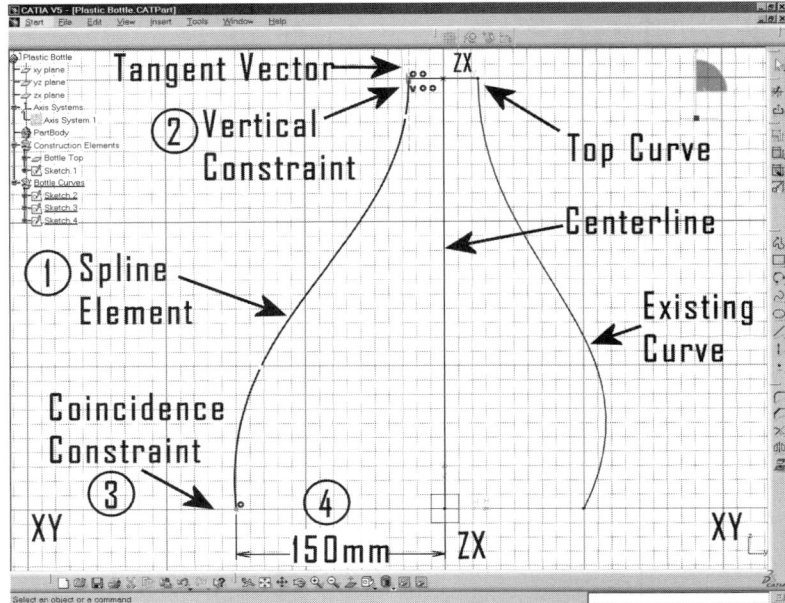

Fig. 7-45.
Curve 3.

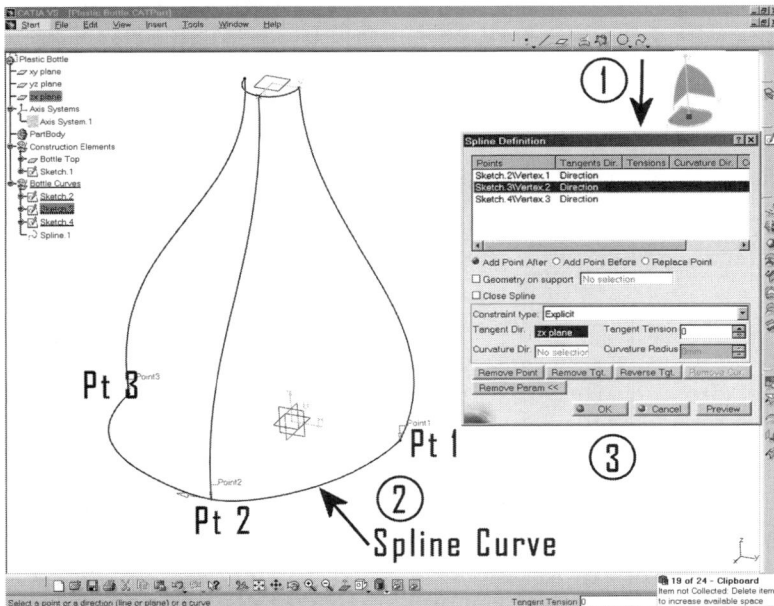

Fig. 7-46. Bottom spline curve.

10 Click on the 3D Spline icon and create a bottom spline curve between the lower three endpoints (1, 2, and 3) of each previously sketched curve. Click on the Add Parameters button and specify tangency directions for each point. Make points 1 and 3 tangent to the YZ plane, and make point 2 tangent to the ZX plane. Click on OK. The bottom spline curve is shown in figure 7-46. Make sure the vector lines are in the same directory.

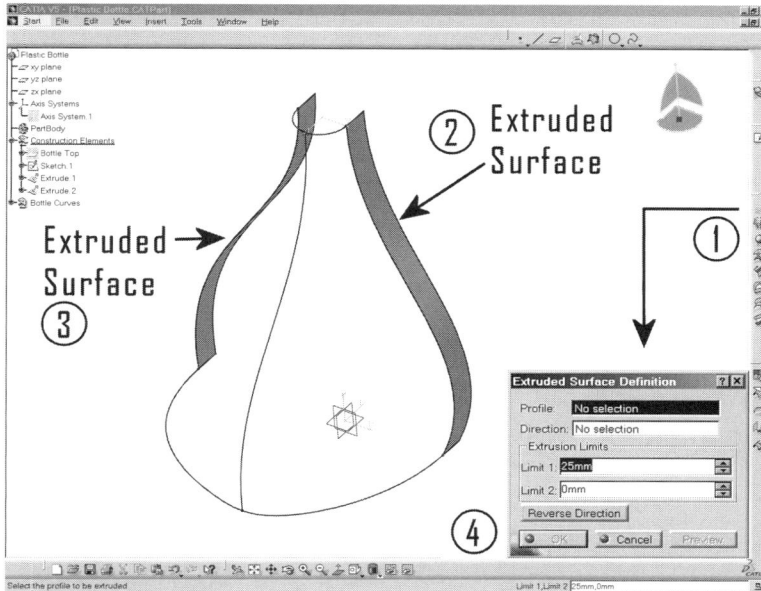

Fig. 7-47. Extruded surfaces

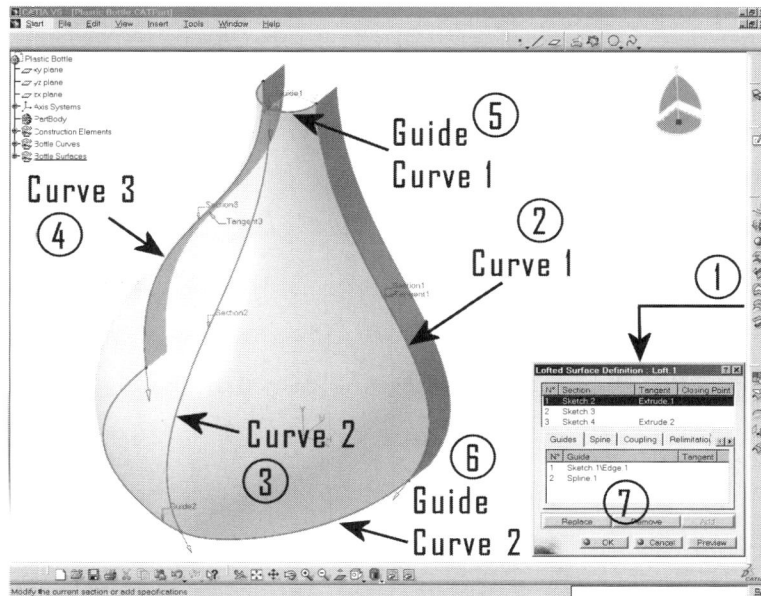

Fig. 7-48. Lofted surface.

11 Create two (2) extruded surfaces 25 mm from the two outer curves that are normal to the YZ plane.

The two extruded construction surfaces should reside in the *Construction Elements* body. These extruded surfaces are shown in figure 7-47.

12 Insert a new open body and name it *Bottle Surfaces.*

13 Create a lofted surface between the three vertical section spline curves. Make sure the vector arrows are pointing in the same direction. Select the top and bottom curves as guide curves. In the Coupling tab, impose tangency at the three edges where the extruded surfaces are located. Click on OK. The lofted surface is shown in figure 7-48.

TIP: *You can hide the two extruded surfaces that were used to hold tangency on the lofted surface.*

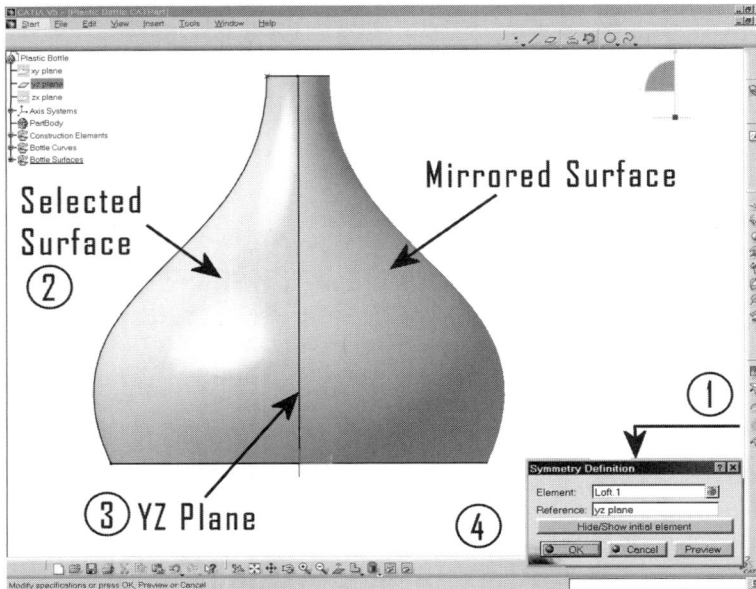

Fig. 7-49. Mirrored surface.

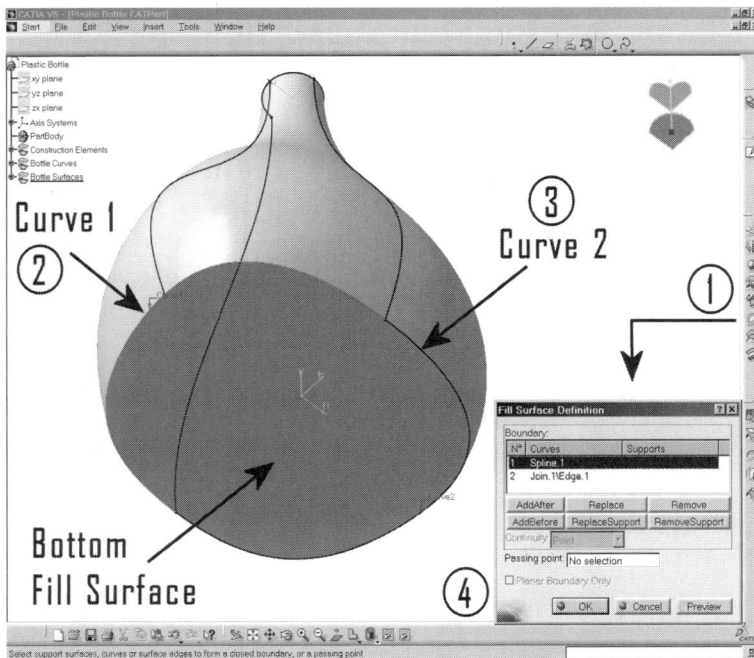

Fig. 7-50. Bottom-filled surface.

14 Mirror the lofted surface about the YZ plane using the Transform Symmetry command. Figure 7-49 shows the mirrored surface.

15 Using the Join command, join the two surfaces to form a single element.

16 Rotate the model to see the bottom of the bottle. Using the Fill command, create a fill surface by selecting the bottom edges of the joined surfaces. Figure 7-50 shows the bottom-filled surface.

17 Using the Join command, join the fill surface with the main bottle surface to form a single element.

18 Insert a new open body named *Handle Curves*. Create a new sketched profile by selecting the YZ plane and then clicking on the Sketcher icon. Sketch a spline curve that begins to outline the handle of the bottle. Exit the Sketcher when complete. Figure 7-51 shows the curve for the handle.

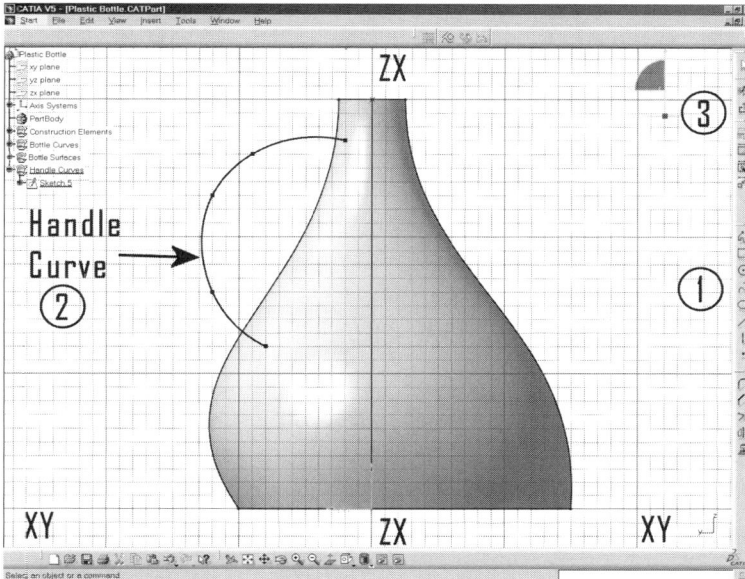

Fig. 7-51. Handle plane.

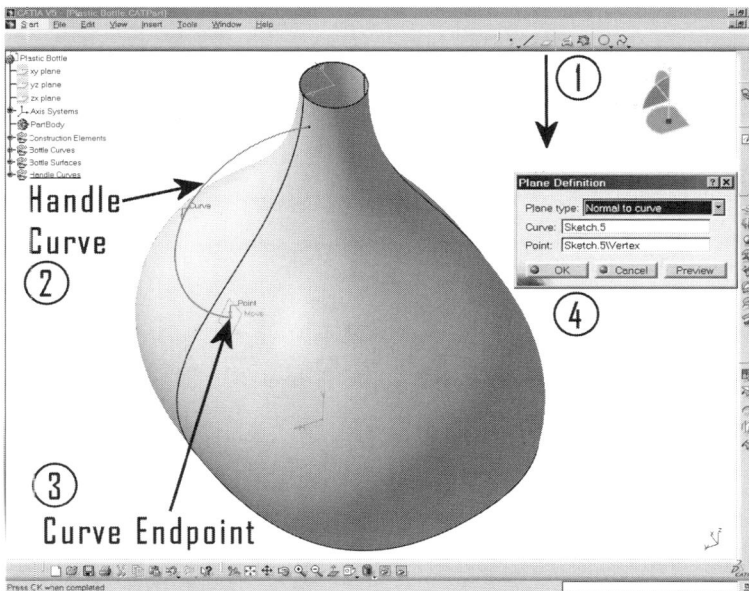

Fig. 7-52. Handle curve.

19 Create a plane at the endpoint of the curve and normal to the curve. Figure 7-52 shows this handle plane.

20 Create a new circular sketched profile using the plane previously created. Constrain the center of the point to the handle spline curve. Create a dimensional constraint and modify the circle diameter to 20 mm. Exit the Sketcher when complete. Figure 7-53 shows the circular handle profile.

21 Insert a new open body named *Handle Surfaces*. Create a swept surface using the handle curve as a guide curve and the circular sketch as the profile. Note that this handle surface completely intersects the other surfaces, as shown in figure 7-54.

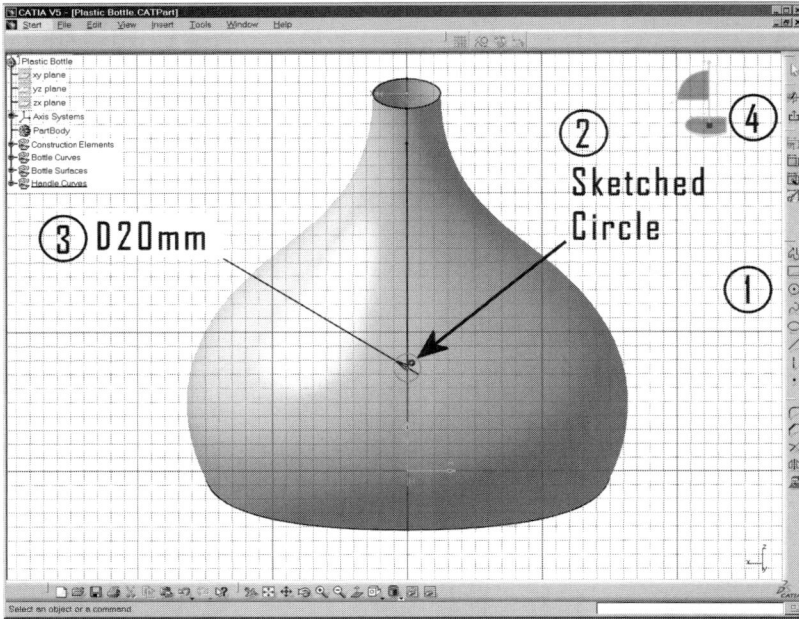

Fig. 7-53. Handle profile.

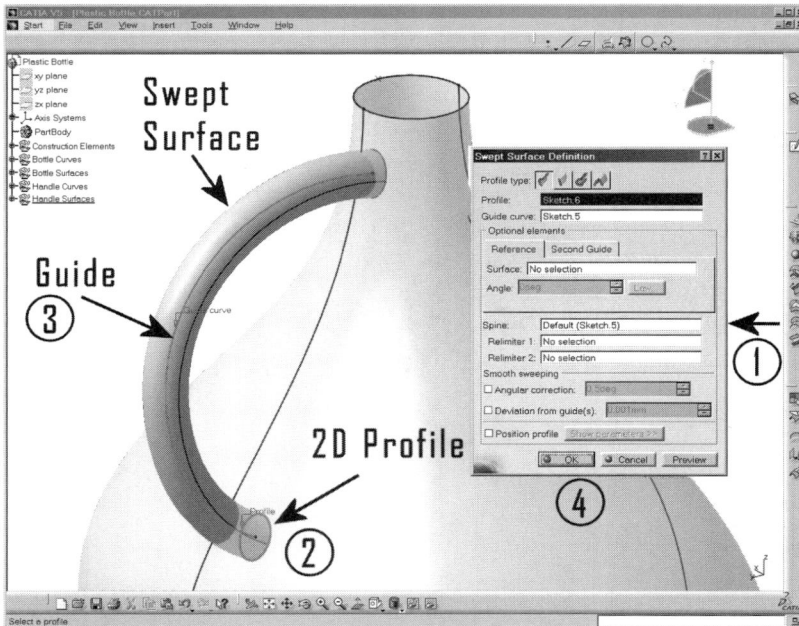

Fig. 7-54. Handle surface.

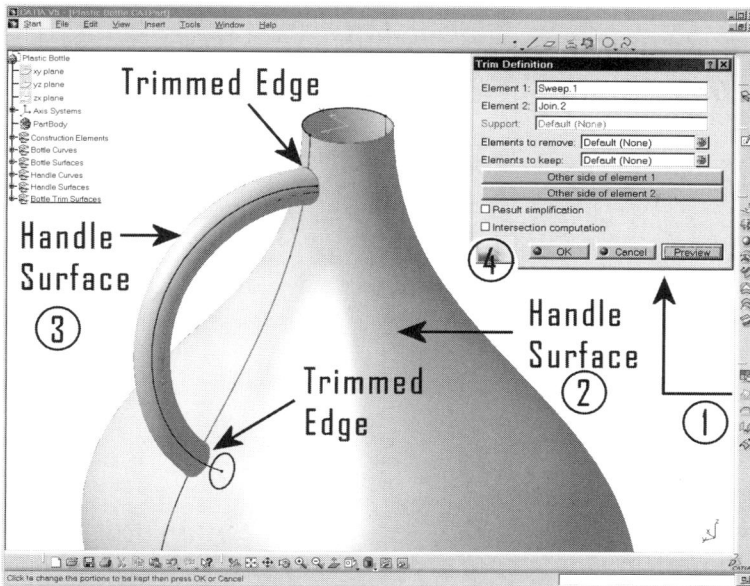

Fig. 7-55. Handle trim surface.

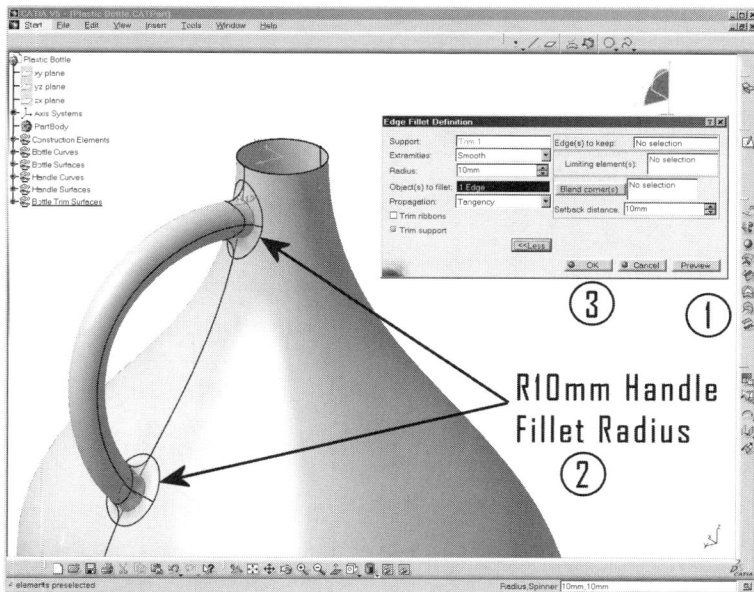

Fig. 7-56. Handle trim surface.

22 Insert a new open body named *Bottle Trim Surfaces.* Using the Trim command, select the swept handle surface with the main surface body of the bottle. Keep the outer surfaces. Figure 7-55 shows this handle trim surface.

23 Create a 10-mm-radius fillet surface at both intersections of the handle with the main bottle surfaces. Figure 7-56 shows these surfaces.

24 Create a 25-mm-radius fillet surface on the bottom edge of the bottle, as shown in figure 7-57.

25 Activate the Part Design workbench. Using the Insert I Surface-Based Features I Surface Thick command, create a solid thin feature from the surface. Use a 3-mm value for the thickness. Figure 7-58 shows the completed solid model design.

The design is now complete. Save the model for future reference.

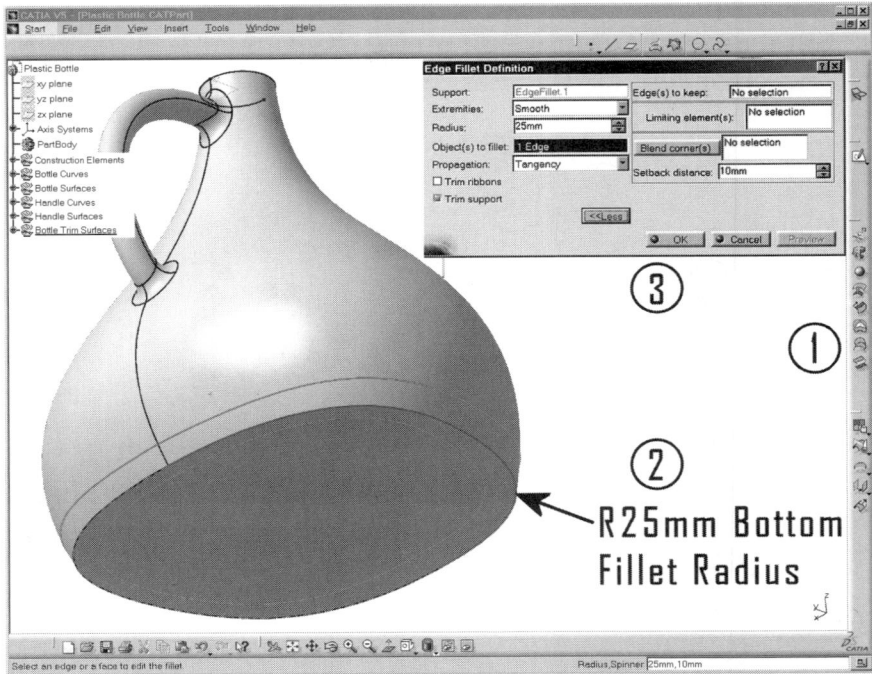

Fig. 7-57.
First bottle
fillet surface.

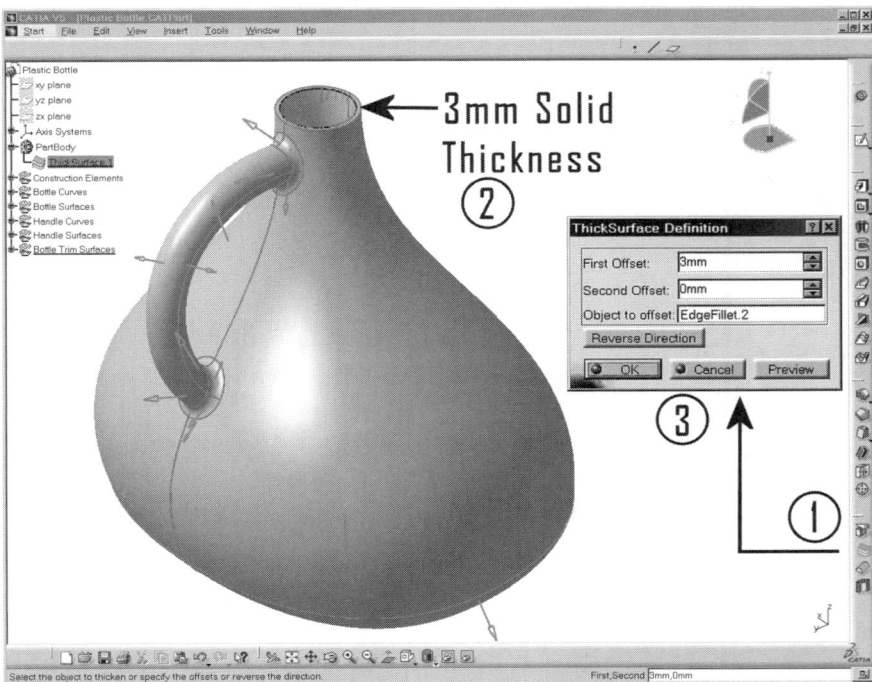

Fig. 7-58.
Second bottle
fillet surface.

Wireframe and Surface Exercise 2: Fan Model

Fig. 7-59. Fan blade.

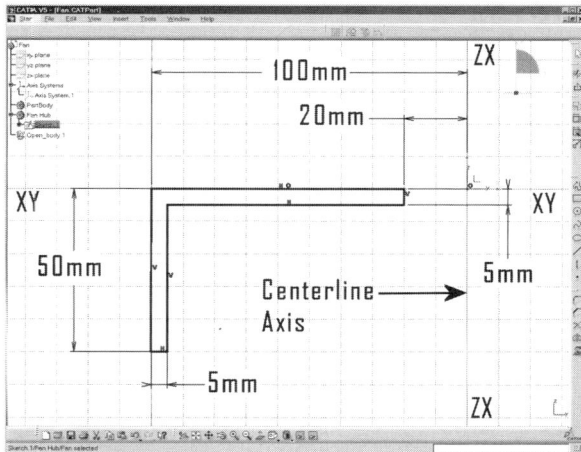

Fig. 7-60. Hub shaft feature.

This exercise utilizes wireframe, surface, and solid features in the construction of a fan model, shown in figure 7-59. The goal of the exercise is to provide you with the opportunity to explore how surface features can enhance the solid modeling experience and provide the detailed interaction of solid and surface features. To create the fan model, perform the following steps.

1   Create a new CATPart and rename the default part name to *Fan Exercise.* Insert a new axis system into the model.

2   Insert a new part body and name it *Fan Hub.*

3   Create a new sketched profile by selecting the YZ plane and then clicking on the Sketcher icon. Sketch the profile shown in figure 7-60, which will define the inner section of the fan hub. Constrain the top line of the hub to the XY plane. Exit the Sketcher when complete.

4   Create a hub shaft feature (revolved solid), shown in figure 7-61, using the previous sketched profile and the vertical Z axis.

5   Activate the Wireframe and Surface workbench.

6   Rename the default *Open Body* to *Blade Construction Elements* by clicking on the Open Body icon using MB3 and modifying the properties.

7   Create a plane that is offset from the YZ plane a distance of 500 mm. Name this plane *Blade Section Plane.*

8 Using the plane named *Blade Section Plane,* create a new sketched profile that describes the inner section of the fan blade hub, shown in figure 7-62. Develop the arc element and constrain the curve upper and lower points to the upper and lower sections of the inner hub solid.

Fig. 7-61. Hub profile.

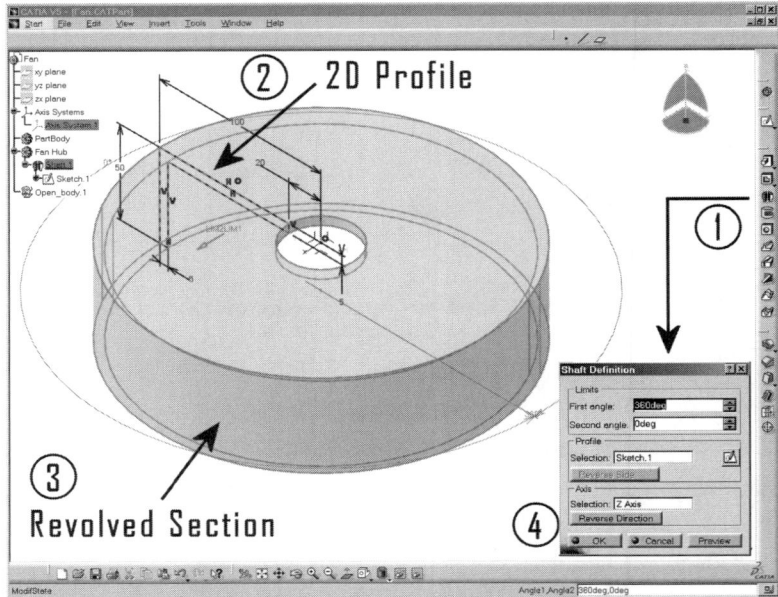

Fig. 7-62. Blade inner hub section.

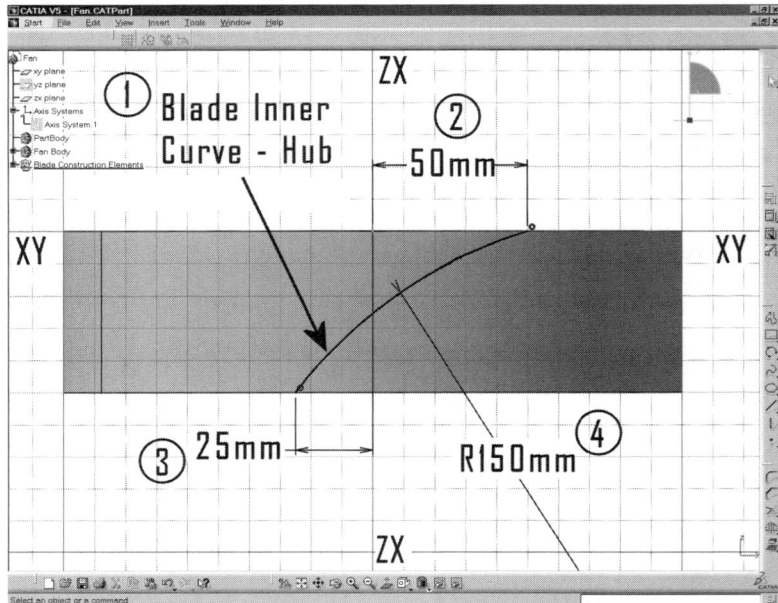

9 Repeat this operation and sketch a new profile that describes the outer section of the fan blade hub, shown in figure 7-63. Develop the arc element and constrain the curve upper and lower points to the upper and lower sections of the inner hub solid. See figure 7-63.

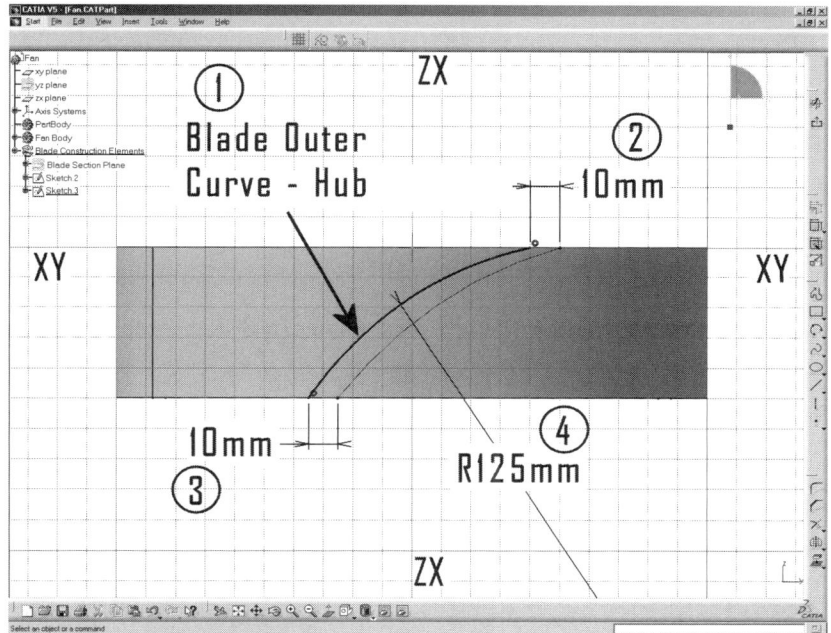

Fig. 7-63. Blade outer hub section.

10 Extrude 550-mm linear surfaces, in one direction, from the inner and outer blade section profiles toward the center of the hub. Figure 7-64 shows these extrusions.

11 Create intersection curves between the two extruded surfaces and the outer face of the fan hub solid. Figure 7-65 shows these intersection curves.

12 Using the hide function, blank off the display of the extruded surfaces and the blade sections.

13 Let's create a new sketched profile. Select the ZX plane and then click on the Sketcher icon. Create a 200-mm arc profile with the top and bottom segment constrained to the top and lower surface of the inner hub. Constrain the center of the arc on the XY plane. Exit the Sketcher when complete. Figure 7-66 shows this outer blade profile.

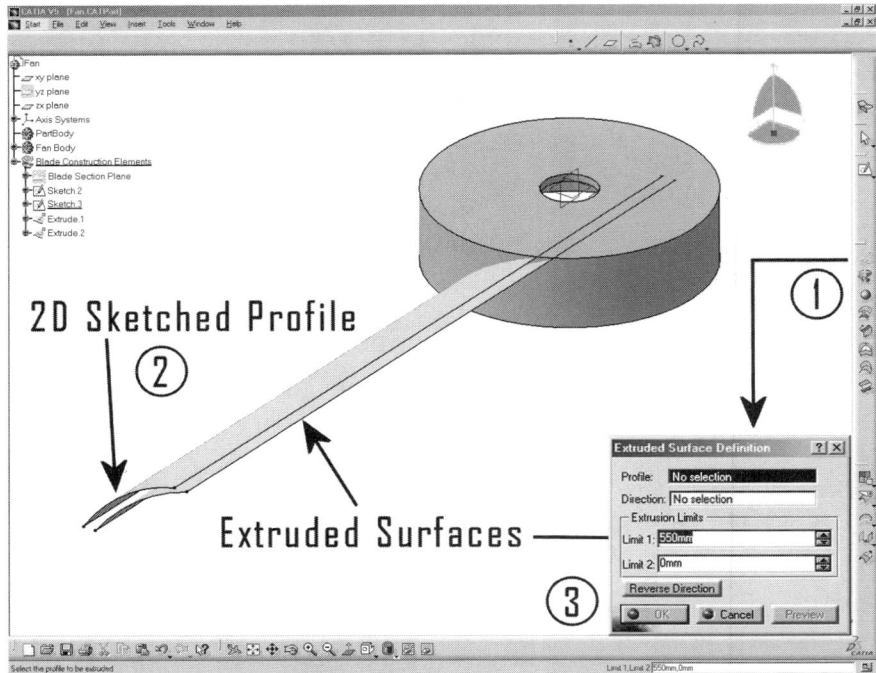

Fig. 7-64.
Extruded 550-
mm linear
surfaces.

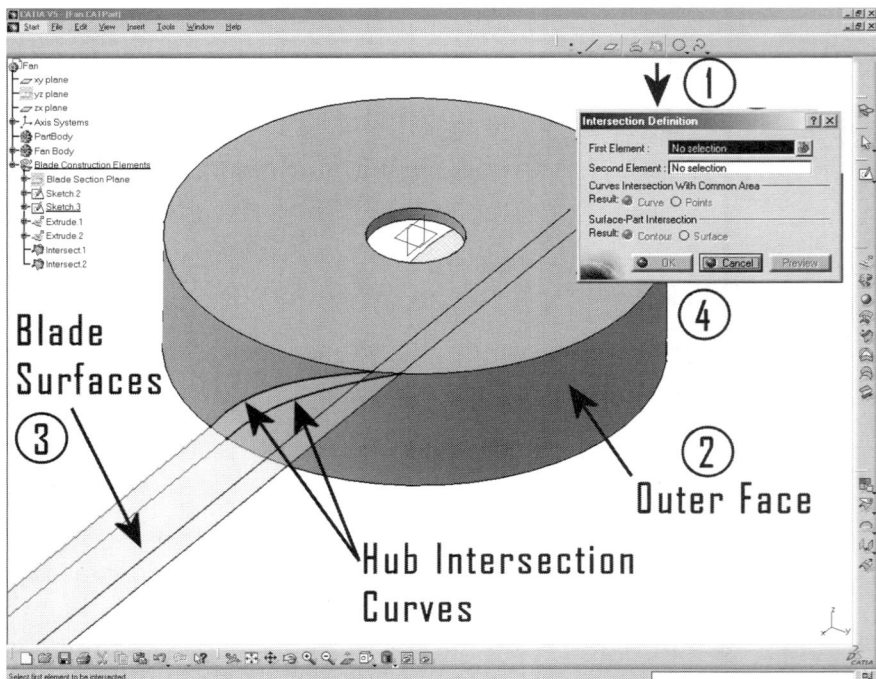

Fig. 7-65.
Intersection
curves for
outer face.

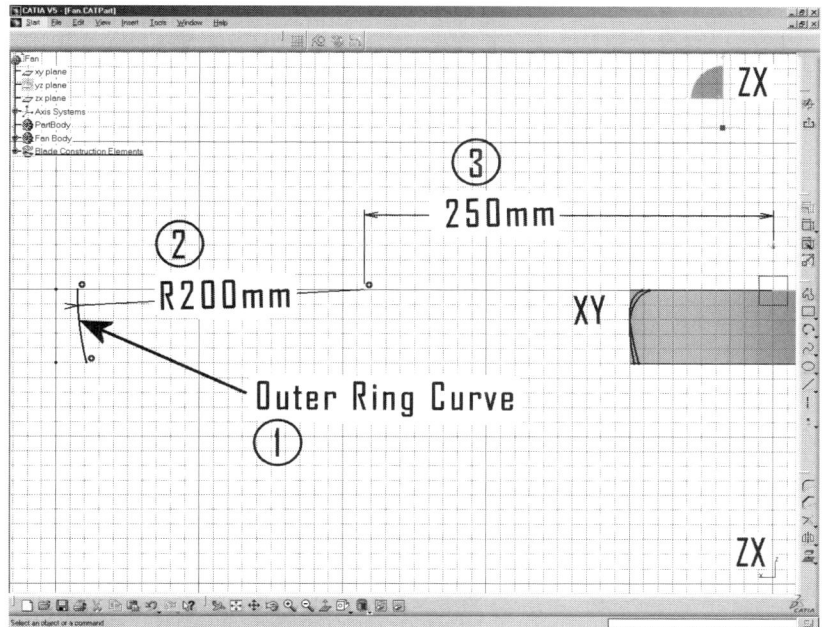

Fig. 7-66. Outer blade profile.

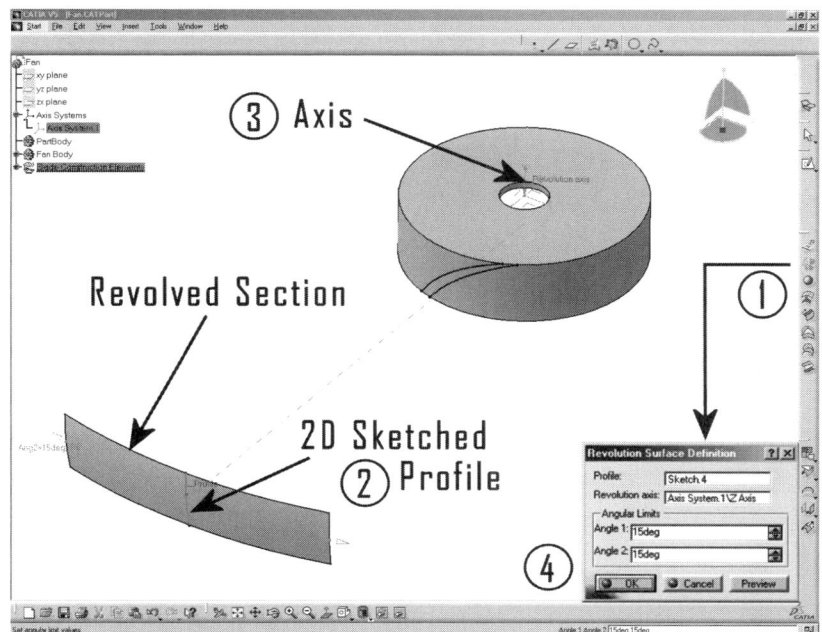

Fig. 7-67. Outer blade surface ring.

14 Create a revolved surface ring using the previously sketched profile and the Z axis. Use 15 degrees for both sides of the angular limits. Figure 7-67 shows this surface ring.

15 Using the plane named *Blade Section Plane,* create a new sketched profile that describes the inner section of the fan blade tip, as shown in figure 7-68. Develop the arc element and constrain the curve upper and lower points to the upper and lower sections of the inner hub solid.

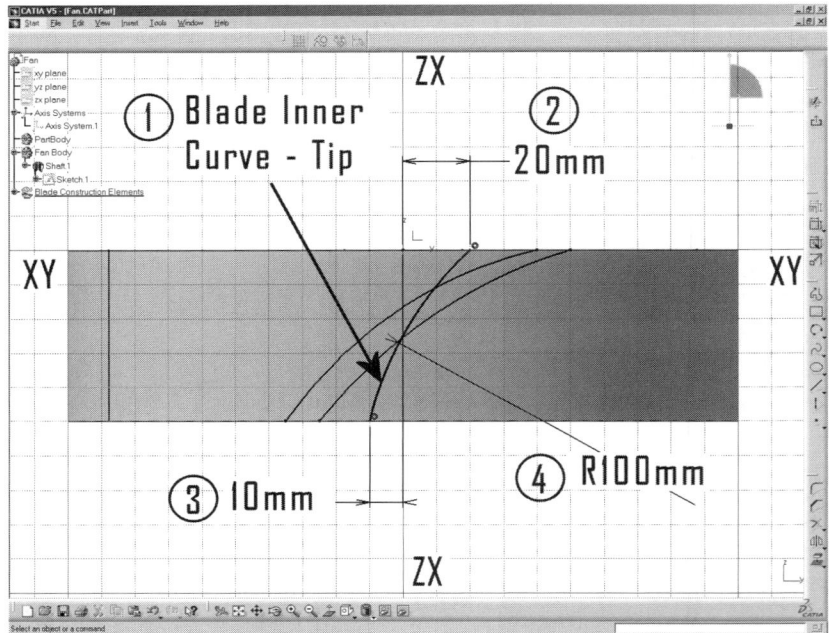

Fig. 7-68. Blade inner tip section.

16 Repeat this operation and sketch a new profile that describes the outer section of the fan blade tip, as shown in figure 7-69. Develop the arc element and constrain the curve upper and lower points to the upper and lower sections of the inner hub solid.

17 Extrude 250-mm linear surfaces, in one direction, from the inner and outer blade tip section profiles that intersect the outer blade ring surface. Figure 7-70 shows these extrusions.

18 Create intersection curves between the two extruded surfaces and the outer blade ring surface, as shown in Figure 7-71.

19 Using the hide function, blank off the display of the extruded surfaces, blade section profiles, and outer ring surface.

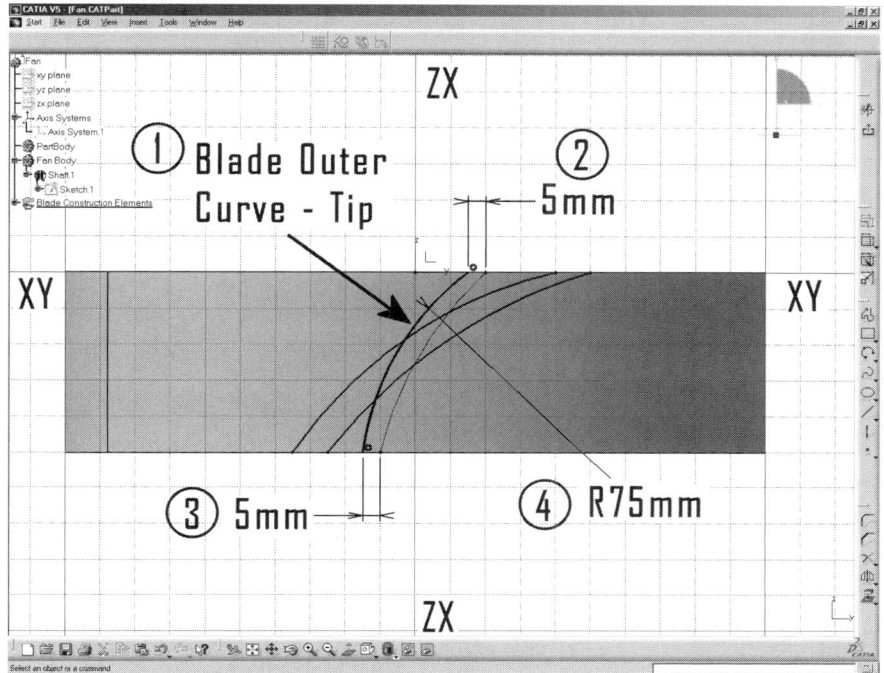

Fig. 7-69.
Blade outer
tip section.

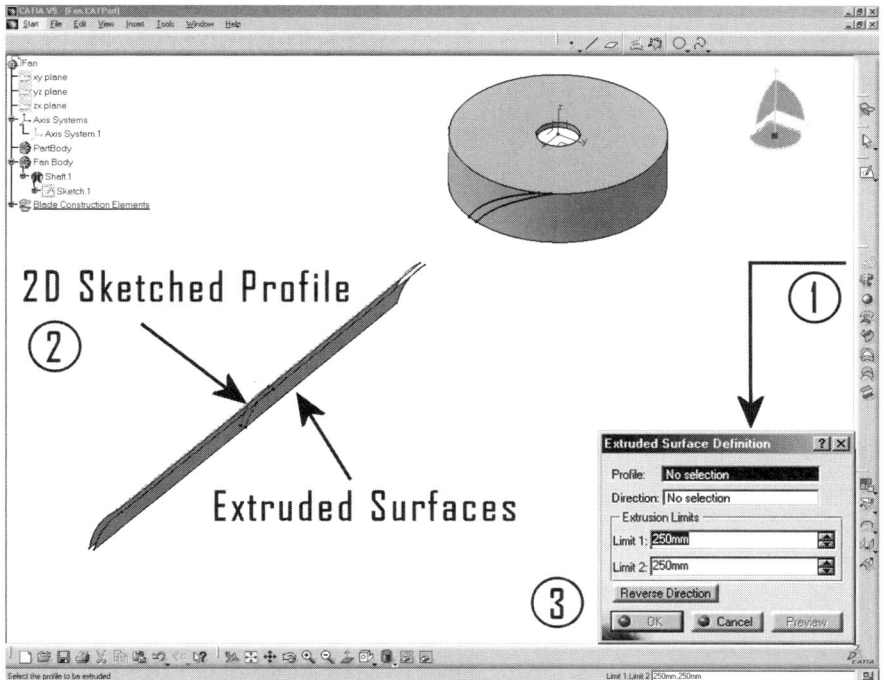

Fig. 7-70.
Extruded
250-mm
linear
surfaces.

20 The design should now look like that shown in figure 7-72.

21 Insert a new open body and name it *Blade Surfaces.*

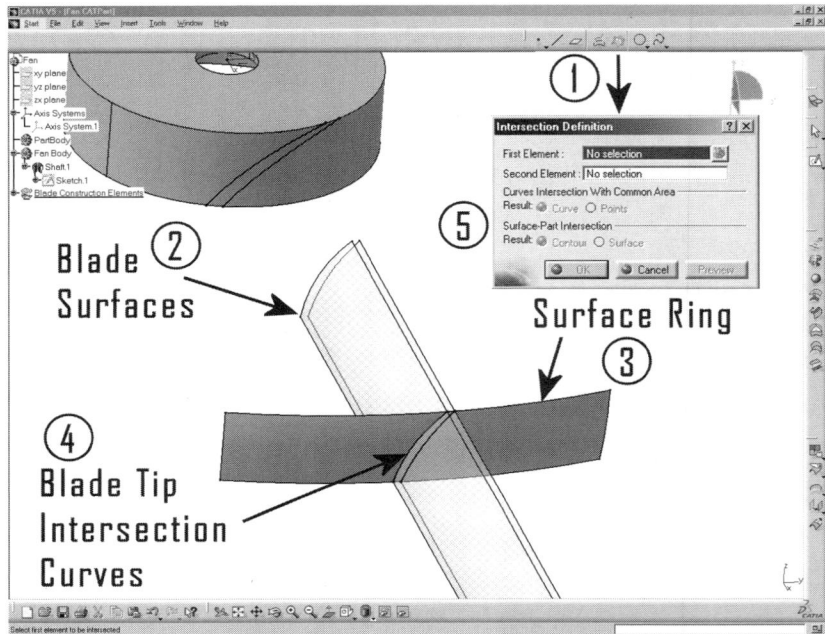

*Fig. 7-71.
Intersection
curves for outer
ring surface.*

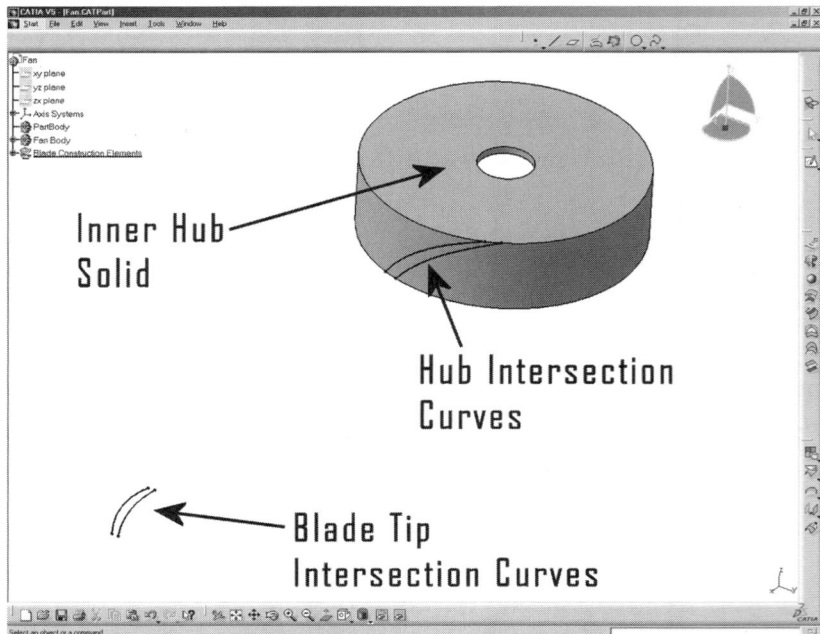

*Fig. 7-72.
Intersection
curves for outer
ring surface*

22 Using the Extract command, create a surface from the outer boundary of the hub face, as shown in figure 7-73.

23 Blank off the display of the *Fan Hub* body, leaving only the extracted surface and the four blade intersection curves, as shown in figure 7-74.

Fig. 7-73. Outer hub face surface extraction.

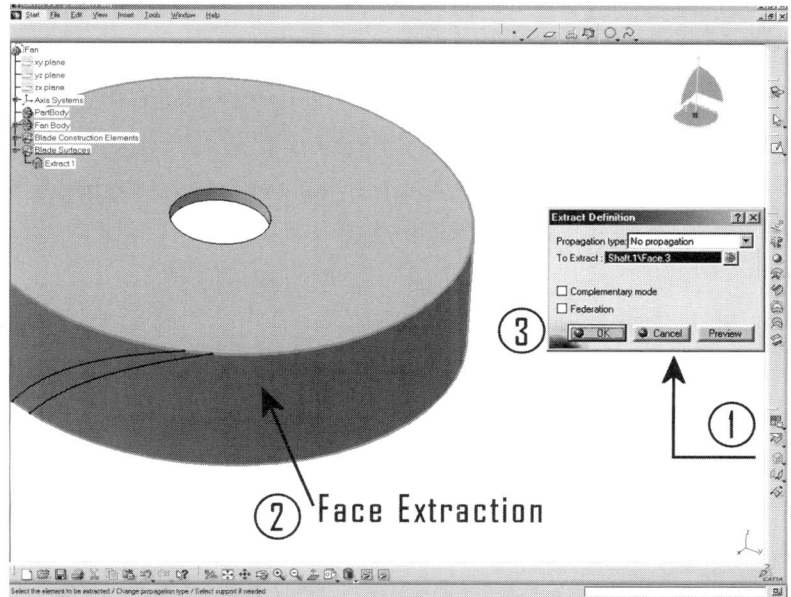

② Face Extraction

Fig. 7-74. Blade definition.

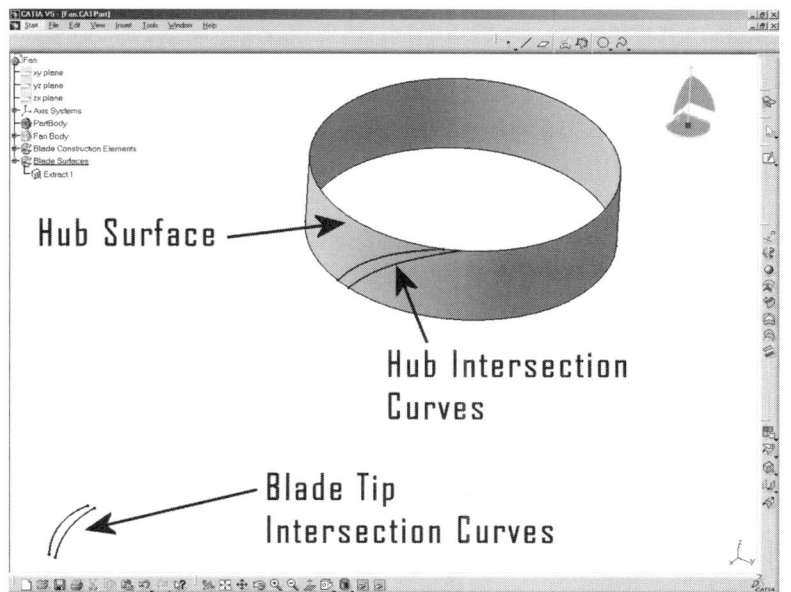

Hub Surface

Hub Intersection Curves

Blade Tip Intersection Curves

24 Using the Blend Surface command, create surfaces that blend the inner and outer sections of the blade, as shown in figure 7-75.

25 Using the Show command, redisplay the outer ring surface.

26 Using the Trim command, trim the two blended inner and outer surface blades with the outer surface ring, as shown in figure 7-76. This will require two operations. See figure 7-76.

27 Using the Trim command, trim the resultant surface with the extracted outer hub surface, as shown in figure 7-77.

28 Using the Fill Surface command, create surfaces that fill the upper and lower open sections of the fan blade, as shown in figure 7-78.

29 Insert a new open body and name it *Blade Surface Join*.

30 Join all of the surfaces to create a complete enclosed surface join element, as shown in figure 7-79.

31 Create a circular pattern of six blade instances of the filleted joined surface blade, as shown in figure 7-80. Use the Z axis for the rotation axis and use a spacing of 60 degrees.

*Fig. 7-75.
Blade surface
blends.*

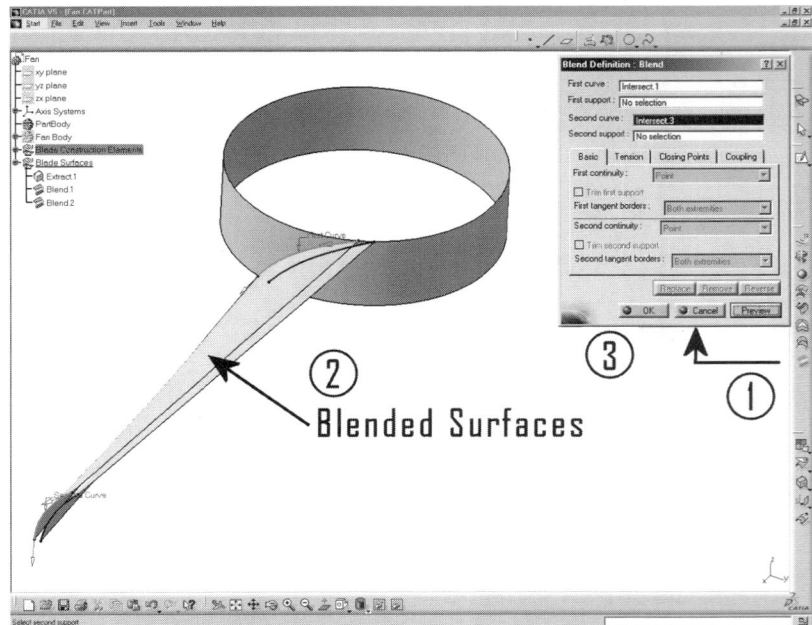

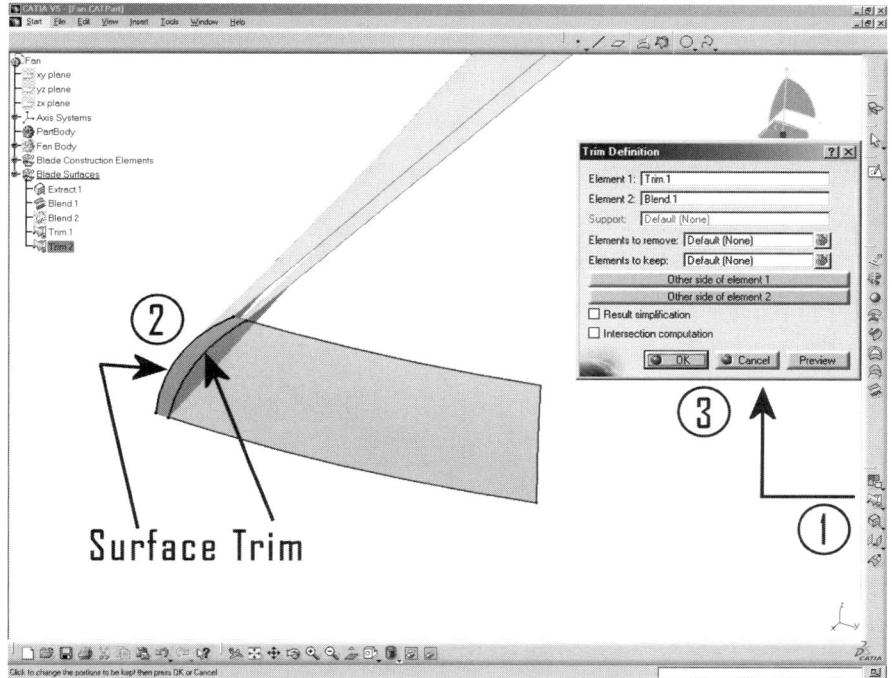

Fig. 7-76.
Outer trim
surface.

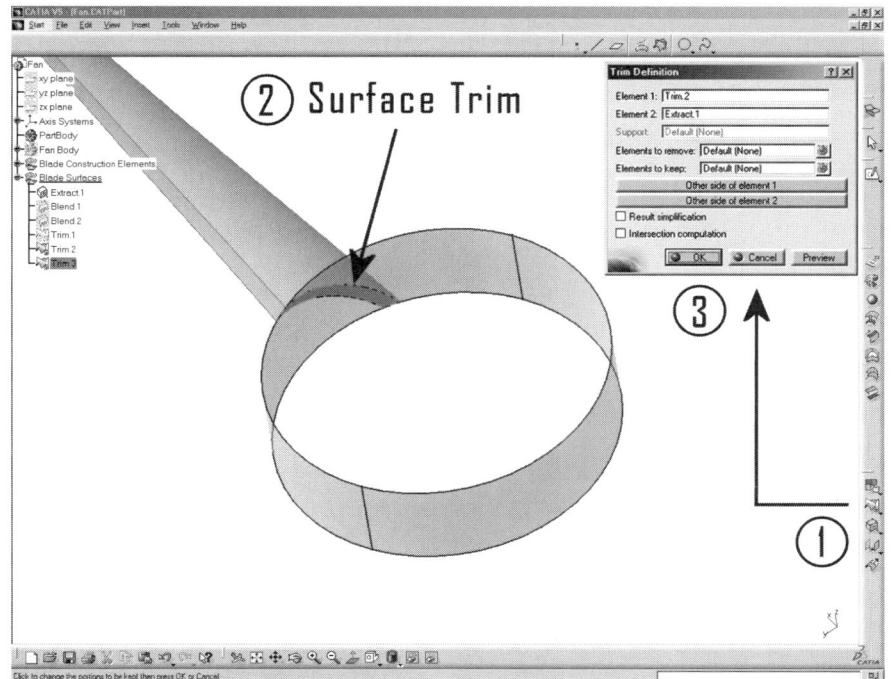

Fig. 7-77.
Hub inner trim
surface.

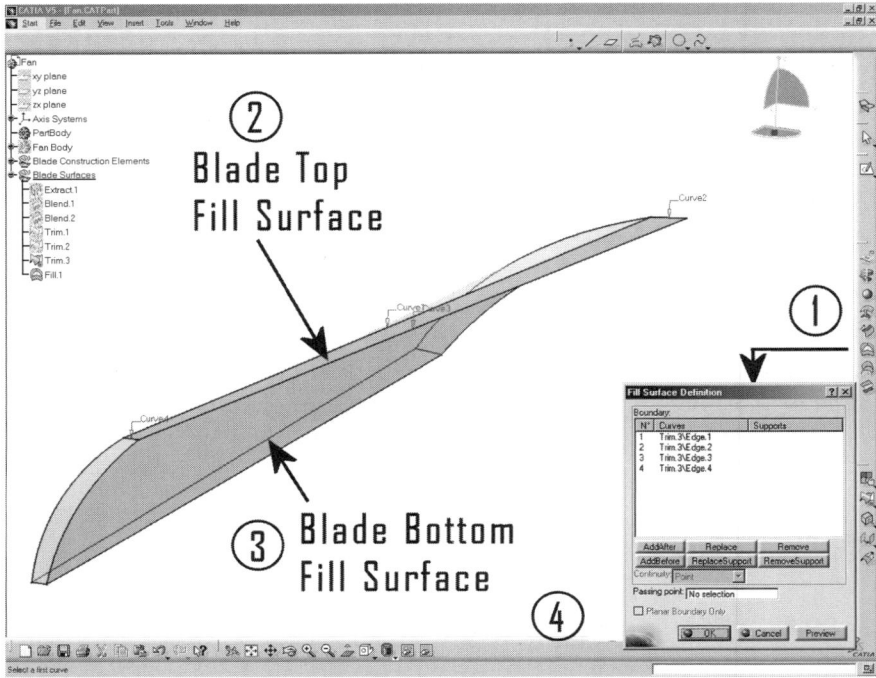

Fig. 7-78.
Fill surfaces.

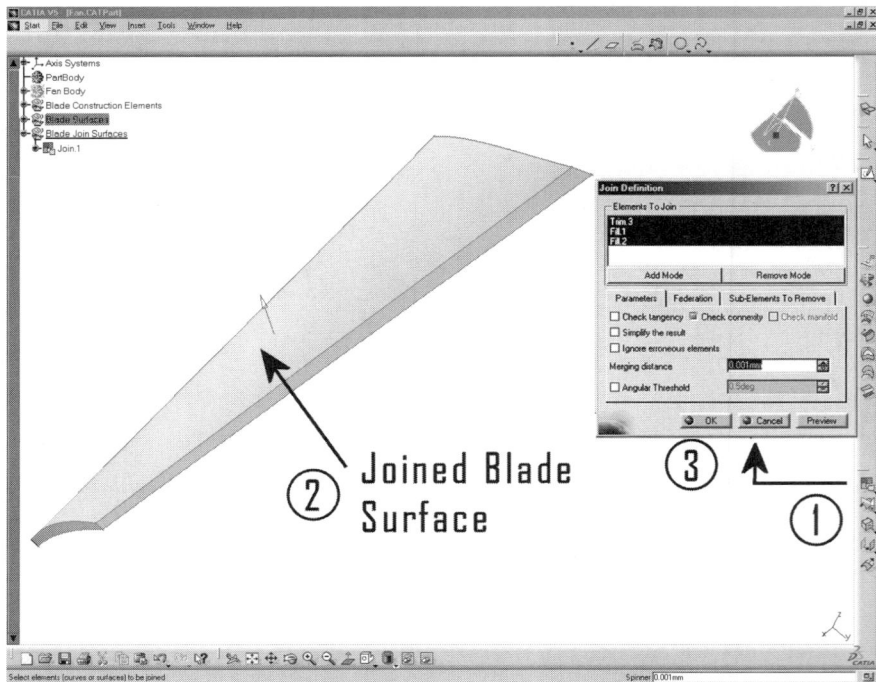

Fig. 7-79.
Surface join.

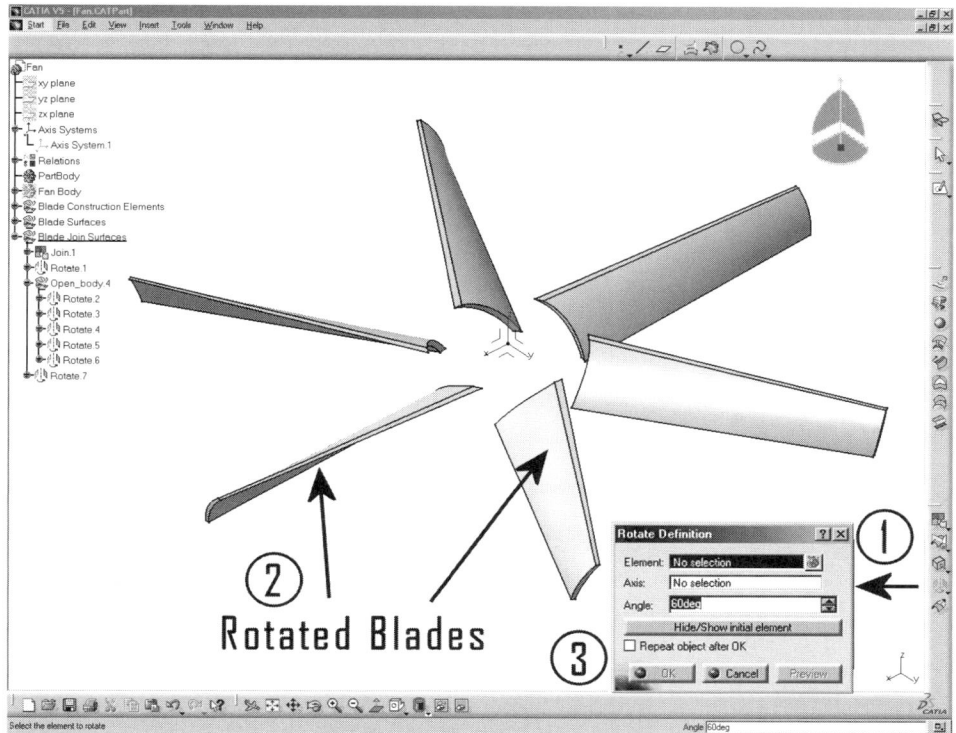

Fig. 7-80. Circular pattern.

32 Activate the Part Design workbench.

33 Insert a new body and name it *Fan Blades*.

34 Create two (2) fan blade solid features using the command Insert I Surface Based Features I Close Surface. Select the original closed surface and one patterned blade to create the solid model fan blades. See figure 7-81.

35 Assemble the two bodies into the default part body by clicking MB3 on the *Fan Hub* part body and select Object I Assemble from the pull-down window. Select *Fan Final* and then click on OK to assemble the two bodies. Figure 7-82 shows this fan assembly.

36 Create 10-mm fillets between each blade edge and the hub face, as shown in figure 7-83.

The design is now complete. Save the model for future reference.

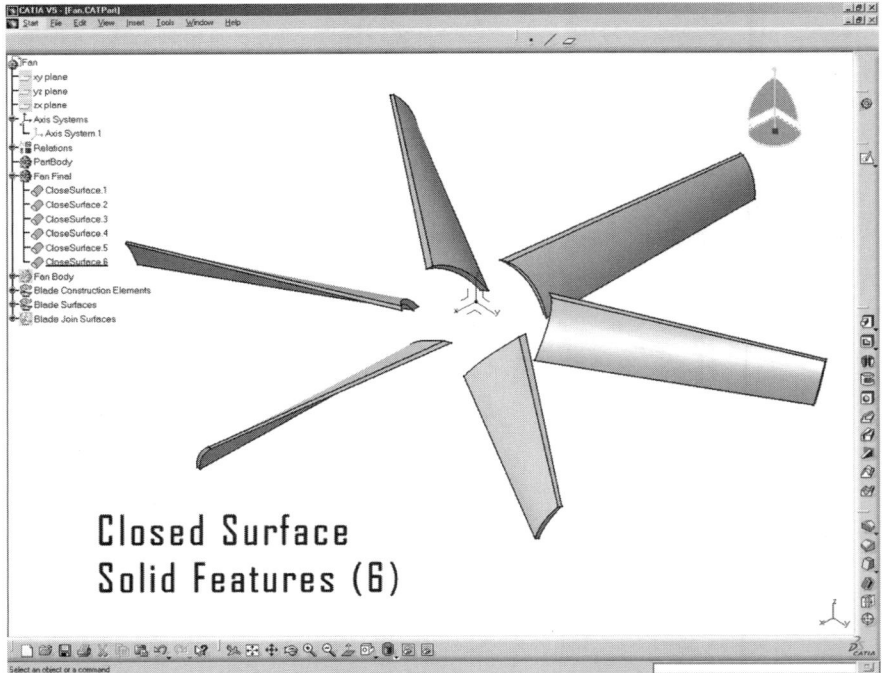

Fig. 7-81.
Closed
surface.

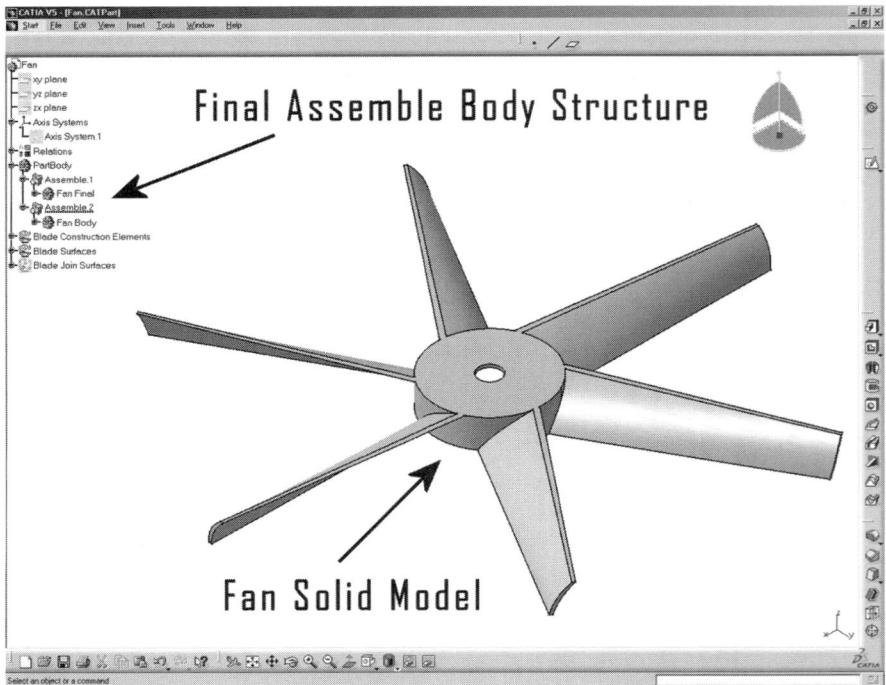

Fig. 7-82.
Fan
assembly.

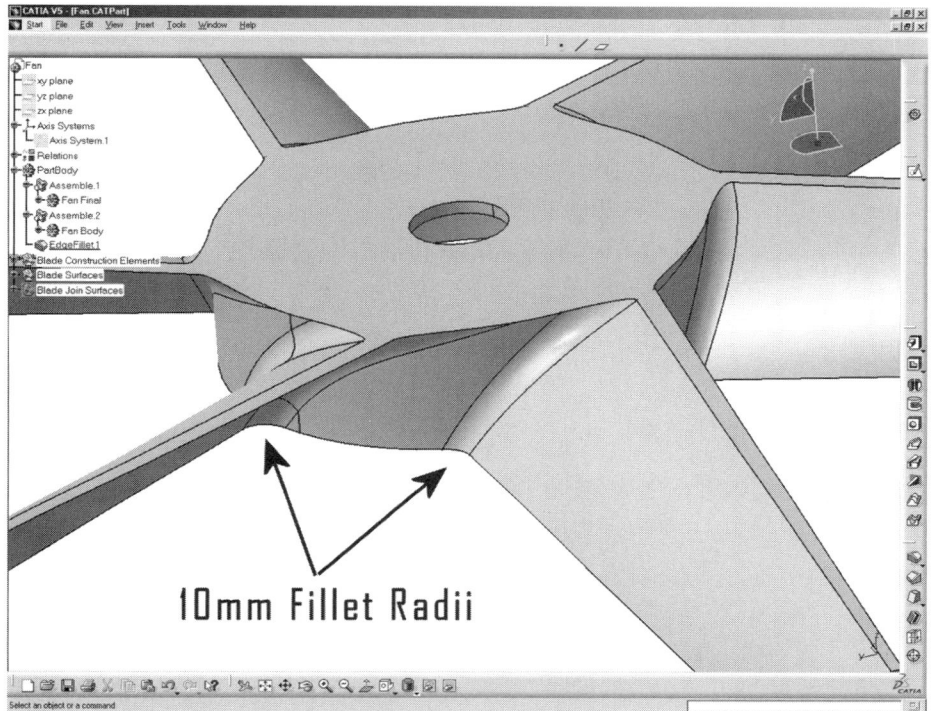

Fig. 7-83.
Fan fillets.

Summary

This chapter introduced you to basic wireframe and surface functionality within CATIA V5. These tools offer a productive and intuitive design environment in which to capture and reuse design methodologies and specifications. Understanding the basics of mixing wireframe, surface, and solid geometry is paramount in building robust models.

The Wireframe and Surface workbench allows you to create wireframe and surface construction elements at any time to enhance the design of solid model parts. These construction elements enrich the capability of a feature-based design and complement the Part Design workbench from a surface-/solids-based hybrid modeling approach. This flexibility also offers the ability to offer post-design and local 3D parameterization toward further capturing part design intent.

Review Questions

1   Name three key differences between surface and solid elements.

2   Why is hybrid modeling more flexible than surface or solid modeling?

3   What types of bodies are associated with part models and surface models?

4   How do surface features enhance the flexibility and efficiency of a solid model part?

5   Name five surface creation types within CATIA V5.

6   What types of operations are available to manipulate wireframe and surface elements within CATIA V5?

CHAPTER 8

THE ASSEMBLY
WORKBENCH

Introduction

THE ASSEMBLY WORKBENCH OFFERS flexible and intuitive tools used to orient parts in relationship to one another. Just as features are combined to form parts, parts can be combined to form assemblies. The Assembly workbench enables you to place component parts and subassemblies together to form finished assemblies. Parts may also be designed within the assembly environment, which allows for proper design in context relationships.

Parts may be modified, analyzed, and reoriented in the assembly environment, depending on defined constraints. Unconstrained part placement methods can also be used in this environment to help accelerate the preliminary packaging of assemblies. In assembly mode, parts can be snapped into position and can be dragged and dropped via various rotation and translation tools. The purpose of this chapter is to introduce basic assembly concepts related to working with the Assembly workbench. (See also the CATIA V5 online documentation for information on advanced assembly functionality.)

Objectives

This chapter covers the following.

- An introduction to the assembly environment
- Assembling components
- Constraining and positioning
- Assembly analysis tools
- Assembly features

The Assembly Environment

Most products and their resultant designs consist of various parts and/or subassemblies. CATIA V5 offers tools that allow for the construction, navigation, and organization of parts and subassemblies to create a finished assembly. These tools allow you to produce a logically organized product design, the components of which are contained within individual, associative documents.

This allows users of the assembly document to navigate from one document to another and to edit designs and create associations between and among documents. Users of such documents can also copy and paste components, add constraints, add new components, load existing models, replace components, modify components, and create dimensional or geometric associative constraints.

Assembly Workbench

There are various methods of creating a new assembly document within CATIA V5, an example of which is shown in figure 8-1. The most common of these methods start from the Start menu, the File menu, or the Workbench icon.

Assembly User Interface

The assembly user interface, shown in figure 8-2, is identical to the part interface except for the configuration tree. Just as parts contain features, assemblies contain parts and other documents.

The configuration tree contains all documents that constitute an assembly. The configuration tree can also contain references to documents that are not in native CATIA V5 formats.

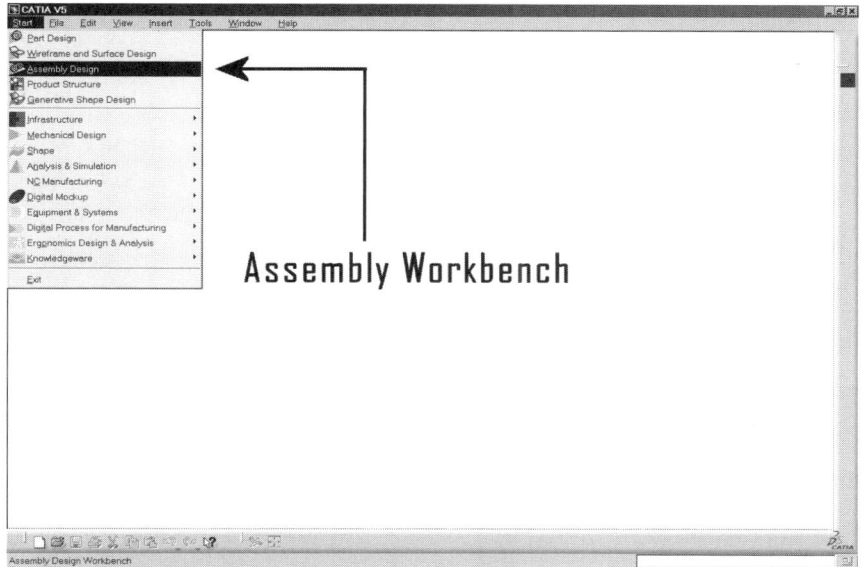

Fig. 8-1. Assembly workbench document.

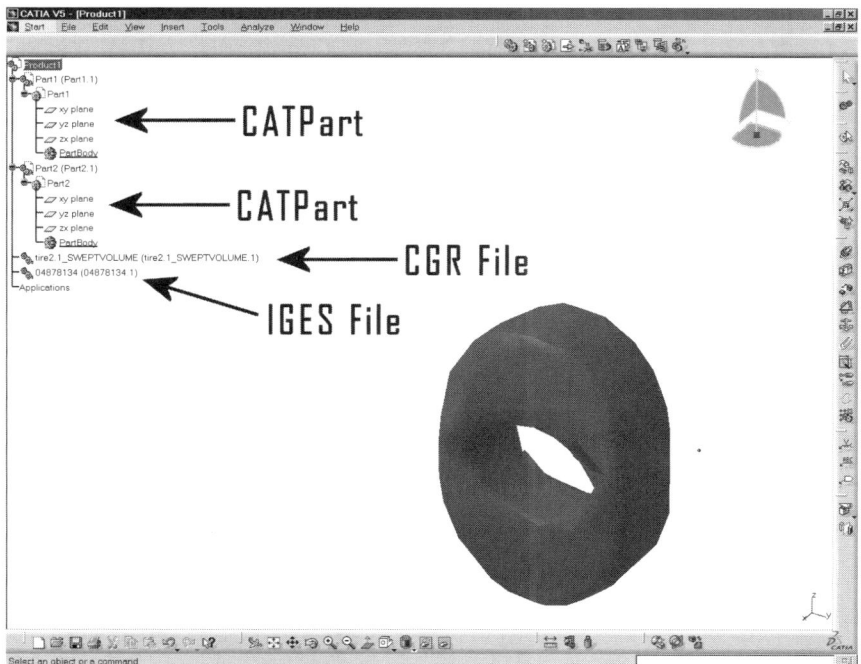

Fig. 8-2. Assembly user interface.

Assembly Documents

Upon creation of a new assembly, CATIA V5 will create a new product structure document (i.e., a CATProduct file), an example of which is shown in figure 8-3. A CATProduct file contains a list of components, constraints, positions, features, applications, and links constituting its assembly structure.

Fig. 8-3. Product structure document.

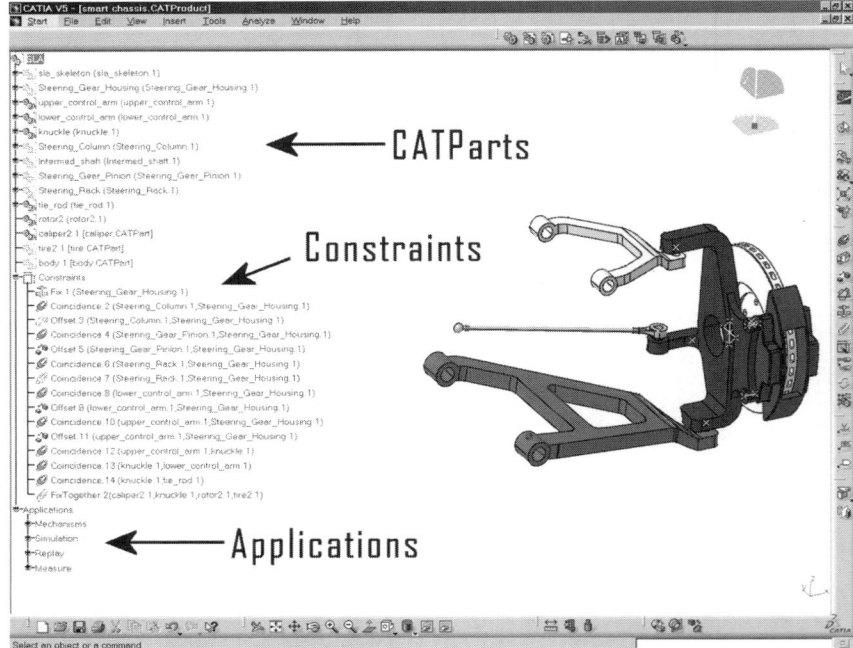

Assembling Components Using the Product Structure Toolbar

Assembling components in CATIA V5 is based on the same principles you would employ were you to physically assemble parts in a prototyping shop. The final assembly is broken down into subassemblies, which may further be broken down into individual parts. The approach to the assemblage design of a mechanical system should follow the same sequence as its physical construction. This is important in ensuring that the actual manufacturing process is reflected in the design and that the required materials are all accounted for. CATIA V5 offers component assembly functionality consistent with the intended assembly sequence of physical components.

The Product Structure toolbar (shown in figure 8-4) within the Assembly workbench is your starting point for introducing components into an assembly. The assembly (or "product structure") environment is designed to support many different types of file formats. These formats may vary depending on the type of CATIA V5 license.

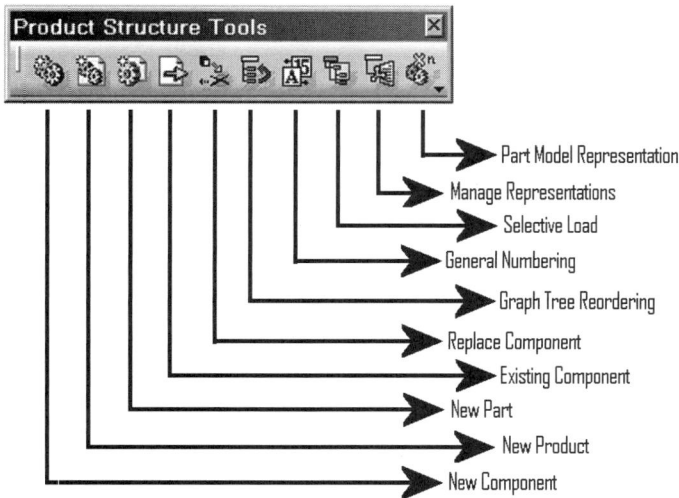

Fig. 8-4. Product Structure toolbar.

The sections that follow describe the options contained on the Product Structure toolbar. Note that whenever you want to select an option from the Product Structure toolbar you must first click on the Product Structure icon in the configuration tree.

New Component

The New Component option, accessed via the New Component icon (shown in figure 8-5), allows for the creation of a new component within an existing assembly, an example of which is shown in figure 8-6.

Fig. 8-5. New Component icon.

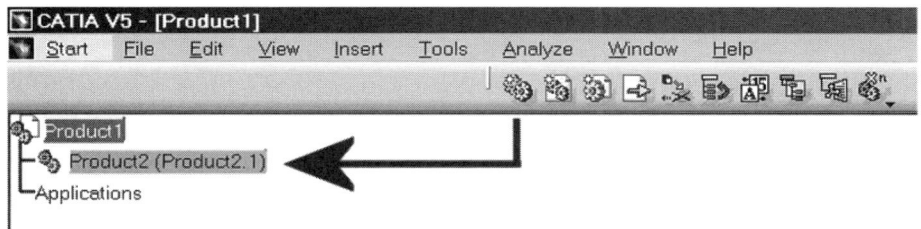

Fig. 8-6. New component created within an existing assembly.

New Product

The New Product option, accessed via the New Product icon (shown in figure 8-7), allows for the creation of a new product within an existing assembly, an example of which is shown in figure 8-8.

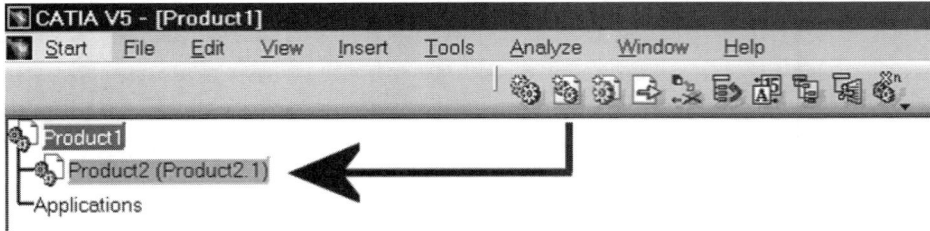

Fig. 8-7. New Product icon.

Fig. 8-8. New product created within an existing assembly.

New Part

The New Part option, accessed via the New Part icon (shown in figure 8-9), allows for the creation of a new part within an existing assembly, an example of which is shown in figure 8-10.

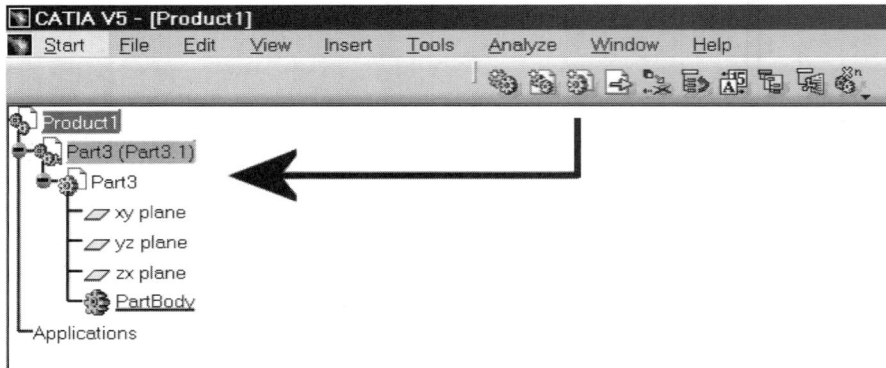

Fig. 8-9. New Part icon.

Fig. 8-10. New part created within an existing assembly.

Existing Component

Fig. 8-11. Existing Component icon.

The Existing Component option, accessed via the Existing Component icon (shown in figure 8-11), allows for the insertion of an existing component into the current assembly. Clicking on the Existing Component icon accesses a pop-up window, which contains file format options for those formats supported for insertion into an assembly. An example of an existing component inserted into an assembly is shown in figure 8-12.

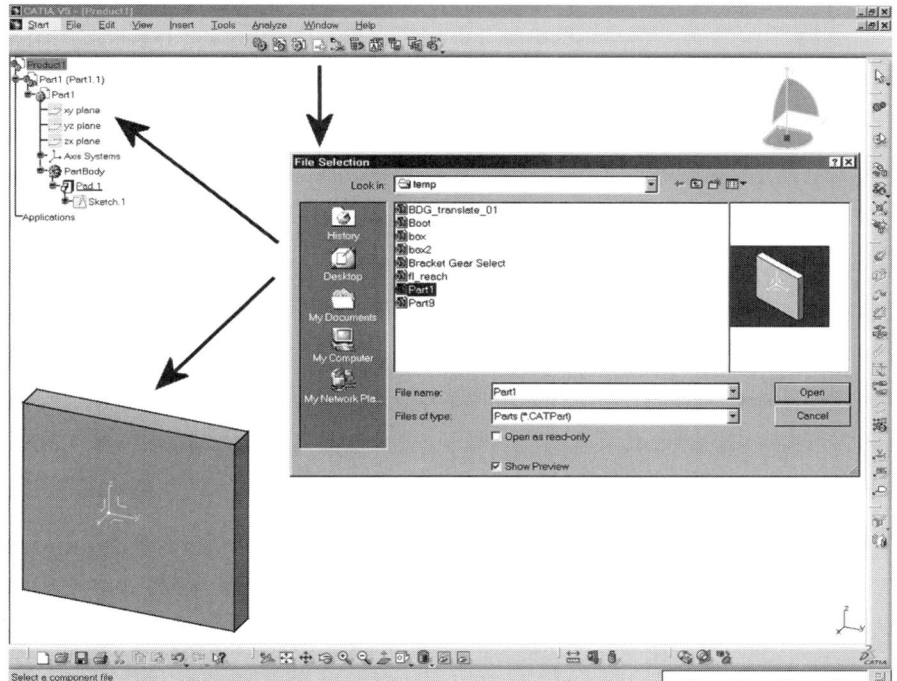

Fig. 8-12.
Existing
component
inserted into
an assembly.

Replace Component

Fig. 8-13.
Replace
Component
icon.

The Replace Component option, accessed via the Replace Component icon (shown in figure 8-13), allows for the replacement of one component with another within the current assembly. Clicking on the Replace Component icon accesses a pop-up dialog window that displays a list of items that will be impacted by replacing one component with another (as shown in figure 8-14), giving you the option in each case of accepting or rejecting the replacement.

Graph Tree Reordering

The Graph Tree Reordering option, accessed via the Graph Tree Reordering icon (shown in figure 8-15), allows for the reordering of existing components within the assembly configuration tree. Clicking on the Graph Tree Reordering icon accesses a pop-up dialog window in which you can reorder components via the Up and Down arrows, as shown in figure 8-16.

Fig. 8-14.
Dialog window
showing list of
items affected
by component
replacement.

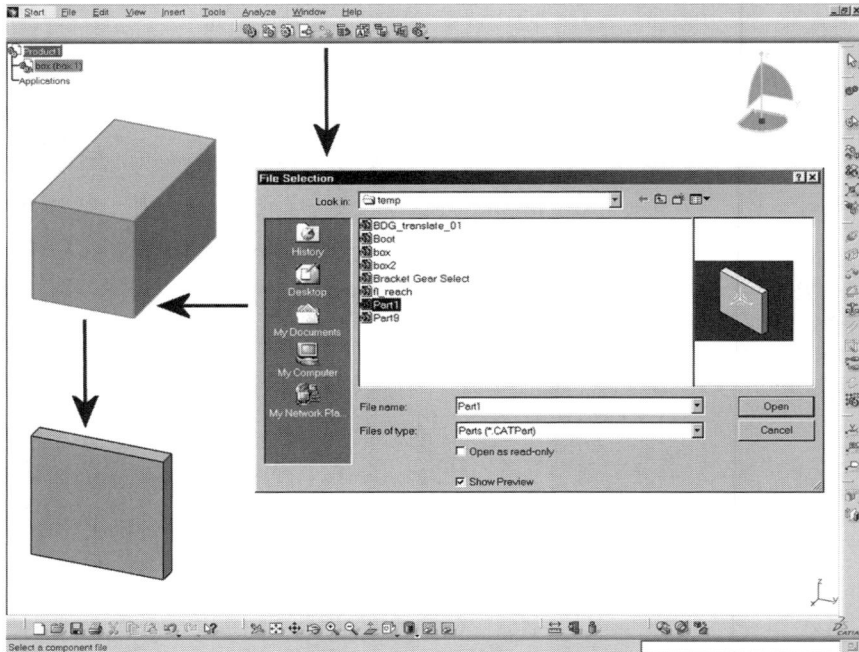

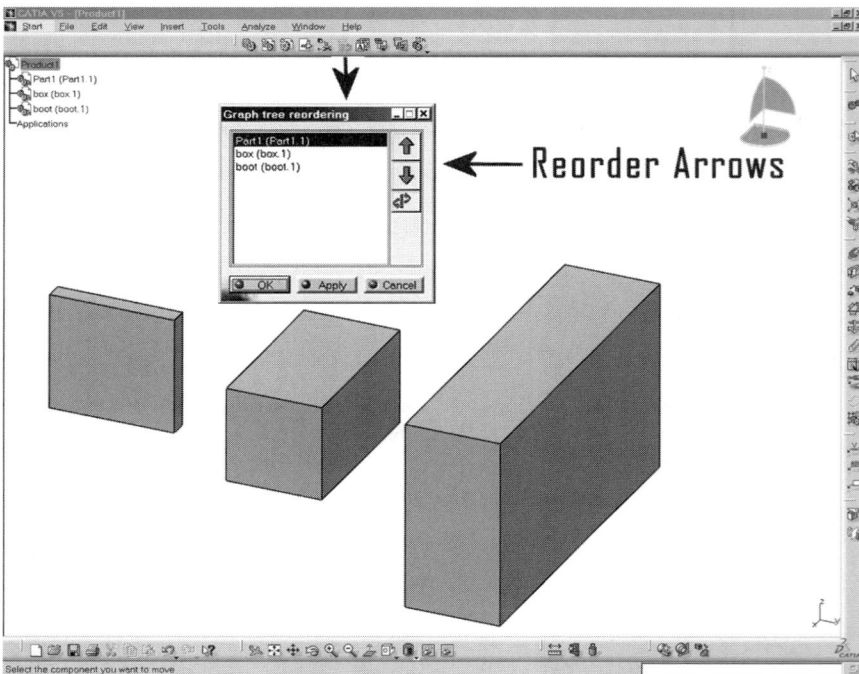

Fig. 8-15.
Graph Tree
Reordering
icon.

Fig. 8-16. Reordering of components via the Graph Tree Reordering option.

General Numbering

The General Numbering option, accessed via the General Numbering icon (shown in figure 8-17), allows for the sequential ordering (per numbers or letters) of current components within an existing assembly. An example of such ordering is shown in figure 8-18.

Fig. 8-17.
General
Numbering
icon.

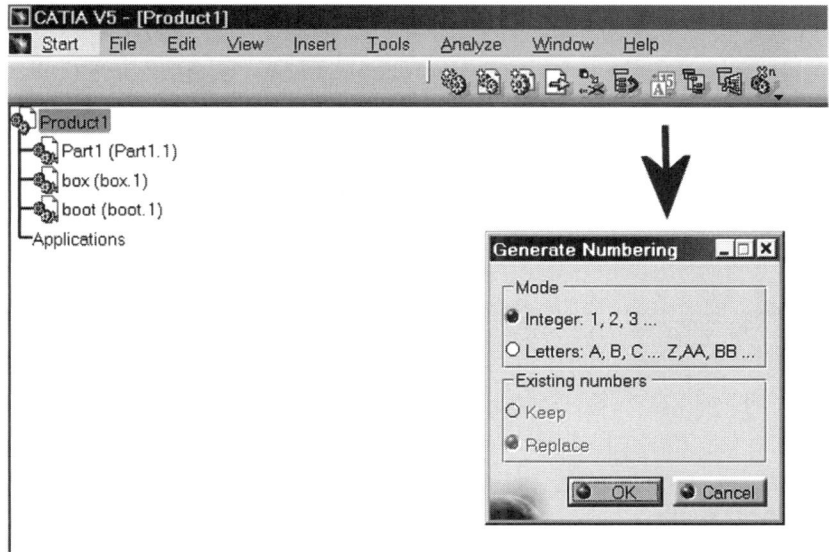

Fig. 8-18. Ordering of components via the General Numbering option.

Selective Load

Fig. 8-19.
Selective
Load icon.

Clicking on the Selective Load option, accessed via the Selective Load icon (shown in figure 8-19), displays a pop-up dialog window that allows you to individually customize the loading of components into an assembly. Individual components may be selected to be loaded, shown, or hidden within the assembly. An example of selective (custom) loading is shown in figure 8-20.

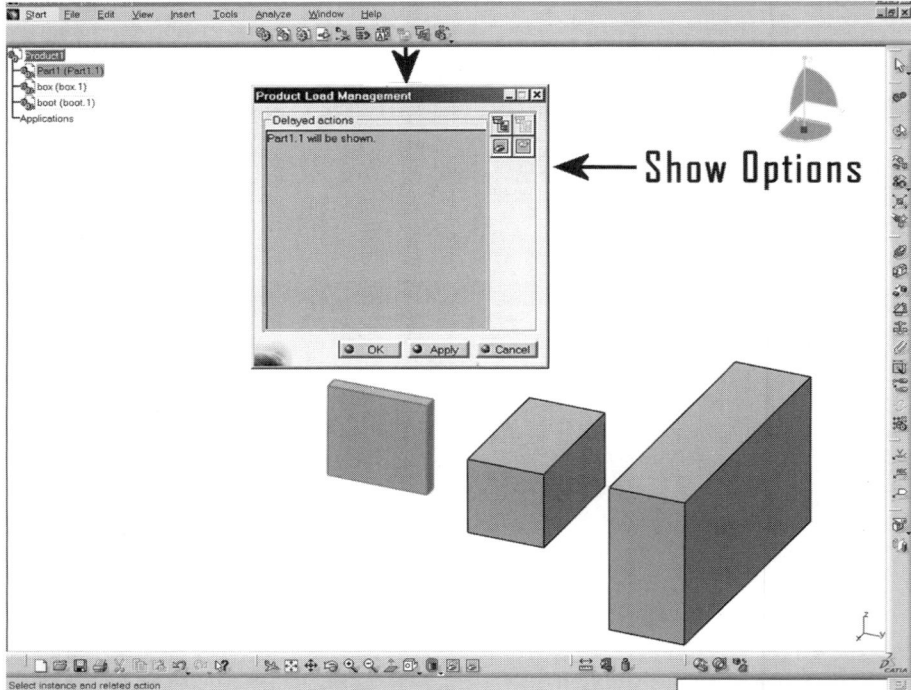

Fig. 8-20.
Selective
(custom)
loading.

Manage Representations

Fig. 8-21.
Manage
Representations
icon.

The Manage Representations option, accessed via the Manage Representations icon (shown in figure 8-21), allows for the management of the type of model representation present within an assembly. Models may be activated, deactivated, removed, replaced, or associated via the pop-up dialog window accessed by clicking on the Manage Representations option. An example of this type of representational management is shown in figure 8-22.

Part Model Representation

Depending on the type of CATIA V5 licensing and configuration options selected (via the Cache Management option), CATIA V5 provides for a lightweight graphical representation of a model within the assembly environment (via the Visualization Mode option). This option is essential to management of large assemblies, whereby you can selectively load a model's full history into a design session.

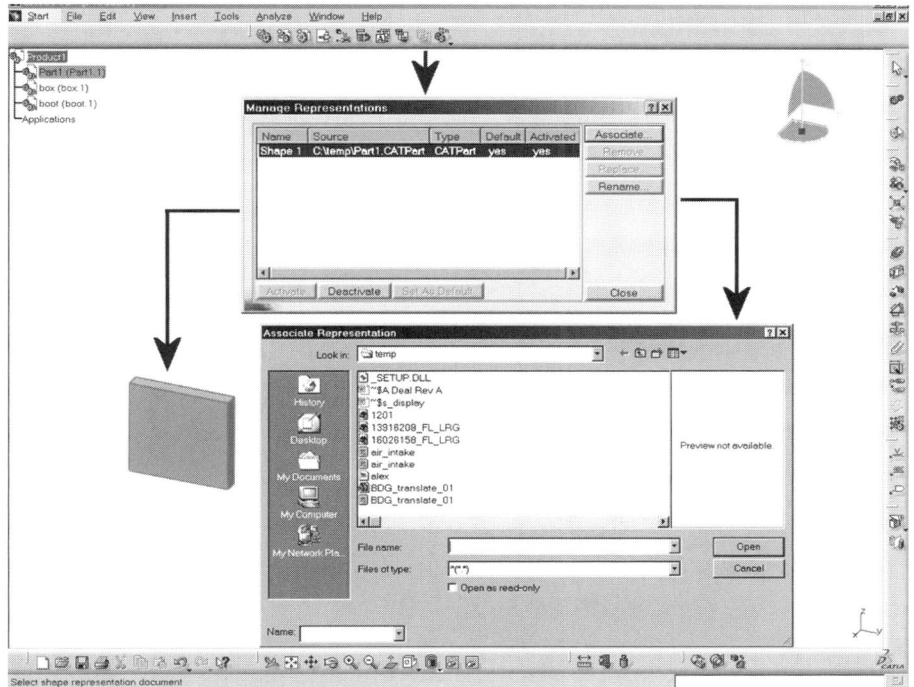

Fig. 8-22.
Representational
management.

This management method utilizes the computer's memory in a more efficient manner and makes large assemblies more manageable and faster to load into session. The graphical representation models are essentially CGR (CATIA Graphical Representation) files (polygon based models) based on DMU technology. (See the CATIA V5 online documentation for more information on model representations.)

TIP: *A model's representation may be also managed by selecting the part and clicking MB3. This displays a contextual menu that allows you to quickly select the desired representation. In addition, double selecting a lightweight visualization model will automatically load the entire model's history into session.*

Constraining and Positioning

CATIA V5 offers an array of assembly tools that fix, position, and constrain components within an assembly. Assembly constraints control the propagation of design modifications from one component to another. Components may be positioned, and constrained or not constrained, depending on the desired relationships of the assembly. However, assembly constraints are often essential to the creation of robust assemblies that capture desired relationships. Assembly constraints are located within the Constraint section of the configuration tree multi instantiate, as shown in figure 8-23.

Fig. 8-23. Configuration tree Constraint section.

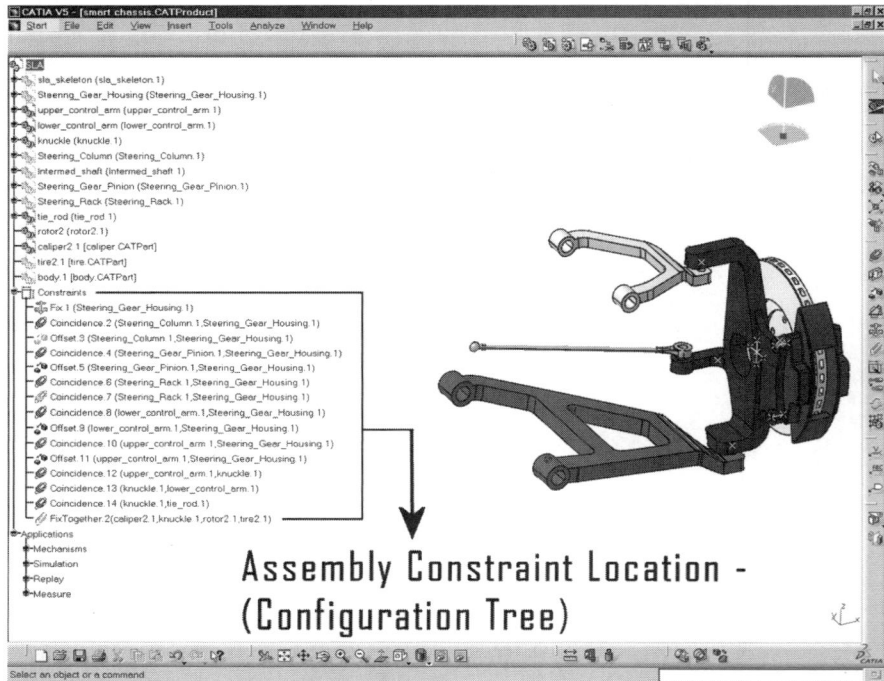

Assembly Constraint Location – (Configuration Tree)

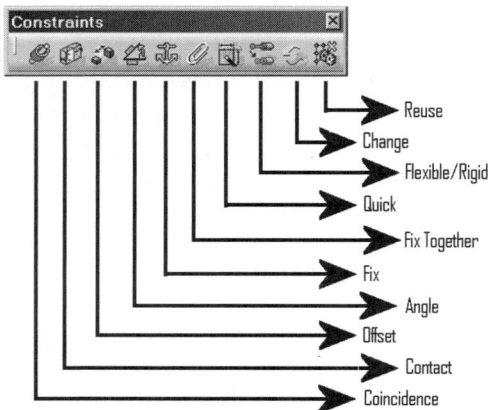

Fig. 8-24. Constraints toolbar.

Constraints Toolbar

The Constraints toolbar contains options for creating assembly constraints between and among components. Other options allow for multi-copies of an existing model within the assembly environment (see figure 8-24).

NOTE: *The order of the component within the configuration tree and the order of selection determine what component will snap to another. A component will always snap to a fixed component, regardless of the order of selection.*

Coincident

Fig. 8-25. Coincident icon.

A coincident constraint is primarily used to align the centerlines between two components. The Coincident option, accessed via the Coincident icon (shown in figure 8-25), is used to create coincident constraints. Various types of coincident constraints can be obtained, depending on which elements of the two components are selected (such as two centerlines). The Coincident option can be used with selection of the elements specified in figure 8-26.

Fig. 8-26. Elements selectable in association with the Coincident option.

	Point	Line	Plane	Planar Face	Sphere Pt.	Curve	Cylinder Axis
Point	X	X	X	X		X	X
Line	X	X	X		X		X
Plane	X	X	X	X	X		X
Planar Face	X		X	X			
Sphere Pt.	X	X	X				
Cylinder Axis	X	X	X				
Curve						X	

A coincident constraint is identified graphically by two green circles. An example of a coincident constraint used to align two centerlines is shown in figure 8-27.

Contact

A contact constraint is primarily used to mate planar faces between components. The Contact option, accessed via the Contact icon (shown in figure 8-28), makes the two selected faces co-planar with respect to each other. The orientation of the co-planar faces is determined by normal vectors that point in opposite directions. Various types of contact constraints can be obtained, depending on the selected elements of the two components. The Contact option can be used with selection of the elements specified in figure 8-29.

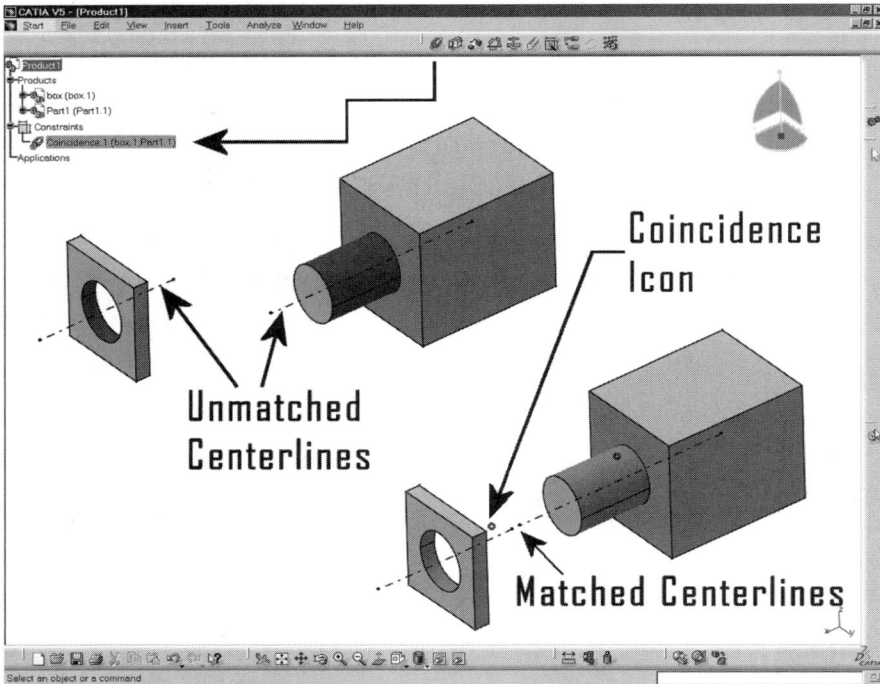

Fig. 8-27. Coincident centerlines.

Fig. 8-28.
Contact
icon.

Fig. 8-29.
Elements
selectable in
association
with the
Contact option.

	Planar Face	Sphere	Cylinder	Cone	Circle
Planar Face	X	X	X		
Sphere	X	X		X	X
Cylinder	X		X		
Cone	X	X		X	X
Circle		X		X	

A contact constraint is identified graphically by two green squares. An example of faces mated via a contact constraint is shown in figure 8-30.

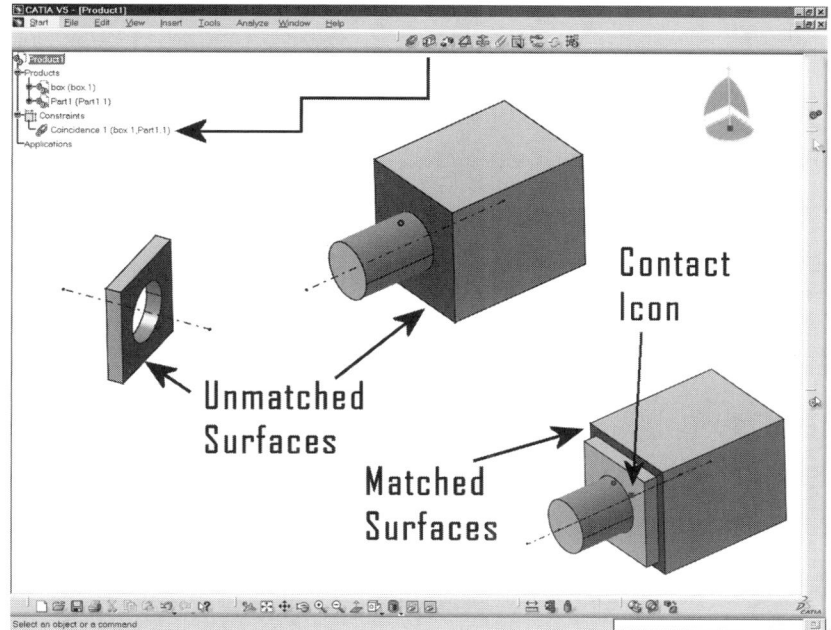

*Fig. 8-30.
Faces mated
via contact
constraint.*

Offset

*Fig. 8-31.
Offset
icon.*

An offset constraint is primarily used to mate planar faces between two components per a specified length (offset distance). The option for this command is accessed via the Offset icon, shown in figure 8-31. Clicking on the Offset option accesses the Offset pop-up dialog window, shown in figure 8-32, in which you establish settings that control the orientation and offset distance of the two components.

*Fig. 8-32.
Offset dialog
window.*

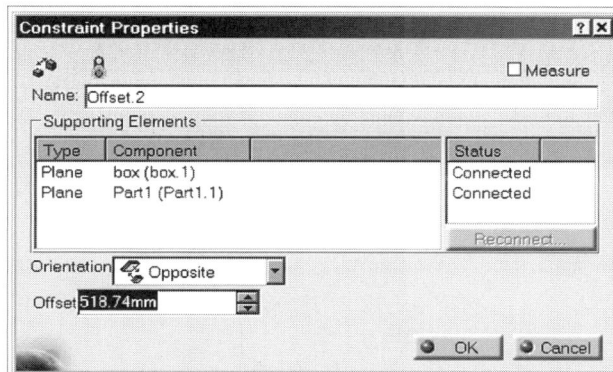

Use of the Offset option makes the two selected faces co-planar with respect to each other (at a specified length, as previously stated). The orientation of the co-planar faces is determined by normal vectors that point in opposite directions. These vectors may be selected to alternate the orientation of the components. Various types of contact constraints can be obtained, depending on which elements of the two components are selected. The Offset option can be used with selection of the elements specified in figure 8-33.

Fig. 8-33. Elements selectable in association with the Offset option.

	Point	Line	Plane	Planar Face
Point	X	X	X	
Line	X	X	X	
Plane	X	X	X	X
Planar Face			X	X

An offset constraint is identified graphically by a green offset dimension between the two faces. An example of faces mated via offset constraint is shown in figure 8-34.

Angle

An angle constraint is primarily used to constrain two planar faces. The option for this command is accessed via the Angle icon, shown in figure 8-35. Clicking on the Angle icon accesses the Angle pop-up dialog window, shown in figure 8-36, in which you establish settings that control the orientation of the two components and the angle defined between them.

Various types of angular constraints can be obtained, depending on which elements of the two components are selected. The Angle option can be used with selection of the elements specified in figure 8-37.

Fig. 8-34. Faces mated via offset constraint.

Offset Dimension

Offset Surfaces

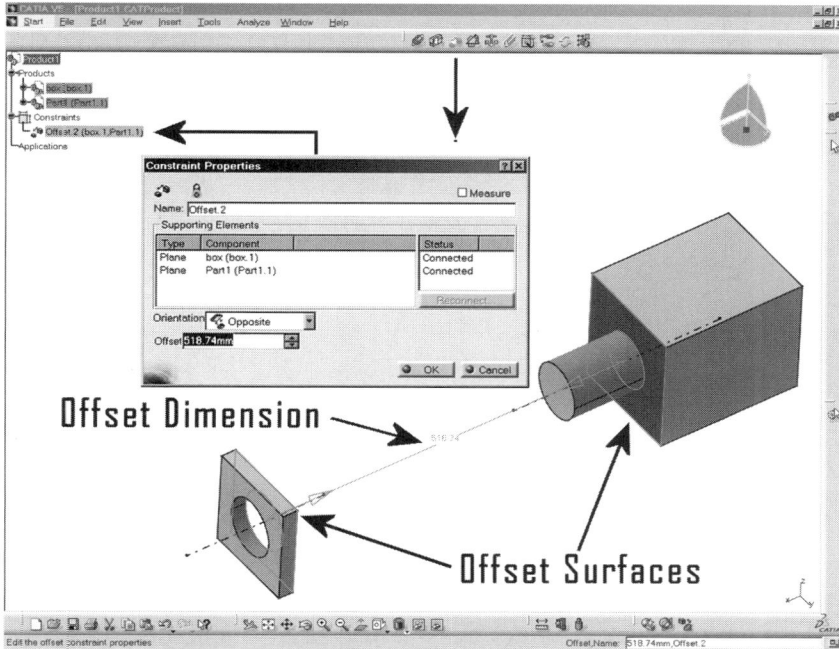
Fig. 8-35.
Angle
icon.

Fig. 8-36. Angle dialog window.

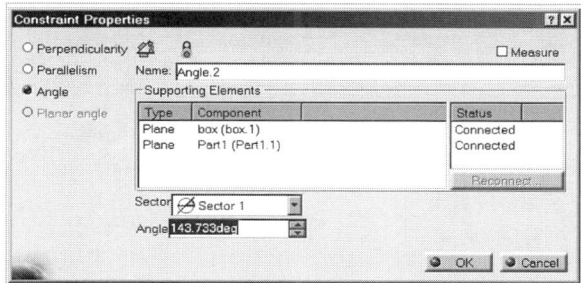	Line	Plane	Planar Face	Cylinder Axis	Cone Axis
Line		X	X	X	X
Plane	X	X	X	X	X
Planar Face	X	X	X	X	X
Cylinder Axis	X	X	X	X	X
Cone Axis	X	X	X	X	X

Fig. 8-37. Elements
selectable in association
with the Angle option.

An angle constraint is identified graphically by a green angle dimension between the two faces. An example of two faces constrained via use of the Angle option is shown in figure 8-38.

Fig. 8-38. Angular constraint produced via the Angle option.

Fix Component

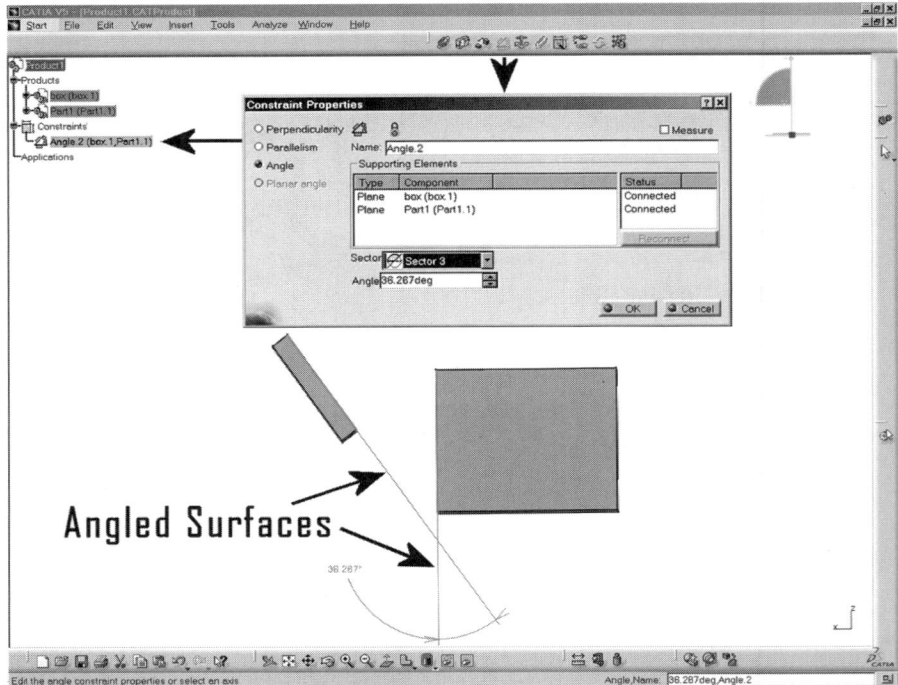

Fig. 8-39. Fix Component icon.

A fixed component constraint is used to fix a component per coordinates in space to prevent it from moving during update of a model. The option for this command is accessed via the Fix Component icon, shown in figure 8-39. This constraint is identified graphically by a green anchor symbol, as shown in figure 8-40.

TIP: *It is good practice to fix at least one component that all other parts can be associated with.*

Fix Together

A fixed-together constraint is used to fix two components jointly per coordinates in space. The option for this command is accessed via the Fix Together icon, shown in figure 8-41. These components will act as one, depending on the manipulation of the lead component.

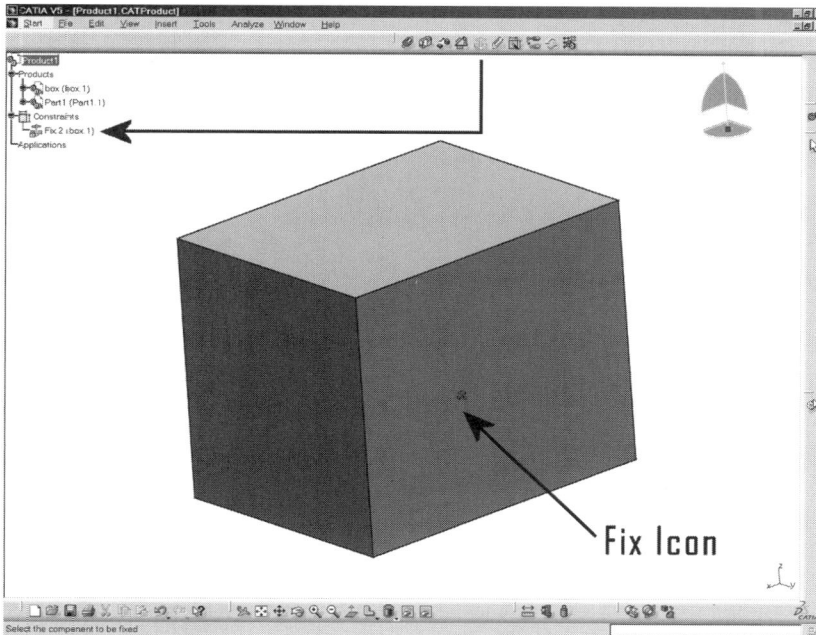

(Left):

Fig. 8-40. Anchor symbol indicating a fixed component constraint.

Fig. 8-41. Fix Together icon.

Quick

Fig. 8-42. Quick icon.

The quick constraint creates the first possible identified constraint depending on the selected elements. The option for this command is accessed via the Quick icon, shown in figure 8-42. This constraint is based on an order of priority determined by the Options section of assembly constraints. (See Chapter 3.)

Flexible/Rigid

Fig. 8-43. Flexible/Rigid icon.

The Flexible/Rigid option, accessed via the Flexible/Rigid icon (shown in figure 8-43), allows you to move flexible and rigid joints within an assembly. (See the CATIA V5 online documentation for more information on flexible and rigid constraints.)

Change

Fig. 8-44. Change icon.

The Change option, accessed via the Change icon (shown in figure 8-44), allows you to change a constraint's type without having to delete the constraint and create a new one. Clicking on the

Change option accesses the Change Constraint pop-up dialog window, shown in figure 8-45, in which you establish settings that determine the type of the selected constraint. Certain constraint-type changes may not be allowed, depending on which elements of a component are selected.

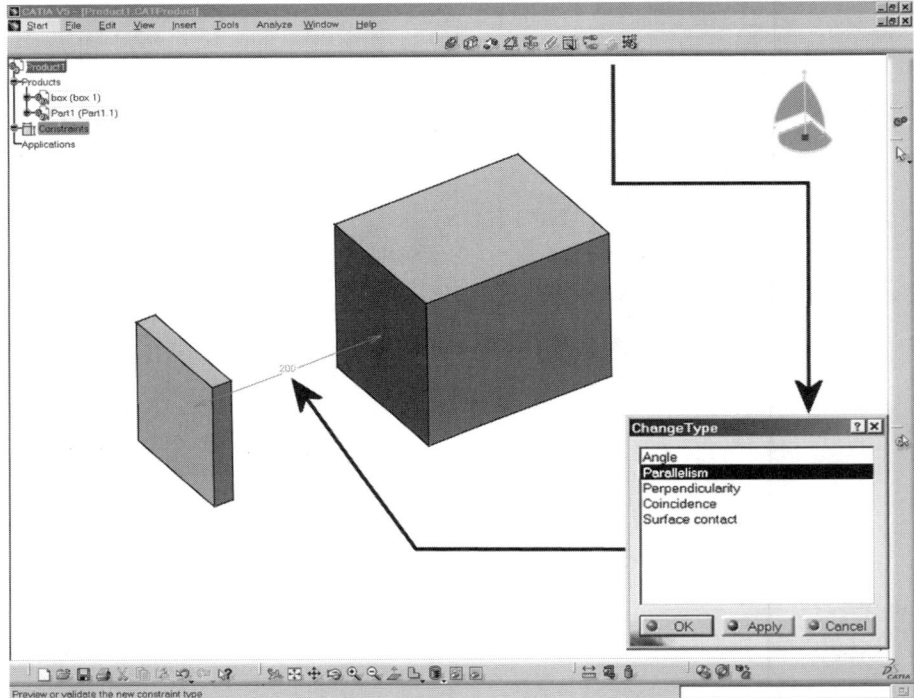

Fig. 8-45. Change Constraint dialog window.

Reuse Pattern

The Reuse Pattern option, accessed via the Reuse Pattern icon (shown in figure 8-46), allows you to repeat a component within an assembly by patterning it in accordance with a pattern residing in the part design. Rectangular, circular, and user-defined patterns can be used for this purpose. (See the CATIA V5 online documentation for more information on assembly pattern use.)

Fig. 8-46. Reuse Pattern icon.

Move Toolbar

The Move toolbar, shown in figure 8-47, offers tools for quickly positioning, translating, rotating, and manipulating components within the assembly environment.

Fig. 8-47. Move toolbar.

Manipulation

Fig. 8-48. Manipulation icon.

The Manipulation option, accessed via the Manipulation icon (shown in figure 8-48), displays the Manipulation pop-up dialog window, shown in figure 8-49, which allows for the mouse movement of any component by any individual degree of freedom. This real-time freehand manipulation is in respect to the selected translation or rotation degree of freedom.

Compass Manipulation

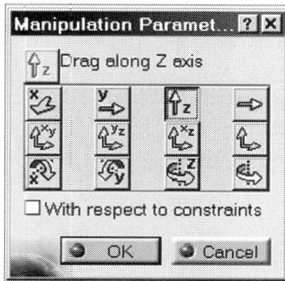

Fig. 8-49. Manipulation dialog window.

In addition to the Manipulation option, the assembly compass may be dragged and dropped onto any component to achieve real-time mouse manipulation movement. As indicated in figure 8-50, once the compass is attached to a component, highlighted in green, translation and rotation movements can be made, depending on the edge, plane, or vector selected within the assembly compass.

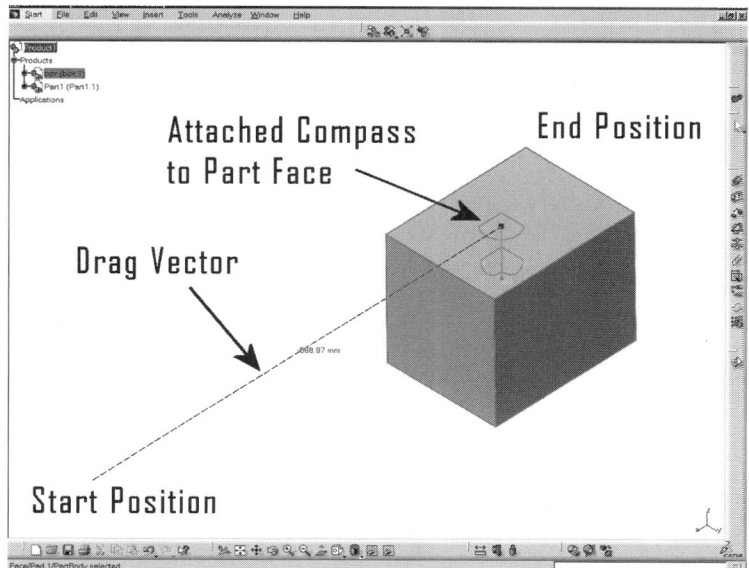

Fig. 8-50. Compass manipulation.

Snap

Fig. 8-51. Snap icon.

The Snap option, accessed via the Snap icon (shown in figure 8-51), snaps the selected element of one component onto the selected component of another component. This option provides as easy means of orienting components within an assembly. The element to be snapped must belong to the active component.

Smart Move

Fig. 8-52.
Smart Move
icon.

The Smart Move option, accessed via the Smart Move icon (shown in figure 8-52), combines the Manipulation and Snap options into one selection. Constraints can be created using the Smart Move pop-up dialog window, shown in figure 8-53.

Fig. 8-53. Smart
Move dialog
window.

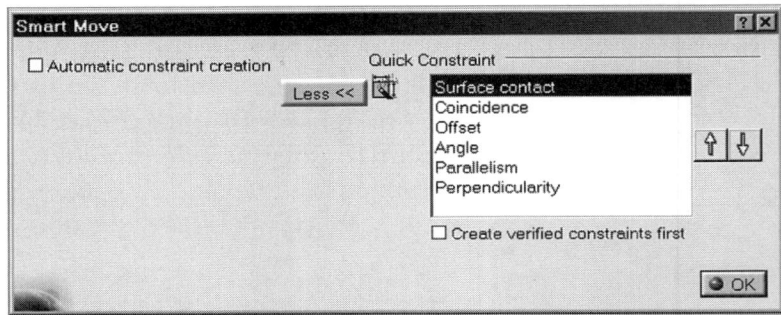

Explode

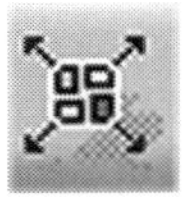

Fig. 8-54.
Explode
icon.

The Explode option, accessed via the Explode icon (shown in figure 8-54), explodes an assembly based on existing constraints within an assembly. The Explode option takes into account only axis/axis and plane/plane coincident constraints. The Explode pop-up dialog window, shown in figure 8-55, offers various tools for manipulating constrained components within the explode environment.

TIP: *It is good practice to anchor (fix) at least one part in the assembly in which the Explode option can center.*

Stop on Clash

The Stop on Clash option, accessed via the Stop of Clash icon (shown in figure 8-56), automatically stops the movement of a component when a clash condition is detected. This occurs during the dynamic movement of a part within an assembly, or during a kinematic simulation.

The Stop on Clash option, accessed via the Stop of Clash icon (shown in figure 8-56), automatically stops the movement of a component when a clash condition is detected. This occurs during the dynamic movement of a part within an assembly, or during a kinematic simulation.

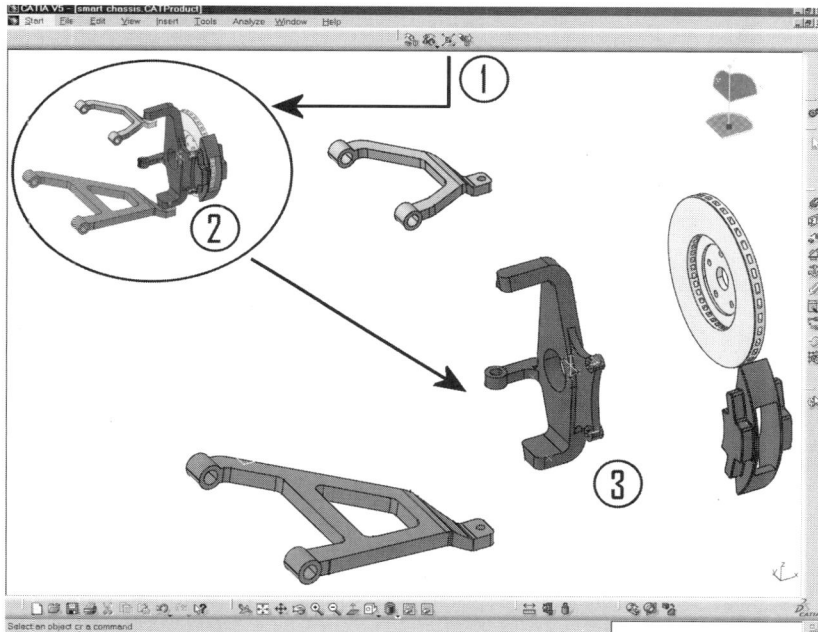

Fig. 8-55. Explode dialog window.

Fig. 8-56. Stop on Clash icon.

Assembly Analysis Tools

Engineers and designers typically perform actions on assemblies once the initial positions have been identified. Different types of analysis provide valuable feedback in order to make design changes based on data-driven results. Assembly tools provide actions that perform interference detections in addition to the functionality they provide for cutting 2D planar cross sections. CATIA V5 offers tools for analyzing an assembly during or after construction. The goal of this chapter is to provide a basic understanding of assembly analysis tools. (See the CATIA V5 online documentation for more information on assembly analysis tools.)

Space Analysis Toolbar

The Space Analysis toolbar, shown in figure 8-57, offers basic functionalities for detecting clearances and clashes, 2D assembly sectioning, and analysis of component distances.

Clash

The Clash option, accessed via the Clash icon (shown in figure 8-58), is used for detecting interferences within the assembly environment. The Clash pop-up dialog window, shown in figure 8-59, offers various options for detecting clashes, contacts, and clearances between and among components.

Fig. 8-57.
Space Analysis
toolbar.

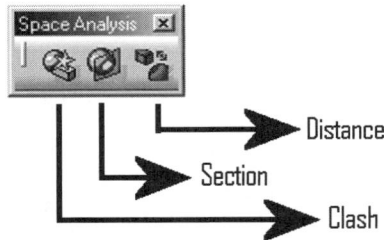

Fig. 8-58.
Clash
icon.

Fig. 8-59.
Clash dialog
window.

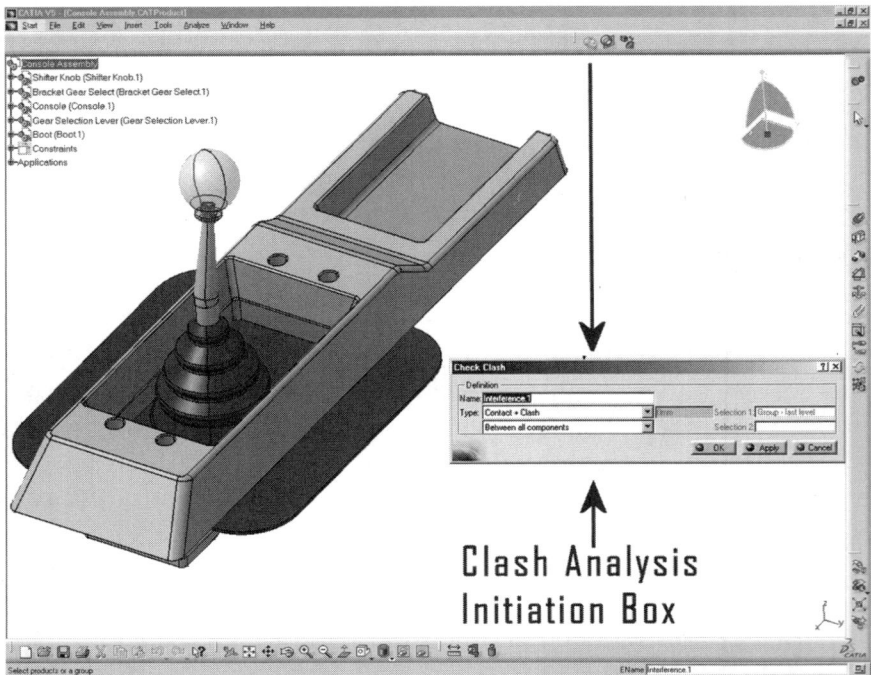

Clash Analysis
Initiation Box

Once a simulation is complete, the Clash Results window (shown in figure 8-60) is displayed, which offers various options for interrogating the results of the simulation.

Fig. 8-60. Clash Results window.

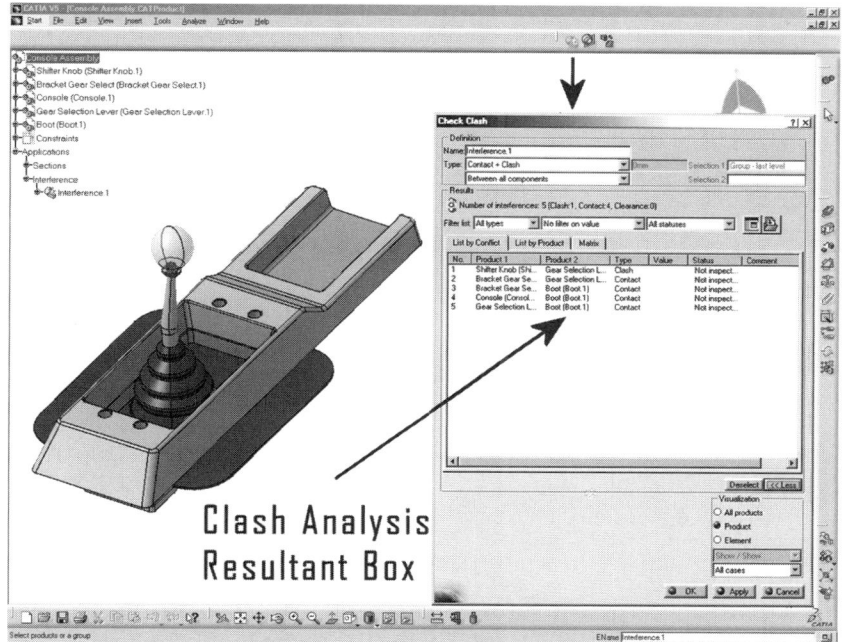

A third window, the Graphical Results window (shown in figure 8-61), is also displayed, which provides further 3D graphical review of individually selected results.

Section

The Section option, accessed via the Section icon (shown in figure 8-62), offers real-time section analysis tools that operate by passing a section plane through each component within the assembly, an example of which is shown in figure 8-63. This is a valuable tool for quickly analyzing 2D cross sections.

Using the mouse, a section plane may be dynamically dragged to update the 2D cross-sectional profile. The 2D cross-sectional profile, an example of which is shown in figure 8-64, may be displayed in a separate pop-up window or within the window of the design assembly.

The Section Definition window, shown in figure 8-65, offers various options for customizing the display of a given cross section.

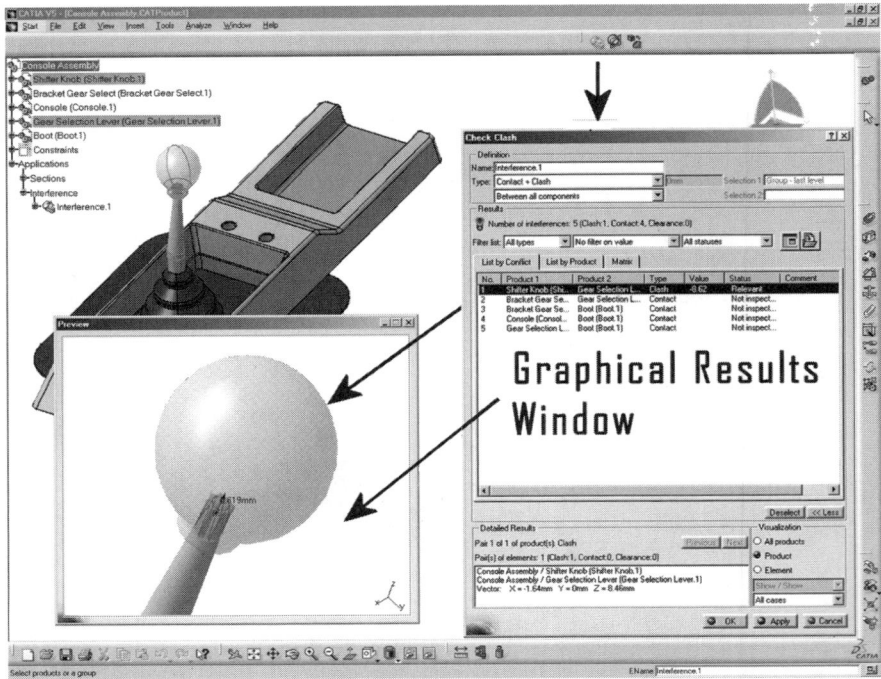

Fig. 8-61.
Graphical
Results
window.

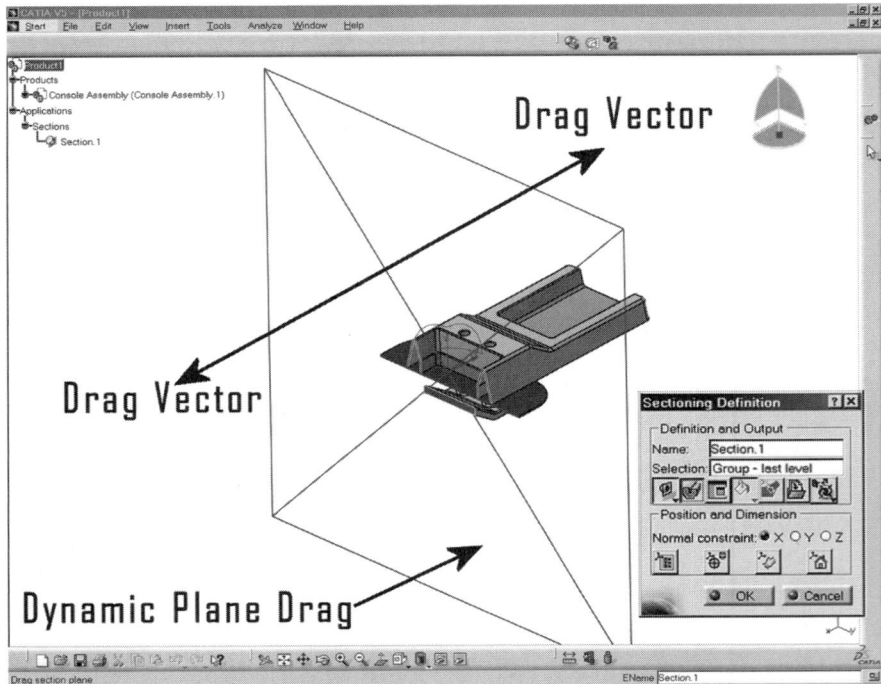

Fig. 8-62.
Section
icon.

Fig. 8-63.
Sectioning
plane.

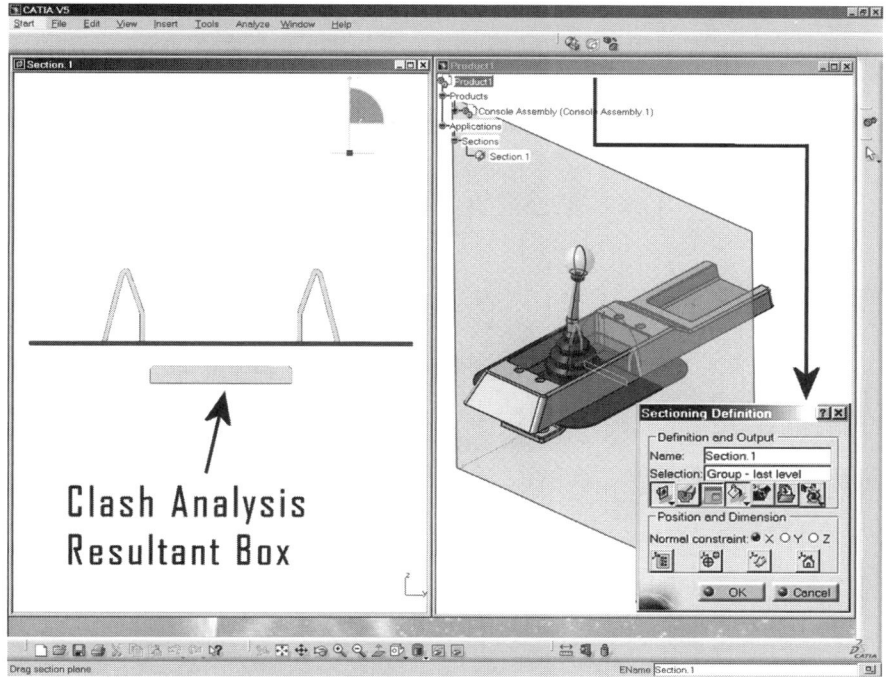

Fig. 8-64. 2D cross-sectional profile.

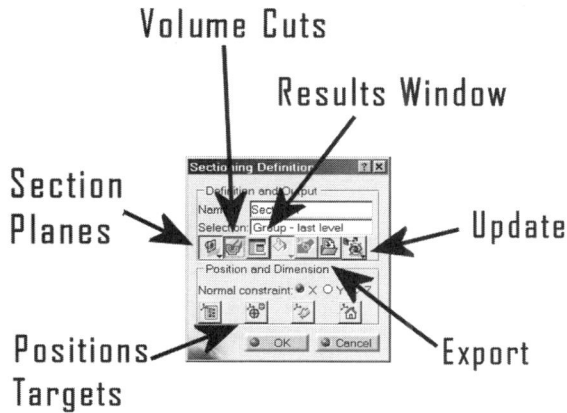

Fig. 8-65. Section Definition window.

Distance

Fig. 8-66. Distance icon.

The Distance option, accessed via the Distance icon (shown in figure 8-66), is used for detecting minimum distances within the assembly environment. The Distance pop-up dialog window, shown in figure 8-67, offers various options for detecting distances between two components or all components.

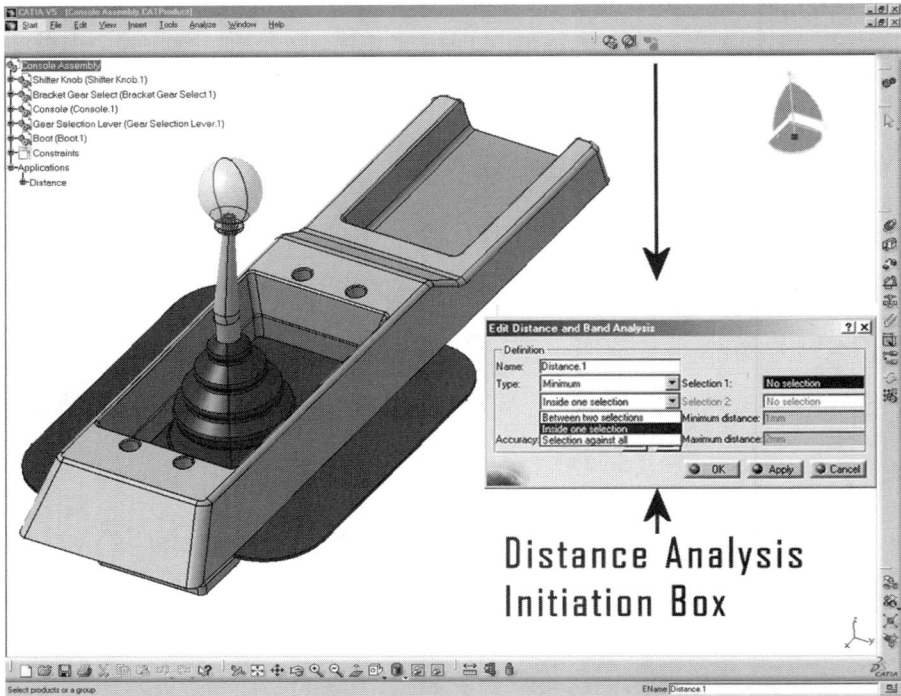

Fig. 8-67.
Distance
dialog
window.

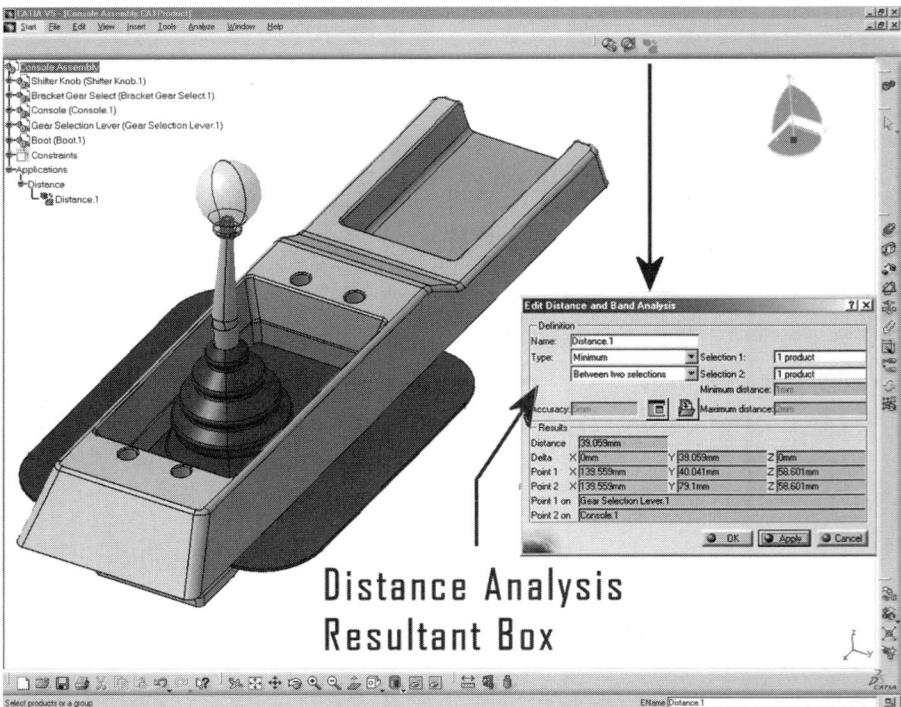

Fig. 8-68.
Distance
Results
window.

Once the simulation is complete, the Distance Results window (shown in figure 8-68) is displayed, which offers various options for interrogating the results of the simulation.

A third window, the Graphical Results window (shown in figure 8-69), is also displayed, which offers further 3D graphical review of individually computed distances.

Fig. 8-69. Graphical Results window.

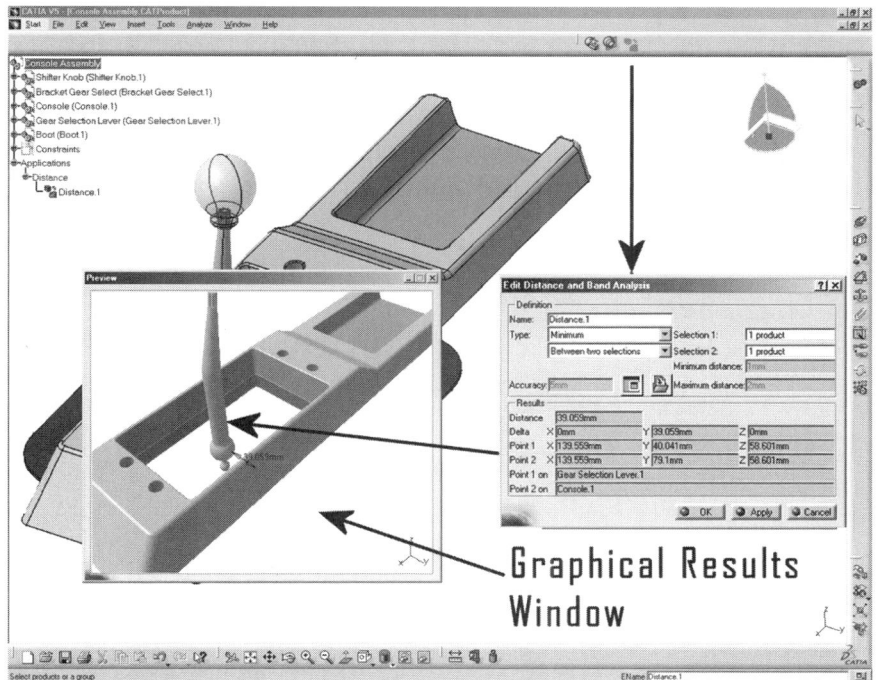

Assembly Feature Toolbar

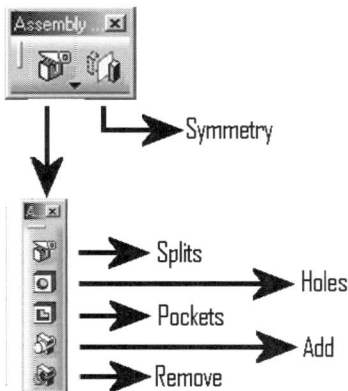

The Assembly Feature toolbar, shown in figure 8-70, allows for the creation of independent features within an assembly. These features exist only in the assembly environment and may affect individual or all components, depending on the options selected. This toolbar allows you to create splits, holes, pockets and symmetrical features, as well as to perform additions and removals.

Fig. 8-70. Assembly Feature toolbar.

Exercise: Using the Assembly Workbench

This exercise takes you through the basics of using the Assembly workbench, including working with tools that control the design. The parts from the previous design exercises are used to demonstrate assembly functions. These parts were designed independently of one another, and thus they "do not know where to go." That is, their places in the assembly have not been defined. The purpose of this exercise is to constrain the parts into the correct relationship within the assembly of the console, shown in figure 8-71. As constraints are added, the parts will move within the assembly workspace and locate to their correct positions. This exercise covers the following concepts and applications.

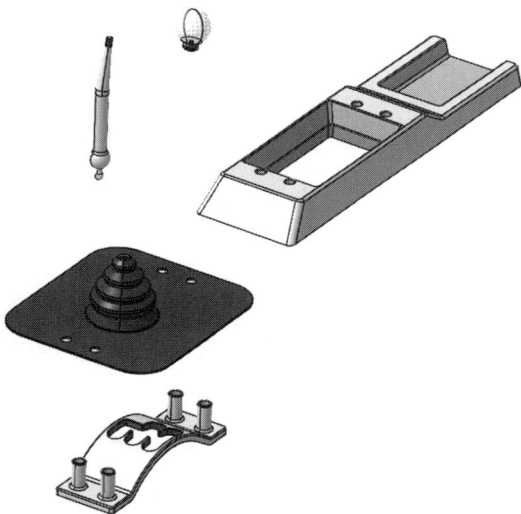

Fig. 8-71. Console to be assembled.

- Creating a product structure and inserting the related CATParts

- Exploding default positions to better access geometry

- Creating a variety of constraints to position and create relationships among the CATPart geometry

- Checking the assembly for clash and interference conditions

- Cutting a section toward analyzing the condition

- Modifying a constraint toward revising a relationship between two CATParts

- Updating a section toward reflecting the modified condition of the assembly

1 Create a new CATProduct and name the default product *Console Assembly*. This structure, shown in figure 8-72, will contain the related parts and constraint relationships.

2 Click on the Insert icon, located on the Product Structure toolbar, and then click on the Console Assembly icon at the top of the Specification tree.

*Fig. 8-72.
Console
assembly.*

This initiates the File Selection window, used to locate and select parts to be brought into the assembly.

TIP: *Parts may also be inserted into an assembly from other documents using copy/paste or drag/drop methods.*

3   Search the file structure until the file that contains the desired parts is located. Multi-select the *Boot, Bracket, Console Gear Lever,* and *Shift Knob* parts from the directory and then click on the Open button to complete the insertion, as shown in figure 8-73.

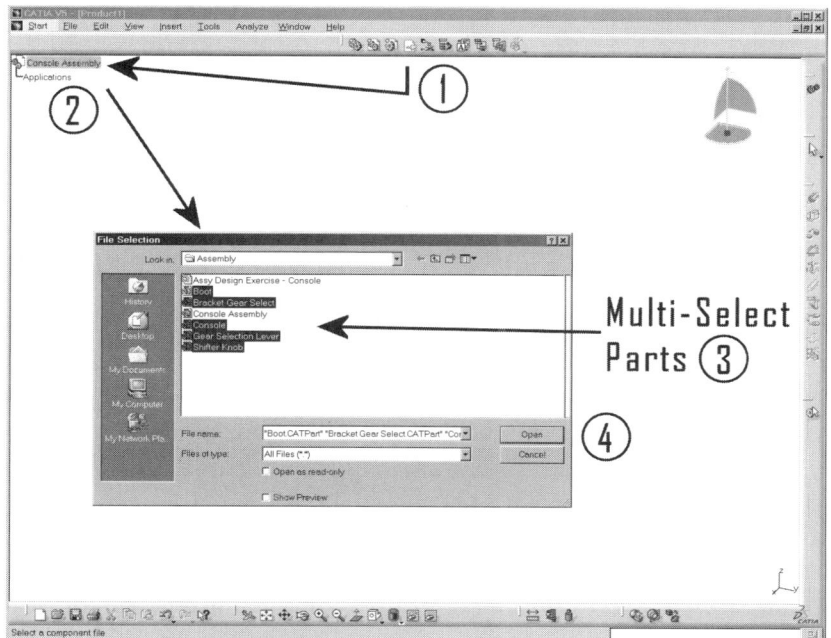

*Fig. 8-73.
Console
assembly
with parts
inserted.*

4   Click on the Explode icon on the Move toolbar. In the Explode window, activate the Fixed Product input box. From the tree, select *Bracket Gear Select* as the fixed part. *Bracket Gear Select*

appears in the Fixed Product space. The bracket gear has been established as the part to remain anchored in position and not move during the operation. Click on OK to accept this setting and to explode the workspace, as shown in figure 8-74.

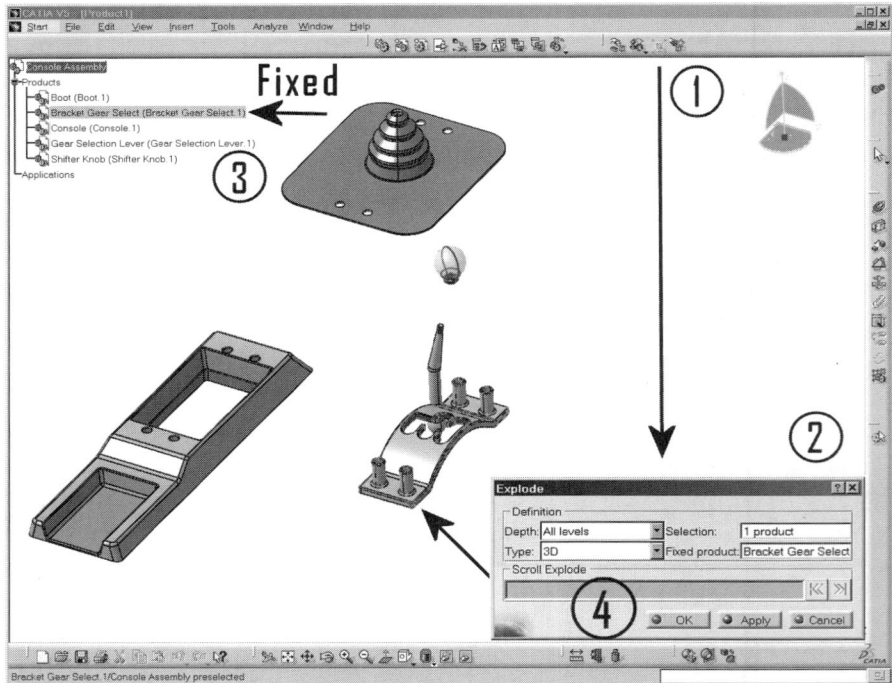

TIP: *Exploding the assembly workspace makes it easier to select the applicable constraints of each part model.*

The next phase involves positioning each part and applying the proper constraints. It is not necessary to move each part into position and then constrain to the final positions. Applying constraints automatically positions the geometry and creates relationships such as coincidence, offset, surface contact, and angle between two parts. It will usually take two or three constraints to position a part. The first constraint that should be applied is one that pertains to a single part. One part will be selected as the anchor of the assembly product. This part will have a "fixed" constraint applied to it for location in space. All other parts in the product will then be constrained in relation to the fixed part.

5 Click on the Fix Component icon on the Constraints toolbar and select *Bracket Gear Select* from the tree or pick the geometry in the workspace that represents the bracket gear.

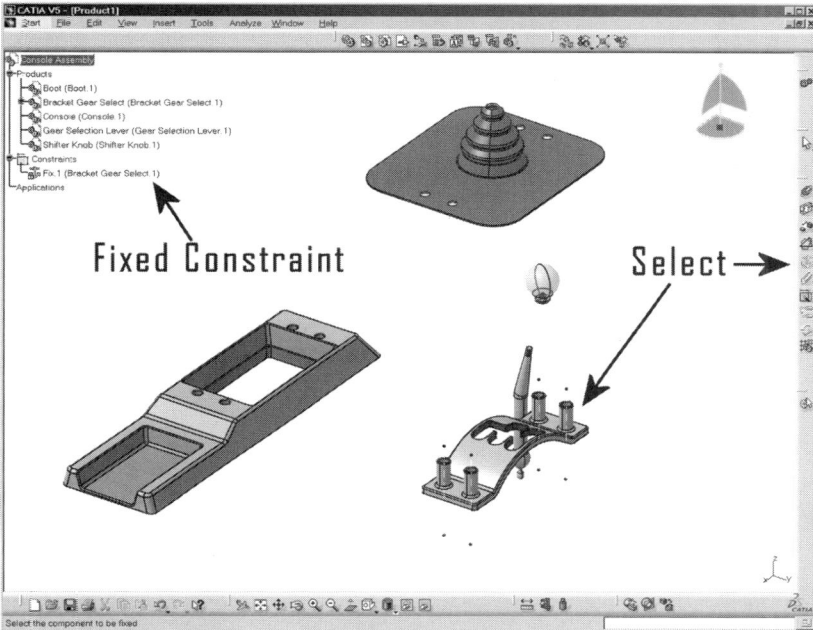

Fixed Constraint

Select ⟶

Fig. 8-75. Parts constrained in relationship to fixed part.

One part in every assembly should be fixed in position. In this assembly, the *Bracket Gear Select* part is anchored and stationary. All other parts will be moved and constrained in relation to the fixed part, as shown in figure 8-75.

NOTE: *The configuration tree now contains a folder named* Constraints. *All constraints common to the assembly will be stored and listed in this folder. The* Bracket Gear Select *part is listed, and bears an anchor symbol, indicating that it is a fixed part. The anchor symbol also appears in the workspace near the* Bracket Gear Select *geometry.*

In the following you will first constrain the *Boot* part. A minimum of two coincident constraints and a surface constraint are required to locate the *Boot* part in relation to the *Bracket Gear Select* part. The *Boot* part was designed 180 degrees out of position, but by applying the constraints it will rotate into the correct alignment with the gear bracket. Continue with the following.

6 Click on the Coincident icon and select the centerline on the Boot part and the centerline on the *Bracket Gear Select* part. Click on the Update icon. The *Boot* part constraint is established, as shown in figure 8-76.

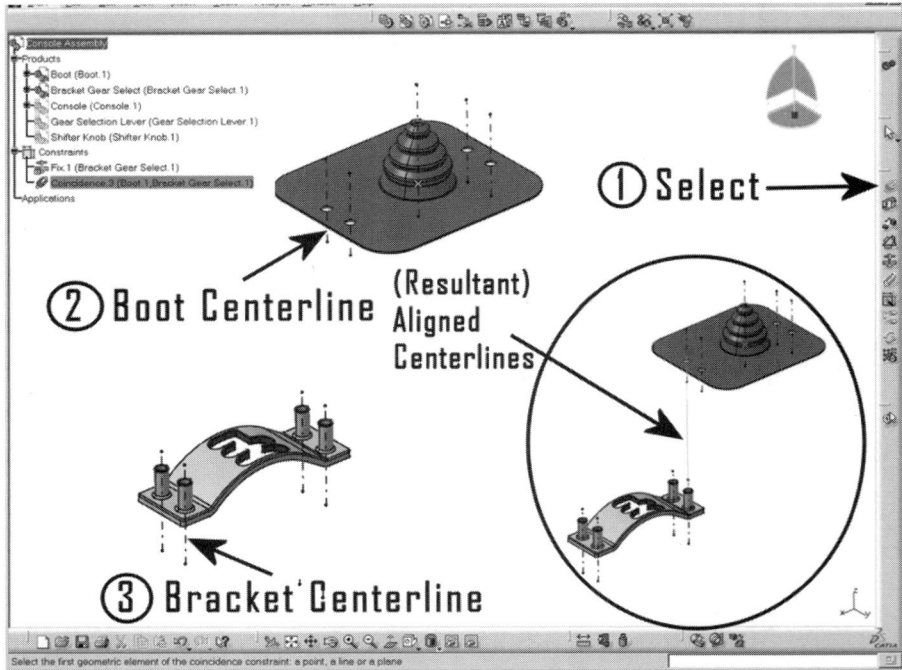

*Fig. 8-76.
Fixed Bracket
Gear Select
part.*

NOTE: *The two centerlines are now aligned and coincident. Listed in the configuration tree are the names of the two parts and the type of constraint that exists between them. The name of the first part selected appears first in the line. Two green circles in the workspace represent the coincident constraints between the parts.*

Next, a second set of centerlines must be selected to align the two parts. The *Boot* part will rotate around the existing coincident constraint and align in the correct position.

7 Select the coincident constraint and then the indicated centerline on the *Boot* part and the indicated centerline on the *Bracket Gear Select* part. Click on the Update icon to effect computation of the move and the constraint, as shown in figure 8-77.

NOTE: *With the addition of a second coincident constraint, the* Boot *part has become aligned with the* Bracket Gear Select *part. Another coincident constraint has been added between the two parts and is listed in the tree. The gap will be closed between the two parts with the addition of a surface contact constraint.*

8 Click on Contact icon and then select the top face of the attachment boss on the *Bracket Gear Select* part and the bottom face of the *Boot* part. The *Boot* part is fully constrained and positioned to the gear bracket Via a boss constraint, as shown in figure 8-78.

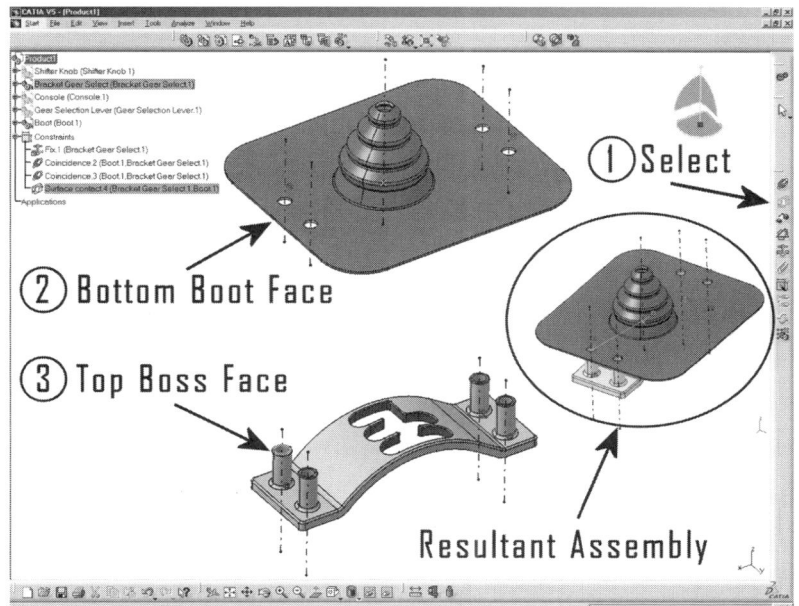

Fig. 8-77. Boot part constraint.

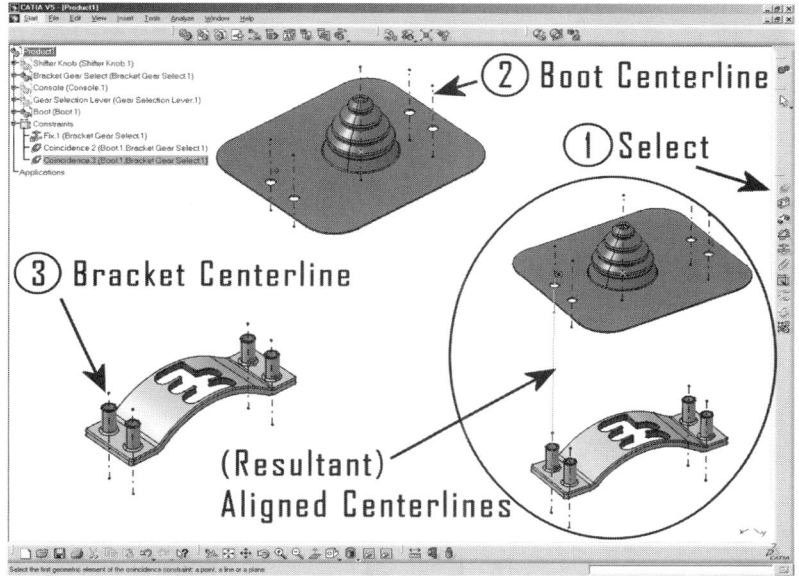

Fig. 8-78. Second set of Boot part centerlines constrained.

The next part to be positioned is the console. It has also been designed 180 degrees out of position, and thus this constraint procedure will be similar to that used for the Boot part. Continue with the following steps.

9 Click on the Coincident icon and then select the indicated centerline on the Console part and the indicated centerline of the *Boot* part. Click on the Update icon to effect computation of the constraint, as shown in figure 8-79.

10 Click on the Contact icon and then select the top face of the *Boot* part and the bottom face of the attachment boss on the *Console* part. Click on the Update icon to update the constraint, as shown in figure 8-80.

Fig. 8-79. Boot part boss constraint.

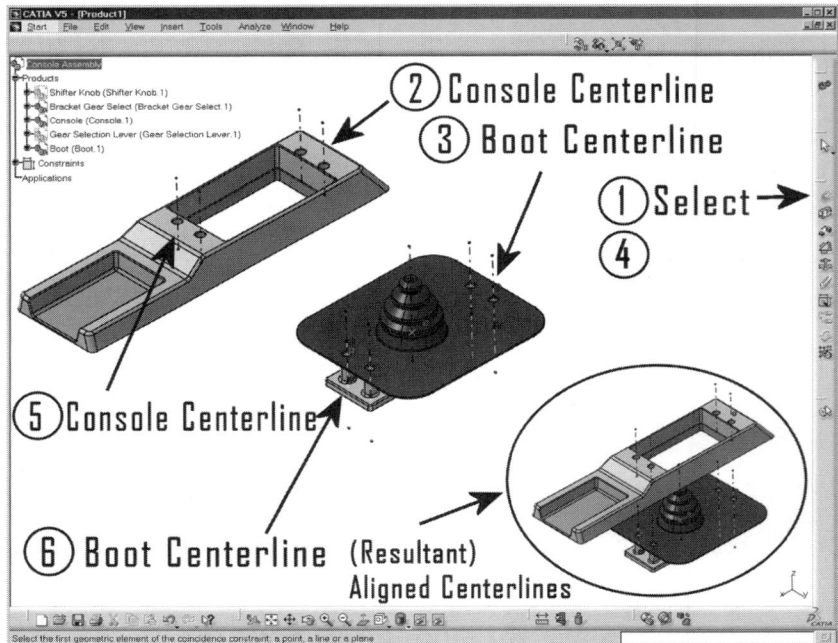

The next two components that will be positioned and constrained are the *Shifter Knob* part and the *Gear Selection Lever* part. After the two are constrained, they will be added to the assembly.

11 Click on the Coincident icon and then select the centerline of the *Shifter Knob* part and the upper angled centerline of the *Gear Selection Lever* part. Click on the Update icon to update the constraint, as shown in figure 8-81.

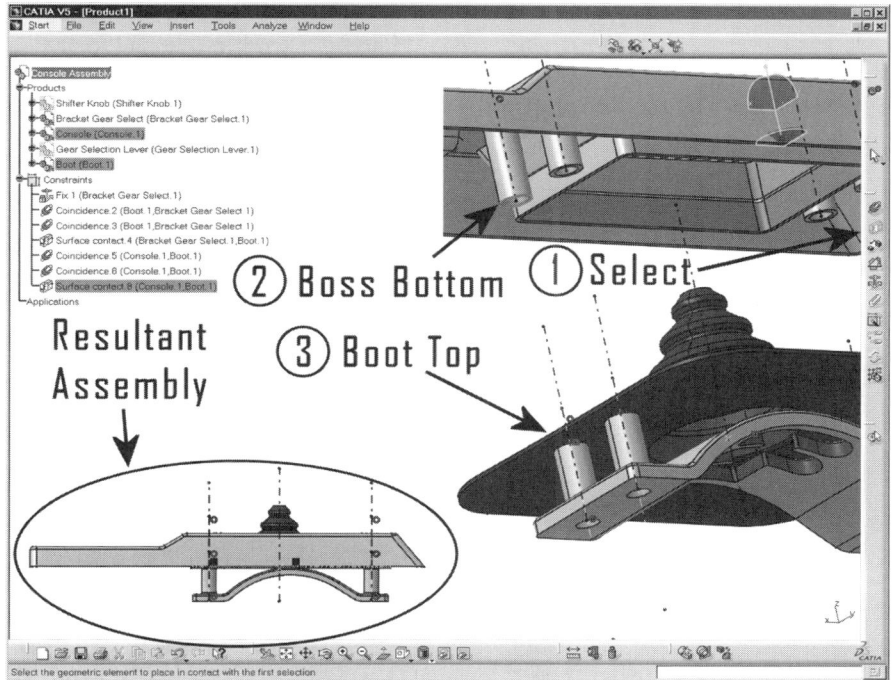

Fig. 8-80.
Centerline
constraint for
Console and
Boot parts.

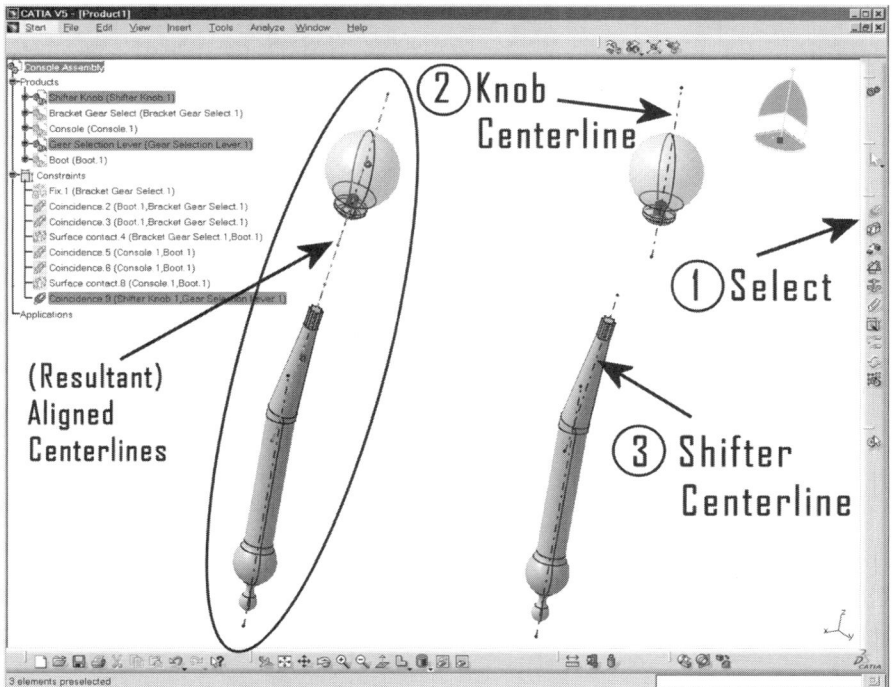

Fig. 8-81.
Contact
constraint for
Console and
Boot parts.

NOTE: *The two centerlines are coincident. The green circle symbol for coincidence is found on both parts in the 3D workspace.*

You now need to create a surface contact constraint between the top mounting faces of the *Gear Selection Lever* part and the inside recessed pocket of the *Shifter Knob* part. Continue with the following steps.

12 Click on the Contact icon and then select the top attachment face of the *Gear Selection Lever* part and the flat face inside the *Shifter Knob* part attachment hole. Click on the Update icon to update the constraint and position the shifter knob, as shown in figure 8-82.

Fig. 8-82.
Shifter Knob
part
constraint.

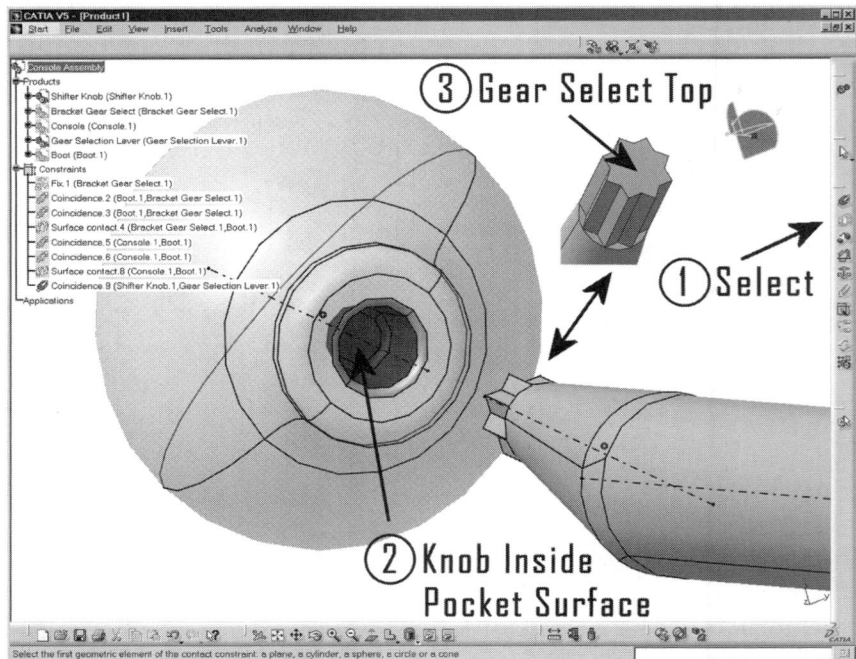

NOTE: *When the* Shifter Knob *part is positioned in the assembly, it will move in accordance with the constraints placed on it.*

13 Select the coincident constraint and then select the main lower centerline of the *Gear Selection Lever* part and the center-line of the *Boot* part. Click on the Update icon to update the constraint, as shown in figure 8-83.

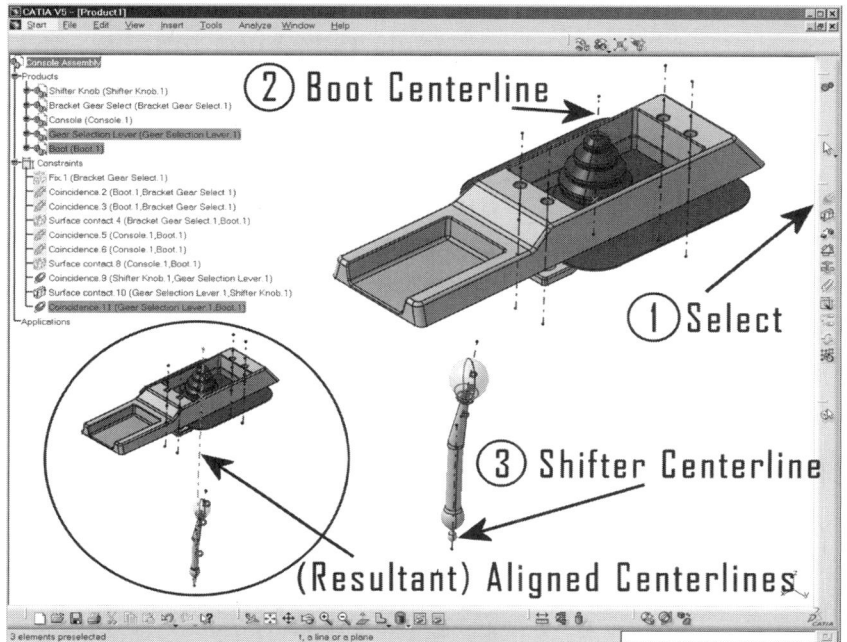

Fig. 8-83.
Shifter Knob
part positioned.

14 Select the coincident constraint and then select the round face indicated on the *Gear Selection Lever* part and the point on the *Boot* part indicated in figure 8-84. Click on the Update icon to update the constraint.

NOTE: *The lever and shifter knob are 90 degrees out of position and need to be constrained in the correct orientation.*

15 Open the *Boot* CATPart and change the ZX plane state from hidden to unhidden. This will be one of the two planes needed to constrain the *Shifter Knob* part to the correct position. Minimize the tree and select the console assembly.

16 Open the *Gear Selection Lever* CATPart and change the state of the YX plane from hidden to unhidden. This will be the second of two planes needed to constrain the *Shifter Knob* part to the correct position. Minimize the tree and select the console assembly.

17 Click on the Coincident icon and then select the *Gear Selection Lever* part plane and the *Boot* part plane. Select Same from the pull-down list in the Constraint Properties window. Click on OK to update the constraint, as shown in figure 8-85.

NOTE: *The assembly is constrained and positioned and the* Gear Selection Lever *and* Shifter Knob *parts have rotated 90 degrees in the correct orientation.*

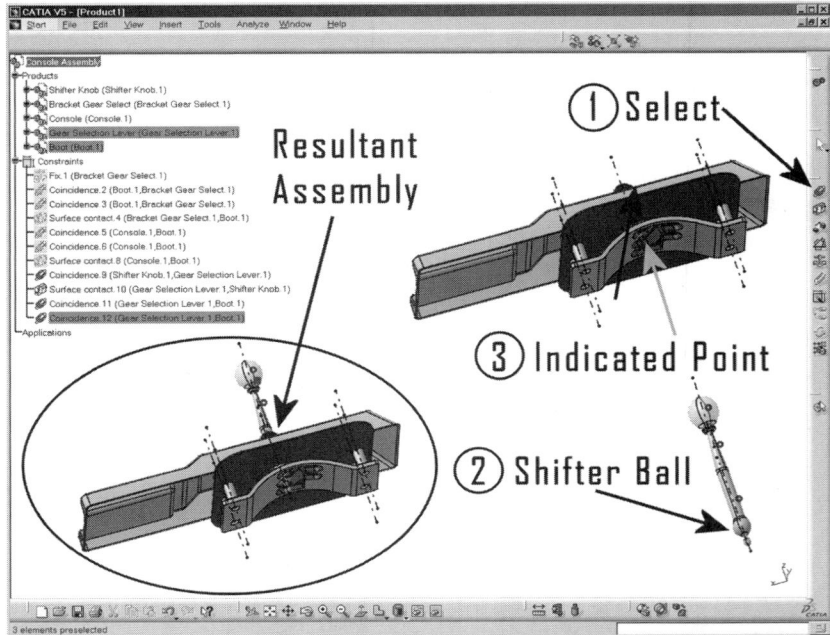

Fig. 8-84. Gear Selection Lever part constraint.

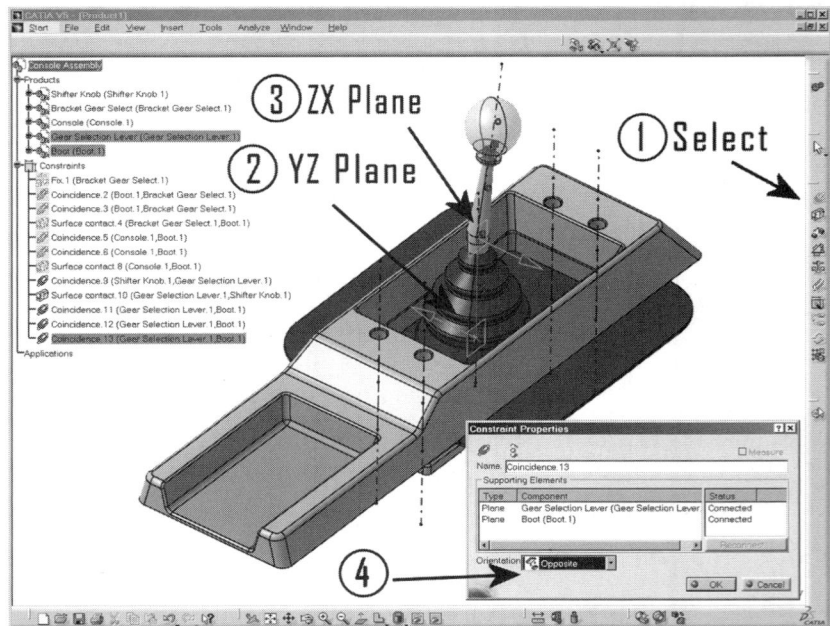

Fig. 8-85. Gear Selection Lever part face/point constraint.

Once an assembly is completed, it is good practice to check for any clashes and interferences. In the following you will analyze the relationship between the *Bracket Gear Select* part and the *Gear Selection Lever* part for a clash condition.

18 Hide all parts in the assembly except for the two that will be analyzed. Click on the Clash icon and in the *Selection.1* window and select *Gear Selection Lever* as 1 Product. Activate the *Selection.2* window and select Bracket Gear Select as 2 Product.

19 Click on OK to process the analysis, as shown in figure 8-86.

The results reveal that a clash condition exists between the two parts, as shown in figure 8-87. The results are also recorded as an interference, in the tree under the Applications branch.

Once a clash has been found, it needs to be investigated. Sectioning is a method of analyzing the visual relationships among parts. Cutting a 2D section enables you to take a closer look at the interference between parts within the assembly.

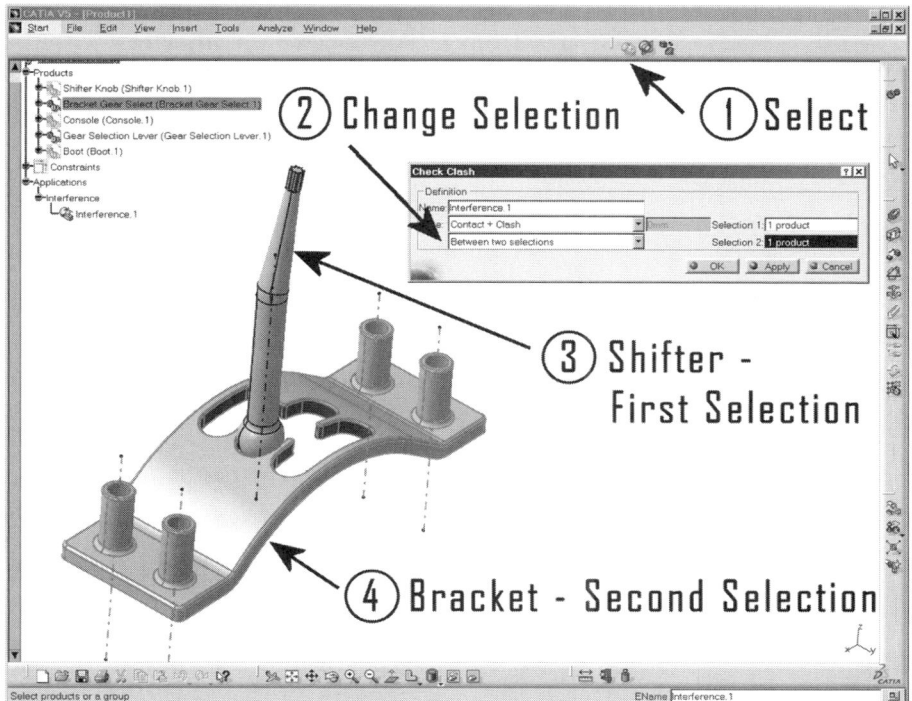

Fig. 8-86. Gear Selection Lever and Boot part planar constraint.

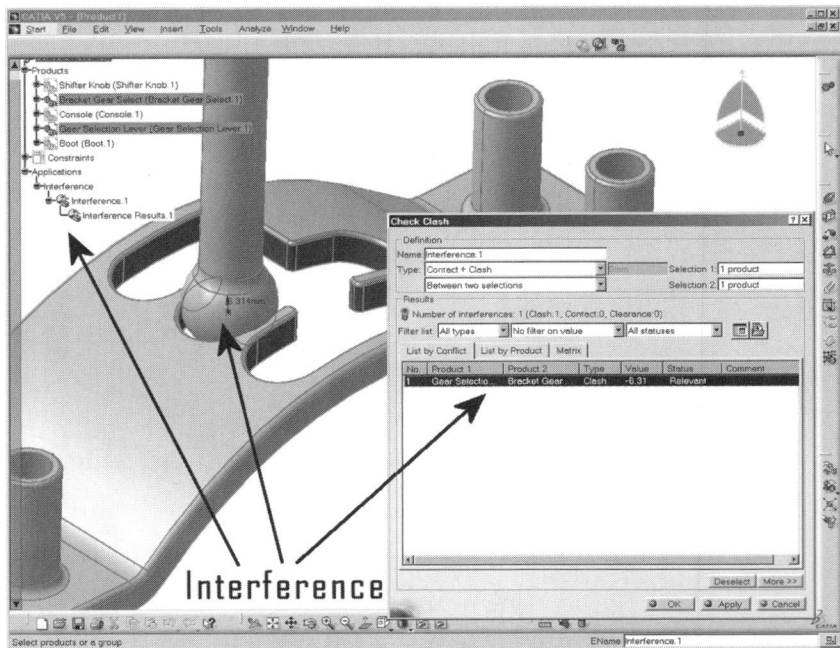

Fig. 8-87. Clash analysis.

20 Unhide all parts so that a section plane can cut through all parts. Click on the Section icon. The default is set for a visible ZX section plane to appear on screen. Pick and hold the arrow on the plane and drag the plane to the center of the assembly, as shown in figure 8-88.

21 Click on the Volume Cut icon to turn the cut from a plane into a volume slice. Click on the Results Window icon.

A second pop-up window displays a planar 2D section of the volume cut. As the volume cut changes, the Results window will update, as shown in figure 8-89. The 2D section in the Results window is interactive and can be zoomed and rotated.

22 Select the indicated green arrow and drag the volume cut to the center of the *Gear Selection Lever* part, as shown in figure 8-91. Click on OK in the window to create the section and list it in the tree. See figure 8-90.

The section reveals a clash between the *Bracket Gear Select* part and the *Gear Selection Lever* part.

23 Zoom in on the Results window to get a close-up of the interference condition, as shown in figure 8-91.

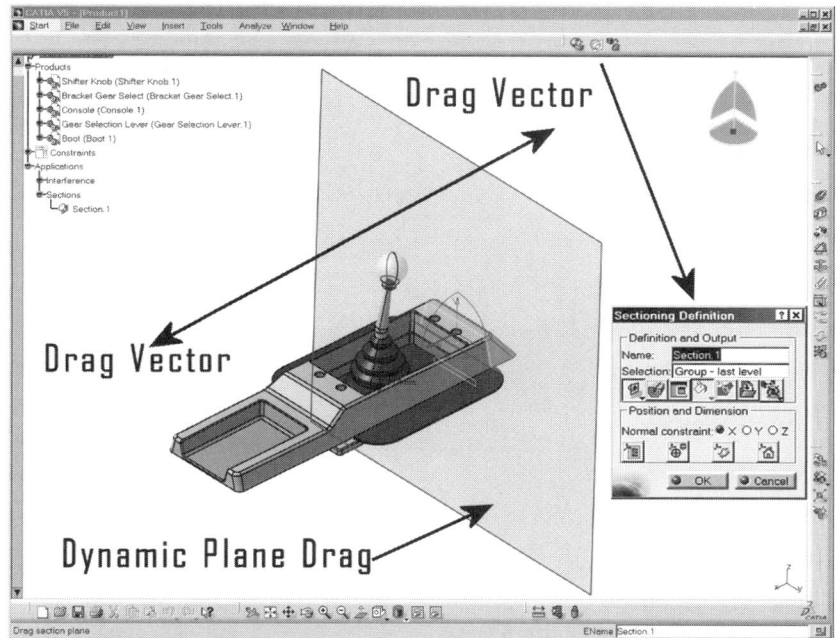

Fig. 8-88, Clash results window.

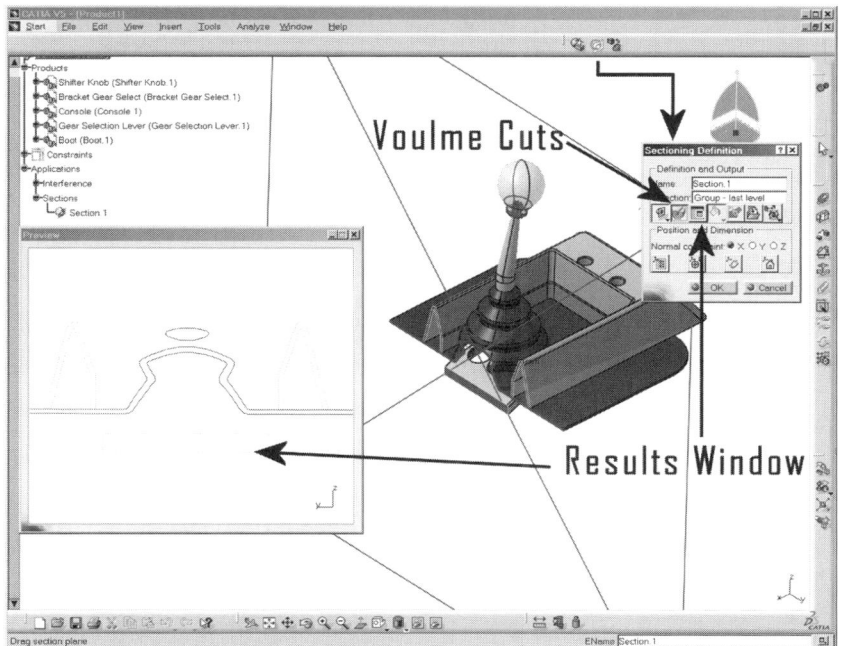

Fig. 8-89. Plane dragged to center of assembly.

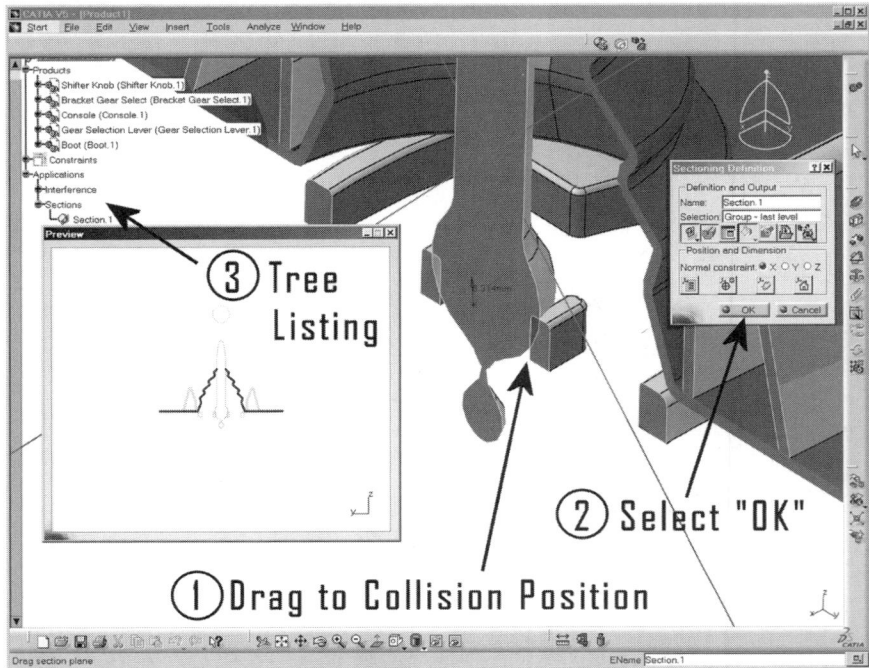

Fig. 8-90.
Updated
Results
window.

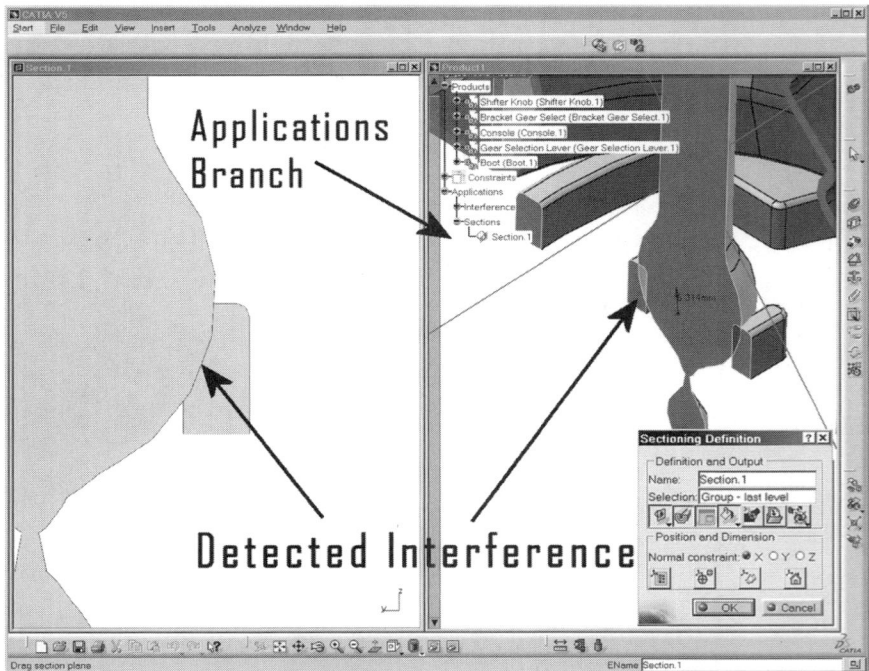

Fig. 8-91.
Volume cut
dragged to
the center of
the Gear
Selection
Lever part.

NOTE: Section.1 *has been added to the Sections sub-branch of the Applications branch in the tree.*

The geometry is correct as designed. The problem lies with the type of constraints used in the assembly. The initial constraint that located the *Gear Selection Lever* part to the *Bracket Gear Select* part was a coincident constraint between two points. It should have been an offset constraint between the two points.

The next phase is to modify the type of constraint from coincident to offset. The section through the part is a living section. When the constraint and the geometry are relocated within the assembly, the section can be updated to reflect the new condition.

24 Place the cursor over the *Cioncidence.11* constraint in the configuration tree and click MB3. The two parts that share the selected constraint highlight in the tree, as shown in figure 8-92.

Fig. 8-92. Close-up of interference condition.

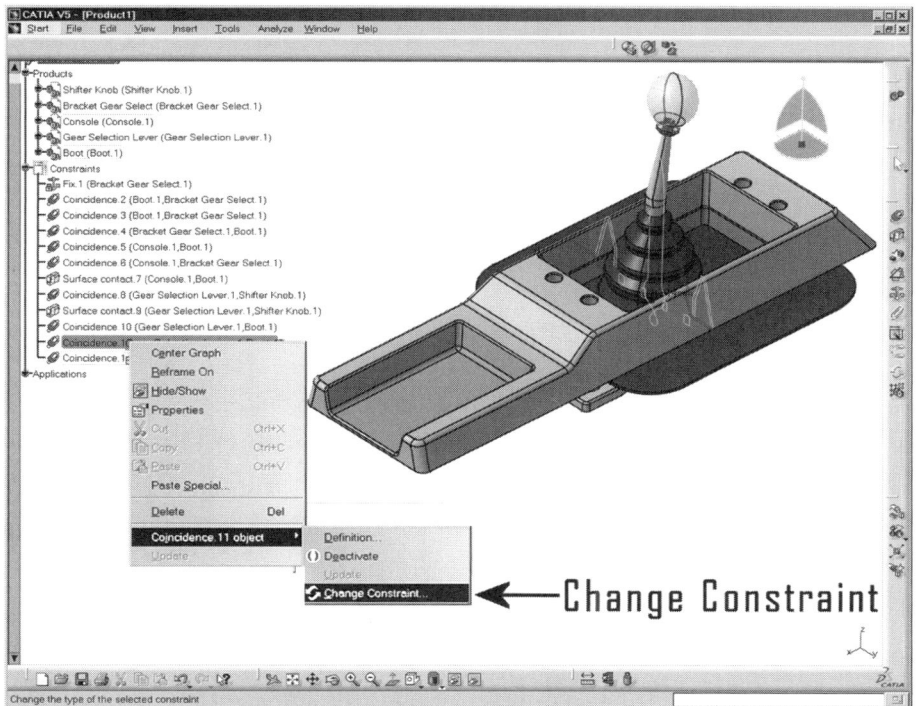

25 In the Change Type pull-down window, shown in figure 8-93, select the *Coincidence.11* object. Click on the Change Constraint icon and within the pop-up window select the Offset option, as shown in figure 8-93.

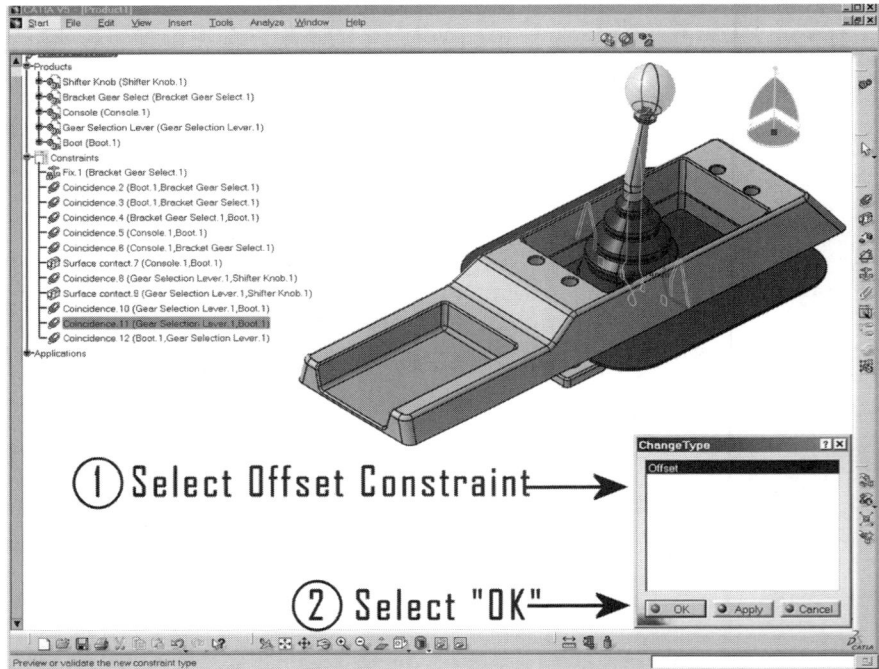

Fig. 8-93. Parts sharing constraint highlighted in tree.

NOTE: *The type of geometry and preexisting constraints determine the options available when changing a constraint type.*

26 In the Constraint Definition window, enter the value *7.2* (mm). The *Gear Selection Lever* part will be raised in the Z direction by the input value, to eliminate the interference between the two parts, as shown in figure 8-94. Click on OK.

The constraint type has been changed from coincident to offset. A new window appears, which displays the values entered for the modified constraint.

27 Click on OK in the window to update the change in constraints, as shown in figure 8-95.

The modified constraint needs to be updated, but the Update icon is not highlighted or available.

Fig. 8-94.
Change
Type
window.

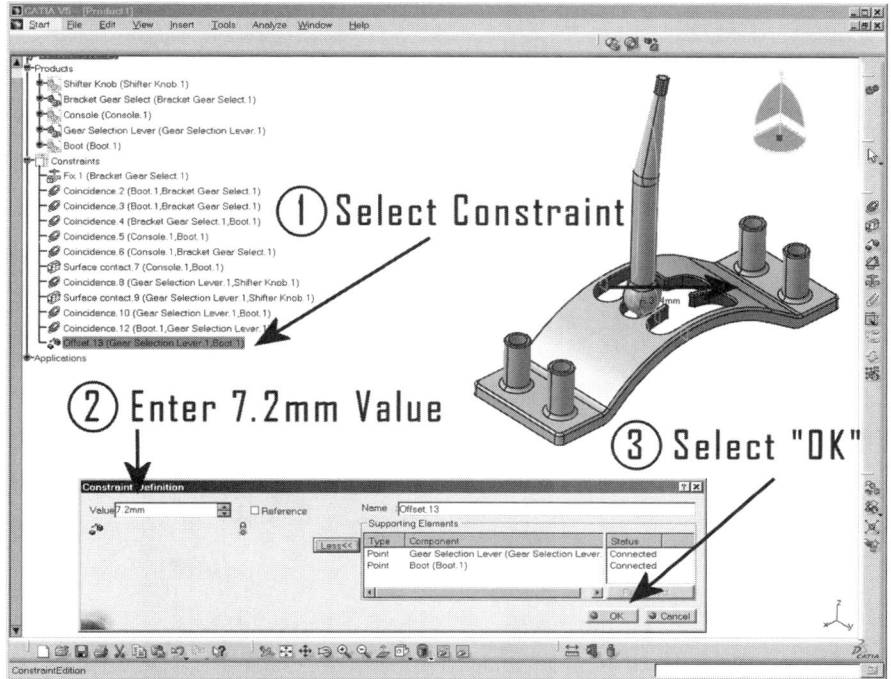

Fig. 8-95.
Offset Z
value.

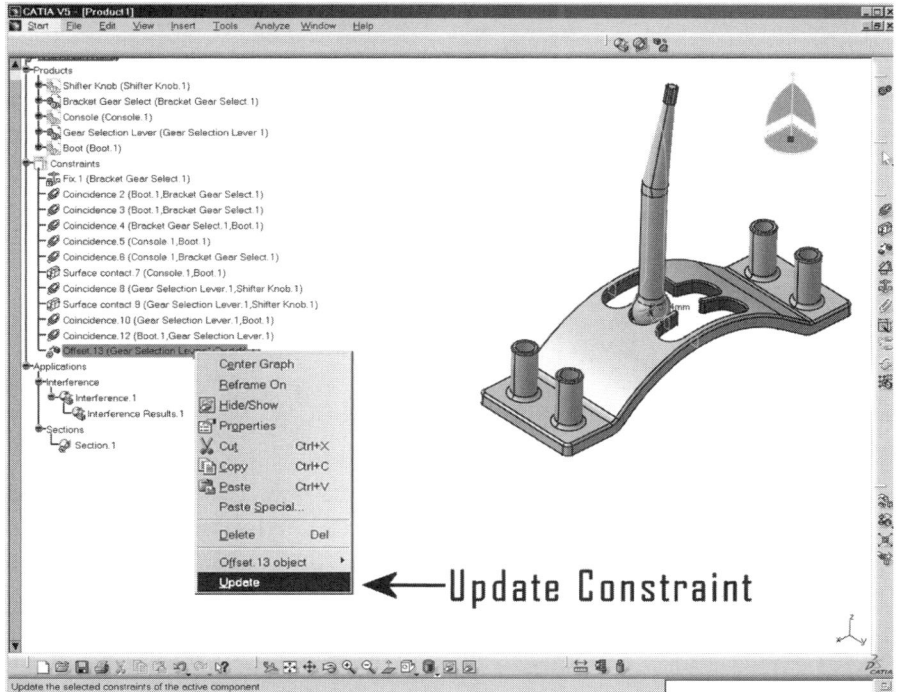

NOTE: *The constraint type and number have changed, and the newly named* Offset.13 *object is listed at the bottom of the tree.*

28 Note that *Offset.13* has an update symbol attached to it in the tree. Click on *Offset.13* and select Update from the pull-down window, shown in figure 8-96.

*Fig. 8-96.
Updated
constraints.*

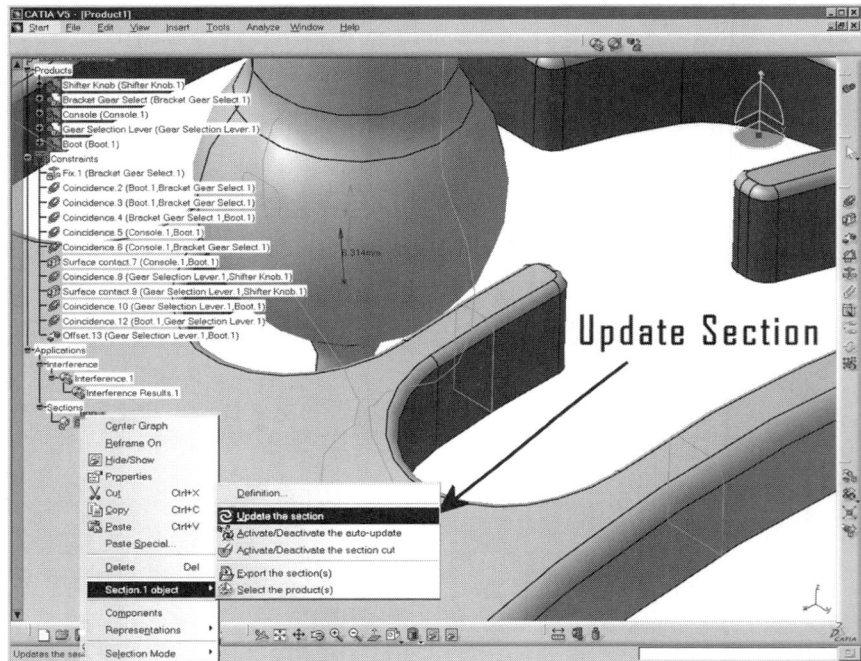

The offset constraint has been modified and updated. The symbol has changed and a value of 7.2 mm has appeared in the workspace between the two points. The *Gear Selection Lever* part geometry has moved in the Z direction as result of the update. The section cut has stayed behind and must be updated to travel with the *Gear Selection Lever* part geometry. These modifications to the constraint are shown in figure 8-97.

29 To update the section, click and hold MB3 on the *Section.1* instance in the tree. From the pull-down window, select the *Section.1* object and then select Update the Section. Select *Section.1* in the tree and observe that the section has updated, following the movement of the 3D geometry, as shown in figure 8-98.

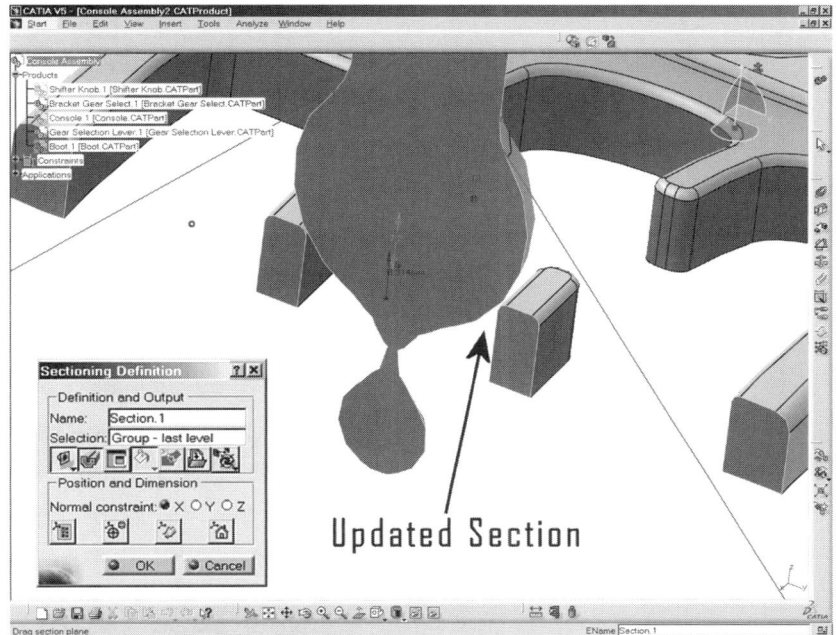

Fig. 8-97. Selecting Update from the pull-down window.

The 3D workspace and the Results window show clearance conditions. More clash analysis may be needed, but for this exercise the lever now clears the bracket, and for visual purposes the assembly is now complete

Summary

The objective of this chapter was to introduce basic assembly tools used for creating and manipulating components. CATIA V5 supports both a top-down and bottom-up approach to assembly design. Components may or may not be constrained to achieve the design in context of the assembly. A full suite of assembly constraints is available to capture the design intent of the product. Analysis capabilities include clash and clearance detection, as well as constraint and dependency analysis tools. CATIA V5 provides basic and advanced functionalities for assembly operations, interactive design, and management of large and complex assemblies within both the Assembly workbench and the Product Structure workbench.

Review Questions

1   What is the difference between a part and an assembly?

2   Why is it important to have assembly functionality?

3   Name five ways to constrain parts within the assembly environ-
     ment.

4   What tools are available for analyzing an assembly?

5   What type of document formats does the Assembly workbench
     support?

CHAPTER 9

THE DRAFTING WORKBENCH

Introduction

THIS CHAPTER IS A CURSORY OVERVIEW of the Drafting workbench, which is used to create detail document drawings. CATIA V5 offers both interactive and generative drafting functionality, depending on the requirements of the design. These two methods of drafting (drawing) are explored in this chapter.

Objectives

This chapter covers the following.

- An introduction to drafting
- Drafting methods
- New drawing creation
- Drafting tools
- An exercise in drafting

An Introduction to Drafting

CATIA V5 offers a full suite of tools for creating 2D detailed drawings based on 3D parts and assemblies. You have the ability to create detailed views that incorporate dimensions, notes, symbols, and all other elements necessary to capture the manufacturing intent of the design. The associative nature of CATIA's drawing modules allows for the automatic update of a drawing based on changes made to 3D parts and assemblies. This is a powerful and time-saving feature in that you do not have to constantly update detail drawings. CATIA V5 offers two types of drafting modes: interactive and generative. These modes (methods) are explored in the sections that follow.

Drafting Methods

The sections that follow describe two types of drafting (drawing) methods (modes) supported by CATIA. Interactive drafting is a 2D standalone drawing mode, and generative drafting is a 2D drawing mode associated with 3D geometry.

Interactive Drafting

This method of drawing creation supports the 2D design world. Views, dimensions, tolerances, text, and so on are manually created, and are not associated to any CATIA V5 3D model. Interactive drafting offers everything you need to create a complete detailed drawing. It can also be used for working with drawings executed in CATIA V4 or other CAD systems. This means that you can view or complete 2D drawings in the CATIA V5 interactive drafting environment. Interactive drafting also lends itself to going from a 2D design process to a 3D associated process.

Generative Drafting

This method of drawing creation supports the 3D design process. In this process, 2D drawing views are created from and associated to the CATIA V5 3D model. There are varying levels of associativity between the 2D and 3D environment. If the CATIA V5 3D model is constructed such that it contains dimensional as well 3D tolerance information, this information can be exploited in generative drafting mode and will appear in a 2D drawing without recreating it.

This process involves a higher level of modeling methodology in that greater care must be taken in building the 3D model so that it may be displayed properly in a 2D drawing. The main idea behind generative drafting is to extract as much information as possible from the 3D model to be displayed in the 2D detailed drawing. The following drawing creation information is based on generative drafting.

Drafting Workbench

There are various starting points for creating a new drawing document within CATIA V5. You can initiate a new drawing from the Start menu, the File menu, or the Workbench icon. Any of these starting points accesses the Drafting workbench, show in figure 9-1.

Fig. 9-1. Drafting workbench.

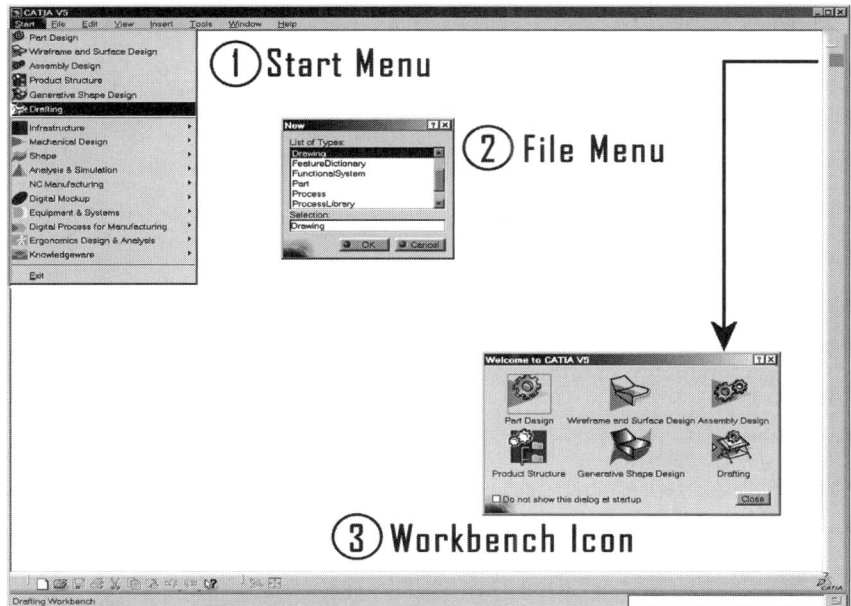

Drafting Interface

The drafting interface, shown in figure 9-2, is similar to the part and assembly interfaces in that there is a configuration tree that identifies the structure of the entire document. One difference, however, is that the drafting interface contains a split screen that separates the configuration tree from the main viewport (window). This is similar to the window configuration accessed via the P1 option.

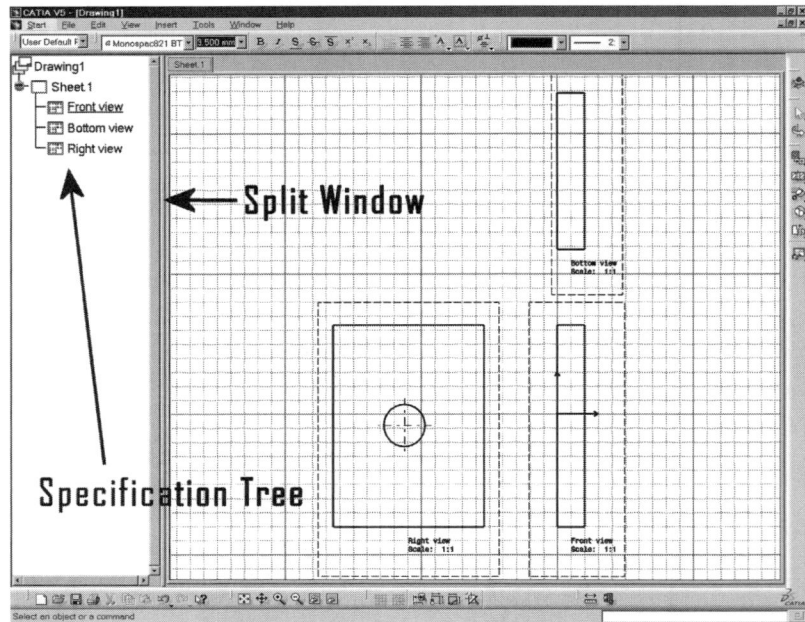

Fig. 9-2. Drafting interface.

Draft Documents

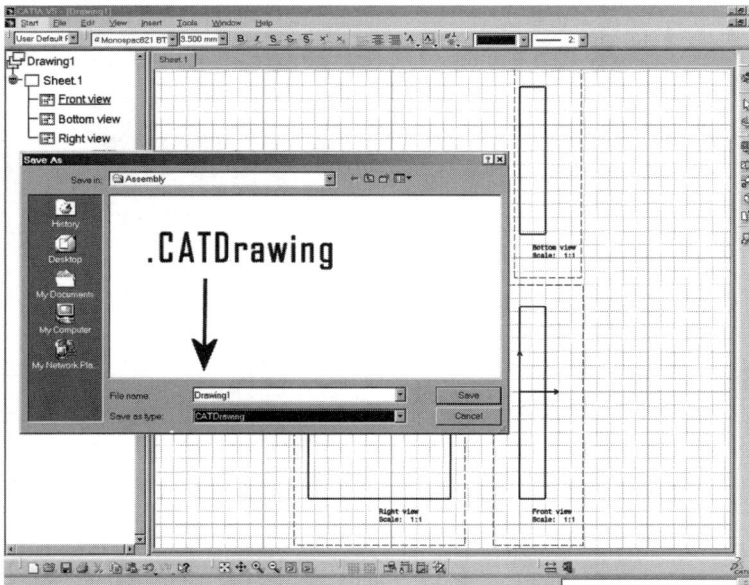

Upon creation of a new drawing, CATIA V5 will create a new drawing document (CATDrawing), an example of which is shown in figure 9-3. A *.CATDrawing* file contains all information that constitutes a drawing.

Fig. 9-3. Drawing document.

New Drawing Creation Process

There are two methods of associating an existing file to a new drawing file. The first method is to first select the Drafting workbench and then open the CATPart or CATProduct intended for the drawing. The second method is just the opposite, in which you first open the CATPart or CATProduct and then select the Drafting workbench. This may seem trivial, but the first method limits the range of options available to you.

"Draw and Then Model" Method

A method known as "draw and then model" is one in which you create a CATDrawing prior to identifying the intended part or assembly. Upon clicking on the Drafting Workbench icon, the New Drawing pop-up window appears (shown in figure 9-4), which contains several options for specifying the parameters of the new drawing. These options are discussed in the sections that follow.

Fig. 9-4. New Drawing window.

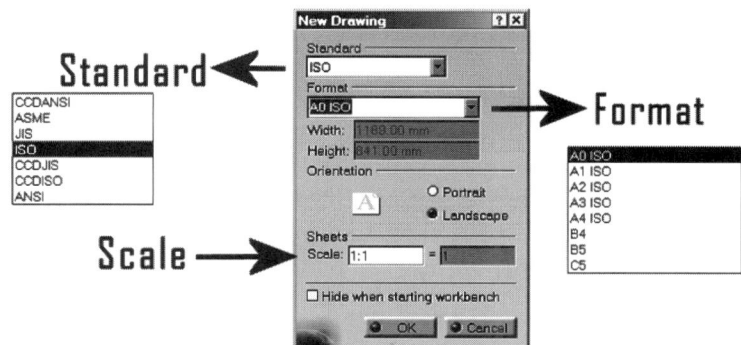

Standards

The Drafting workbench supports international standards (such as ISO, JIS, ANSI, and ASME) and customized standards, toward meeting corporate requirements. CATIA's Standards editor allows you to specify the values and measures used in the creation of drawings toward achieving uniformity and consistency.

Formats

Depending on the standard selected from the Standards editor, the Formats drop-down list offers predefined formats as options. Custom formats can be added to the list.

Orientation

The Orientation option allows you to specify a portrait or landscape orientation for a drawing.

Scale

The Scale option allows you to set the initial scale of a drawing. This establishes the scale of the part model as it relates to the drawing format. Once the desired criterion is selected, a new blank CATDrawing document is created. The next step is to open the intended CATPart or CATProduct and begin creating views, dimensions, notes, and other elements that complete the drawing.

"Model and Then Draw" Method

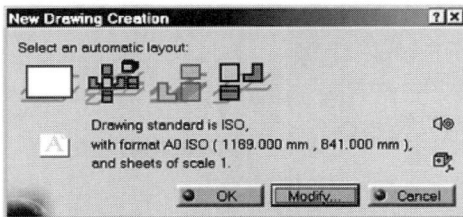

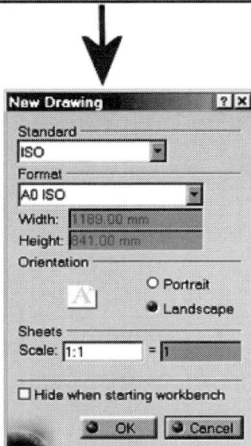

This method calls for the CATPart or CATProduct to be open in session prior to entering the Drafting workbench. Clicking on the Drafting Workbench icon accesses the New Drawing Creation window, shown in figure 9-5.

This window contains automatic layout configuration options. Empty Sheet produces the same results as the previous method. The other layout options are All Views, Front Bottom and Right, and Front Top and Left. Clicking on the Modify button accesses the New Drawing window, in which you can customize the selection.

Fig. 9-5. New Drawing Creation window.

The "model and then draw" method offers a real time-savings benefit when initially creating the drawing using the standard view scheme. The views created with this method are fully modifiable once the CATDrawing document is created.

Drafting Tools

Once a CATDrawing document is created, via either blank or automatic view creation, CATIA V5 offers a full suite of tools for completing the drawing. Various toolbars offer basic and advanced functions for creating, manipulating, and customizing the entire drawing document. This section reviews the basic tools required in the creation of views, dimensions, and notes. (See the CATIA V5 online documentation for more information on advanced drafting and drawing functionality.)

Views

One of the very first steps in creating a detailed drawing is to create the appropriate views. These views allow you to display a part model from various perspectives, toward reflecting the nature of the part or assembly. Views may be displayed in a variety of ways, toward highlighting feature details to lend clarity to the construction and manufacturing processes.

Views Toolbar

The Views toolbar, shown in figure 9-6, offers all functionality required to create any type of view in support of a drawing. This toolbar is essential for creating drawing views when using the "draw first and then model" method. The Views toolbar is divided into six basic categories: Projections, Sections, Details, Clippings, Break, and Wizards.

NOTE: *The active view is framed in red, whereas inactive views are framed in blue. The active view is always underlined in the configuration tree. New views are generated from the active view.*

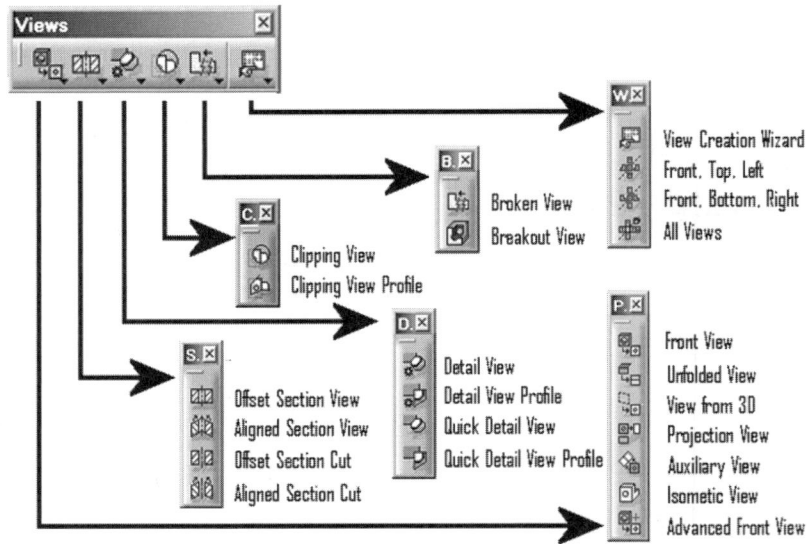

Fig. 9-6. Views toolbar.

View Properties

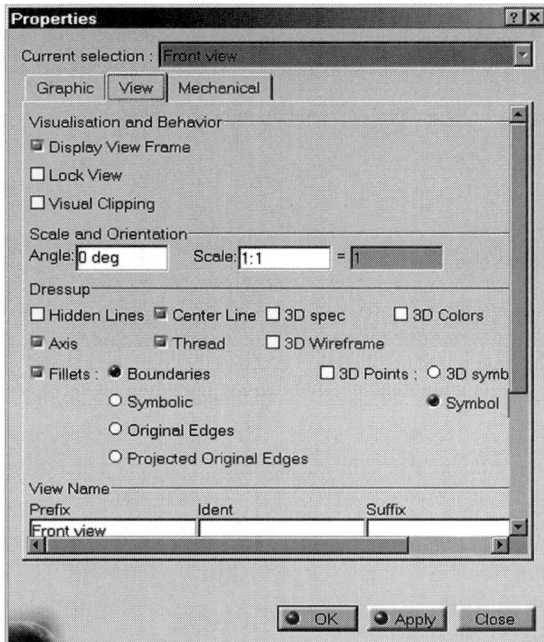

Fig. 9-7. View property settings.

Views can be customized to fit the needs of particular drawings. You customize a view by establishing its property settings, accessed from the specification tree (by right-clicking in the tree with MB3). The Properties window accessed in this manner contains the following settings: Visualization/Behavior, Scaling/Orientation, Dress-up, Naming, and Generation Mode. These settings are shown in figure 9-7.

Graphic Properties Toolbar

The Graphic Properties toolbar, shown in figure 9-8, offers modification options for quickly altering the appearance of elements within a view. Color, line weight, and line type are a few items that may be modified.

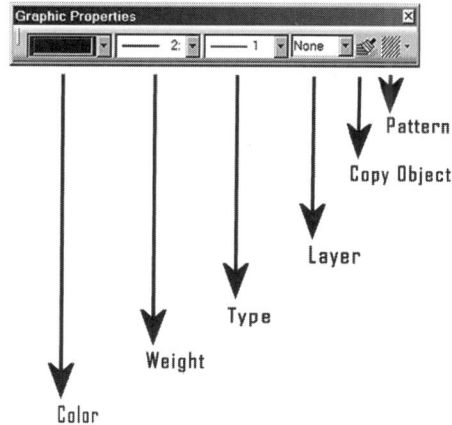

Fig. 9-8. Graphic Properties toolbar.

Dimensioning

There are several ways to generate dimensions within a drawing. One method is to generate dimensions from the 3D features that reside within the part model (i.e., display the dimension from the features that were used to construct the part model). These dimensions are bi-directional in that a dimensional modification, within the drawing, will propagate into the 3D part model.

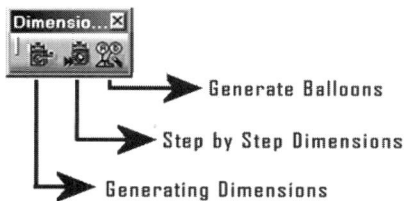

Fig. 9-9. Generation toolbar.

Generation Toolbar

The Generation toolbar, shown in figure 9-9, offers the functionality to generate dimensions from a 3D part model. The Generation toolbar offers automatic dimension generation, step-by-step generation, and balloon generation.

Dimensioning Toolbar

The Dimensioning toolbar, shown in figure 9-10, offers the functionality to create dimensions on 2D elements. The Dimensioning toolbar is divided into three categories: Dimensions, Extension Line Interruptions, and Tolerancing.

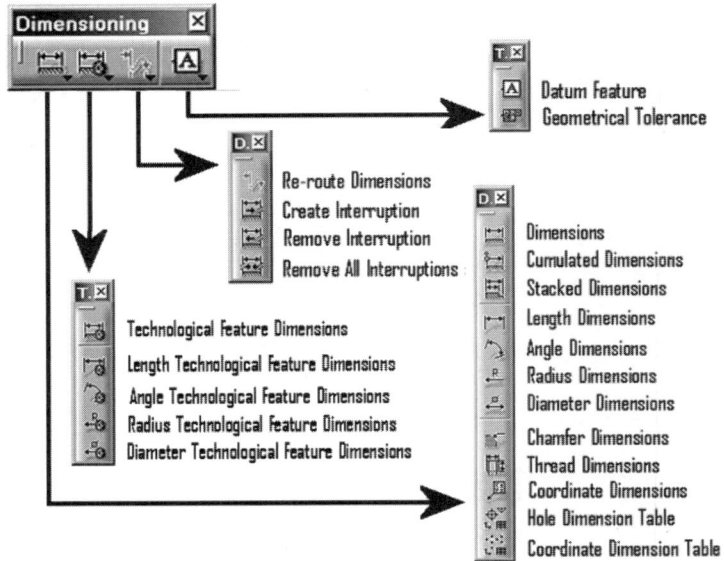

Fig. 9-10.
Dimensioning
toolbar.

Dimension Properties Toolbar

The Dimension Properties toolbar, shown in figure 9-11, offers modification options for quickly altering the appearance of dimensions within a drawing. Dimension Styles, Tolerances, Height, and Decimal (number of decimal places) are a few of the options that may be modified. Use these toolbars to create and modify dimensions and tolerances that correctly define the part.

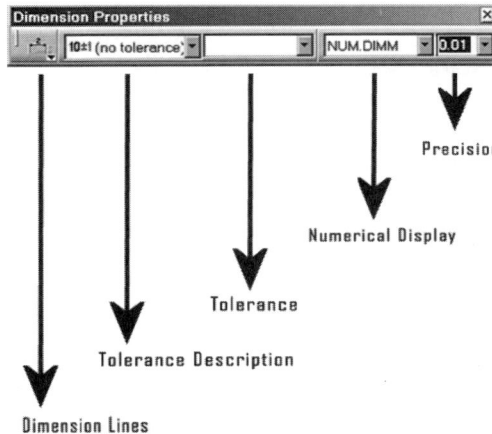

Fig. 9-11. Dimension
Properties toolbar.

Annotations

The creation of annotations within a drawing further describes a part model where dimensions do not such information. Annotations typically take the form of text and symbols.

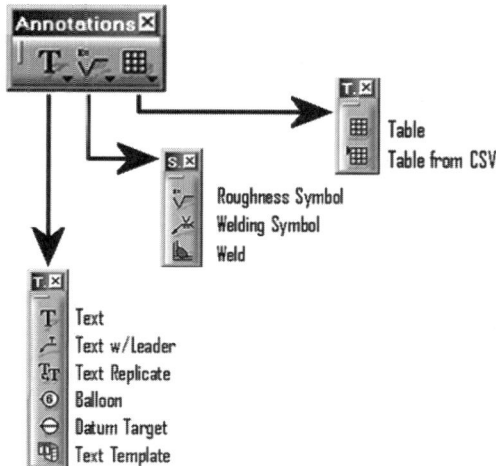

Fig. 9-12. Annotations toolbar.

Annotations Toolbar

The Annotations toolbar, shown in figure 9-12, provides the functionality that supports text callouts and symbols. The Annotations toolbar is divided into three categories: Text, Symbols, and Tables.

Text Properties Toolbar

The Text Properties toolbar, shown in figure 9-13, offers modification options that allow you to quickly alter the appearance of text within a drawing. Text Font, Tolerances, Height, Bold, Italic, and Justification are a few of the options that may be specified.

Fig. 9-13. Text Properties toolbar.

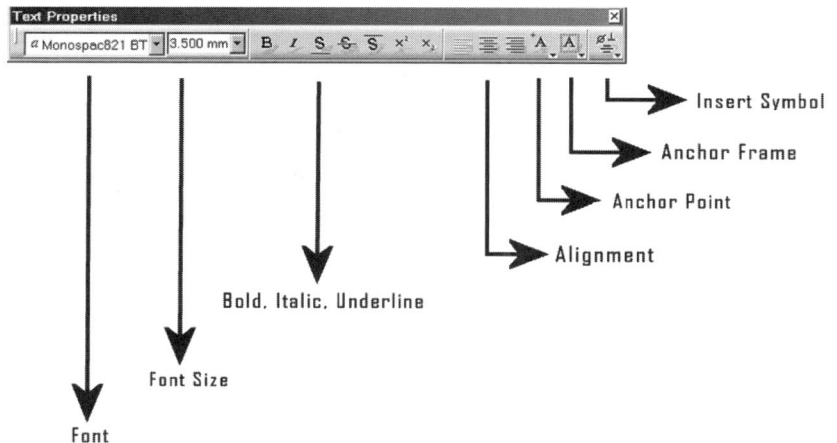

2D Geometry Creation

In some instances, 2D geometric elements may be required in a drawing. CATIA V5 offers the ability to create and modify these 2D elements, which are attached to specific drawing elements or views.

Geometry Creation Toolbar

The Geometry Creation toolbar, shown in figure 9-14, contains options used to create a variety of 2D geometry elements. These options are similar to the element creation options found in the Sketcher workbench.

Fig. 9-14. Geometry Creation toolbar.

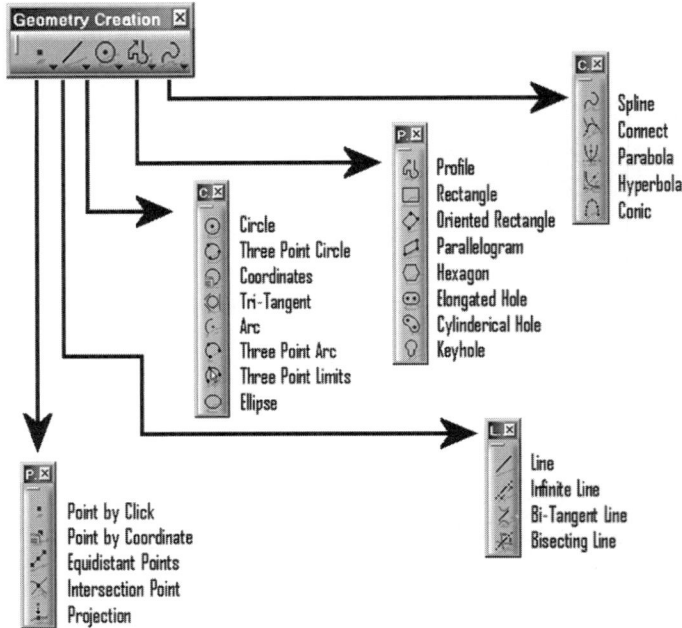

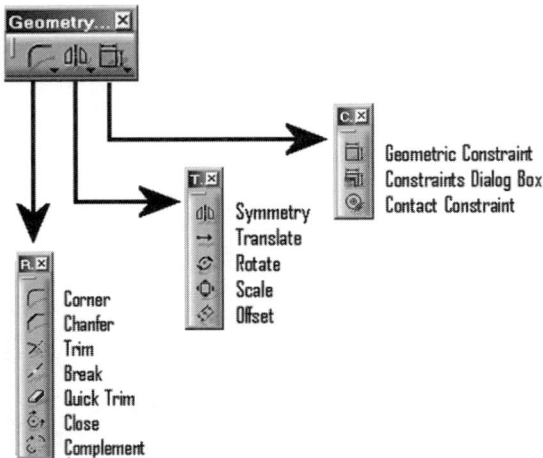

Geometry Creation

Circle
Three Point Circle
Coordinates
Tri-Tangent
Arc
Three Point Arc
Three Point Limits
Ellipse

Profile
Rectangle
Oriented Rectangle
Parallelogram
Hexagon
Elongated Hole
Cylinderical Hole
Keyhole

Spline
Connect
Parabola
Hyperbola
Conic

Line
Infinite Line
Bi-Tangent Line
Bisecting Line

Point by Click
Point by Coordinate
Equidistant Points
Intersection Point
Projection

Geometry Modifications Toolbar

The Geometry Modifications toolbar, shown in figure 9-15, contains options used to create a variety of features that support the 2D geometry elements of a drawing. These options are similar to those found in the Sketcher workbench.

Fig. 9-15. Geometry Modifications toolbar.

Geometry...

Corner
Chanfer
Trim
Break
Quick Trim
Close
Complement

Symmetry
Translate
Rotate
Scale
Offset

Geometric Constraint
Constraints Dialog Box
Contact Constraint

Drawing Exercise: Working with the Bracket Gear Select Part

This exercise takes you through the process of creating a detail drawing from a part. You will use the *Bracket Gear Select* part (shown in figure 9-16) of Chapter 8 to perform basic drawing and drafting functions. This exercise covers the following concepts and applications.

Fig. 9-16.
Bracket Gear
Select *part.*

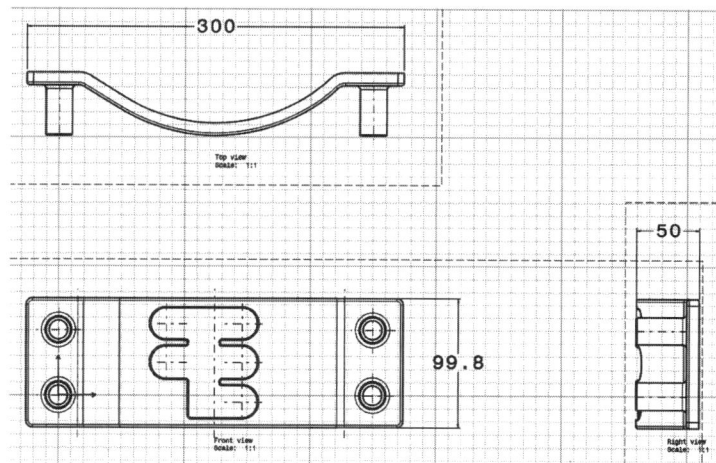

- Creating a new drawing document (CATDrawing)
- Creating views and dimensions
- Creating notes
- Modifying elements
- Making 3D changes
- Updating a drawing

To work with the *Bracket Gear Select* part, perform the following steps.

1   Start a CATIA V5 session and create a new drafting document by clicking on the Drafting Workbench icon.

2   In the New Drawing window, shown in figure 9-17, select an ANSI D Size format, 1:1 Scale, and Landscape orientation for

the new document. Click on OK to accept the creation of the
new blank drawing document (see figure 9-18).

3   Open the *Bracket Gear Select* part created in Chapter 8.

*Fig. 9-17.
Selections for new
drawing setup.*

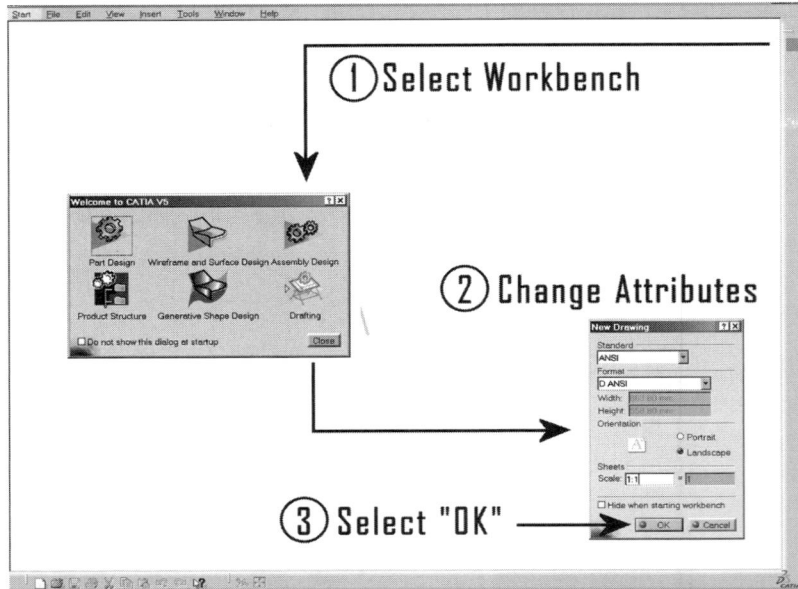

*Fig. 9-18. New
Drawing window.*

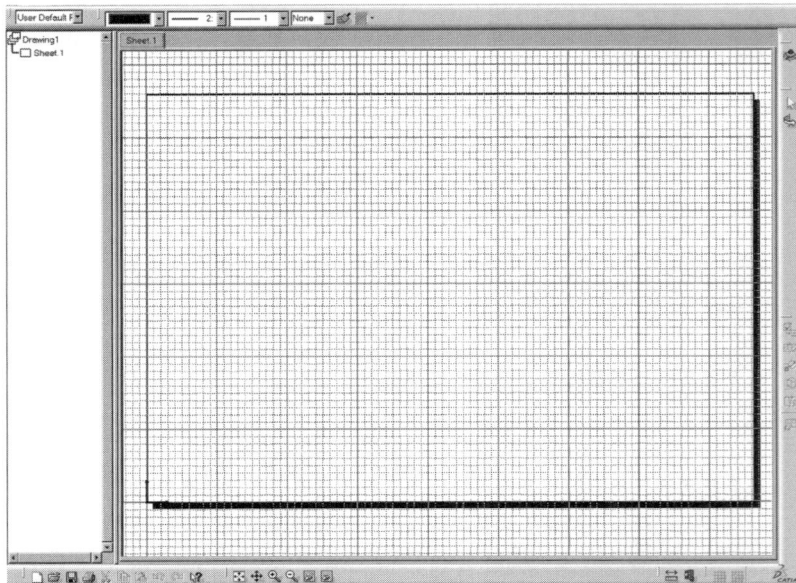

4 In the top menu bar, select *Window/Tile Horizontally* to position the windows for easy access to both documents, as shown in figure 9-19.

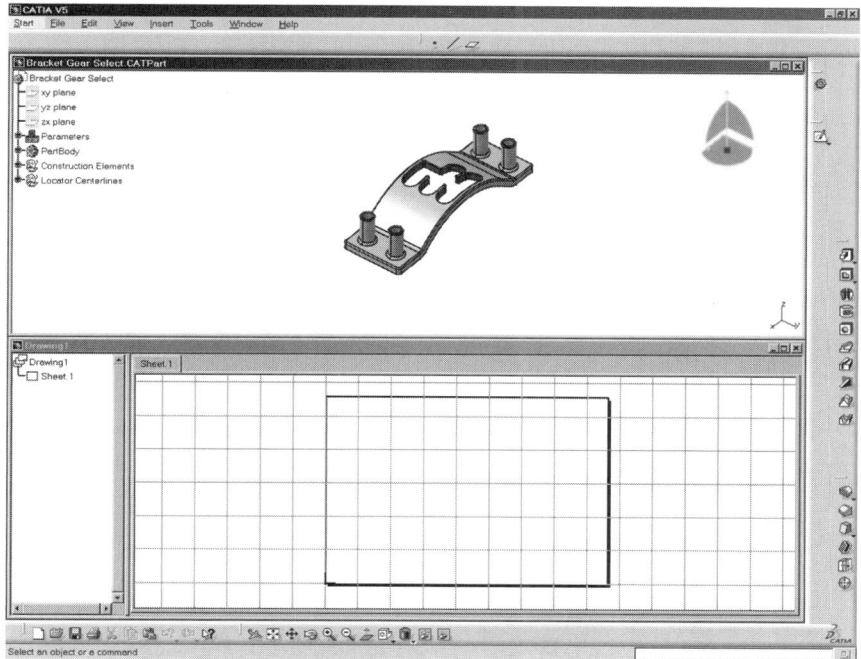

Fig. 9-19.
Window tiling.

5 Make the drawing document active and then click on the Front View icon (shown in figure 9-20) on the View toolbar.

6 Within the model document, select the front planar surface of the bracket to be the reference plane of orientation for the creation of the front view, as shown in figure 9-21.

The front view of the part, surrounded by a dashed green frame, will appear in the CATDrawing. This green frame means that this view is active. To the right of this view is a view manipulation compass tool. By clicking on the arrows you can change the orientation of the active view to obtain the view you want to start with.

7 Click on the center of the compass to accept and finalize the desired position, as shown in figure 9-22.

NOTE: *The completed front view will be surrounded with a red dashed frame.*

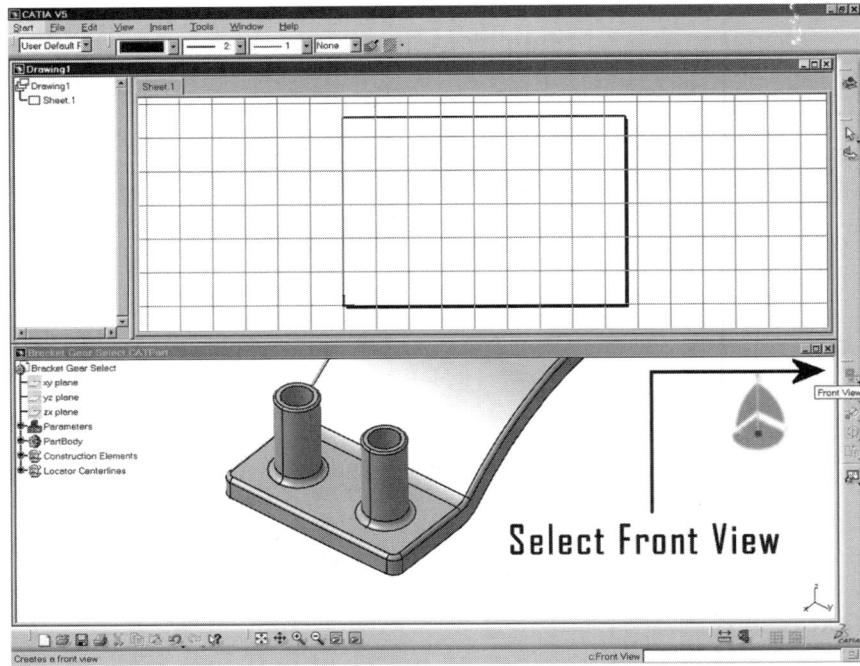

Fig. 9-20.
Front View
icon.

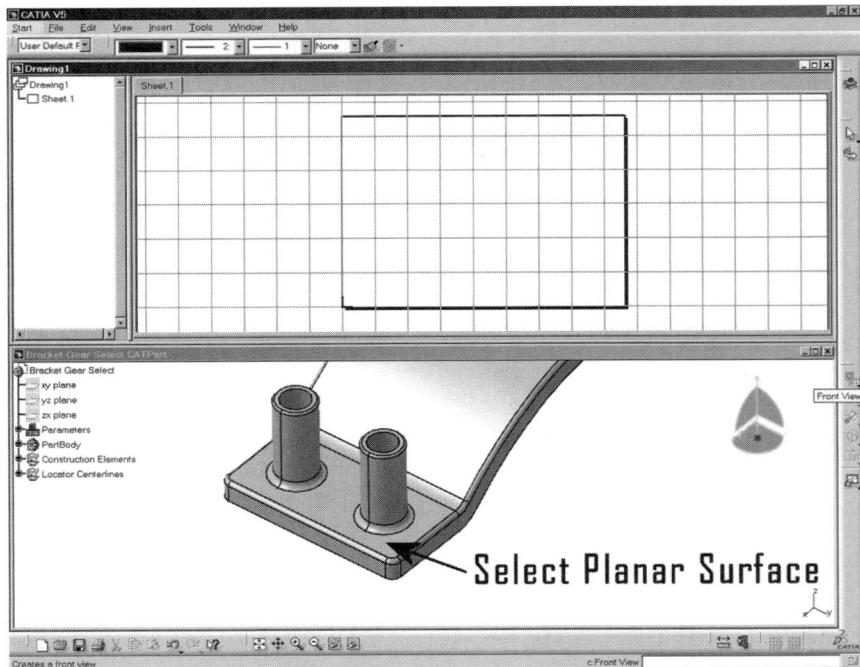

Fig. 9-21. Final
position in front
view.

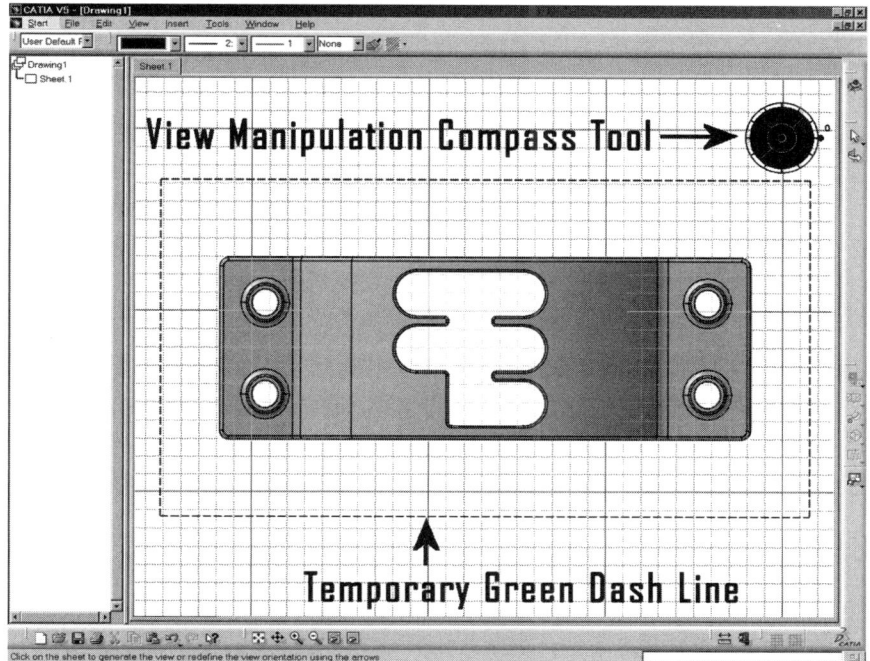

*Fig. 9-22.
Final
position in
front view.*

8 Maximize the drawing document to full screen.

The front view is now complete. However, it appears that the format seems to be too small to contain the part. To correct this, continue with the following steps.

9 Select Front View in the specification tree, and then change the scale from 1:1 to 1:2 in the Properties window.

10 Click on the Project View icon, located on the Views toolbar. Begin rotating the cursor around the active view to quickly display the various projection views that can be created from this view. Click on the top display view, shown in figure 9-23, to accept the completion of the projection view.

Note that the nonactive views are surrounded by a blue frame. A view may be turned active by double clicking on the view or selecting a view option in the specification tree.

11 Create a right-side view of the Create a right-side view of the bracket using the Projection View option. Using the mouse cursor, select and move each view to finalize the desired positions within the drawing. The drawing should now look like that shown in figure 9-24.

Fig. 9-23.
Top view.

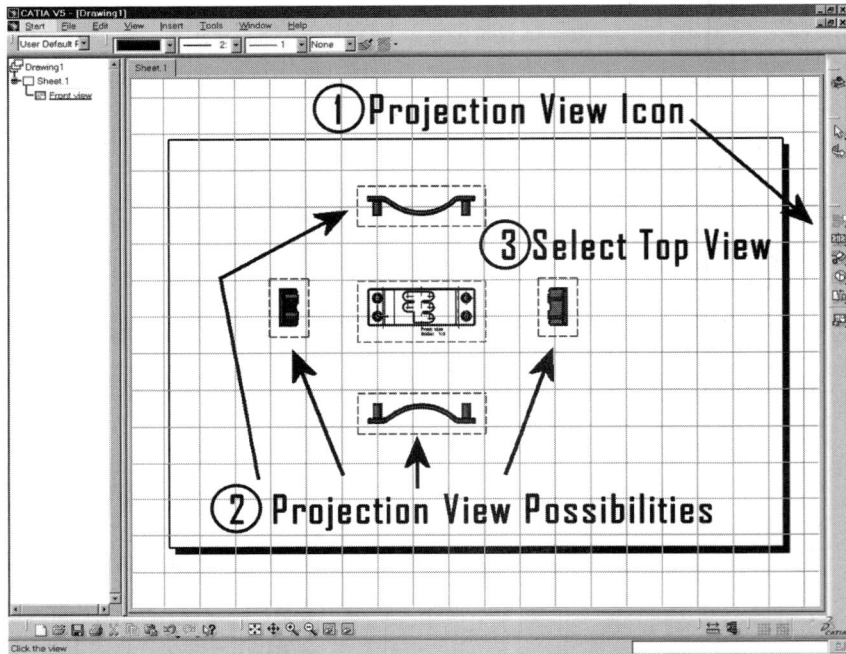

Fig. 9-24.
Drawing
views.

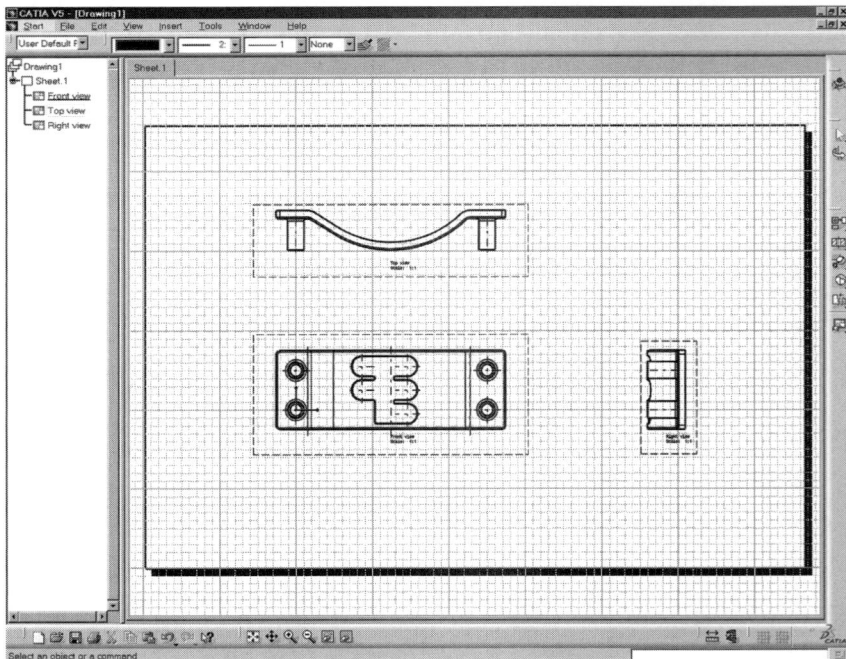

NOTE: *Projection views move along the line of projection from the parent view. Moving the parent view also moves the associated projection views.*

12 Click on the Default Dimension icon on the Dimensions toolbar. While holding down the Ctrl key, select the outer vertical edges of the bracket from the top view. Place the overall horizontal length of the bracket above the part, as shown in figure 9-25.

NOTE: *CATIA V5 will create a dimension from the initially selected element until another element is selected, via multi-selection using the Ctrl key.*

*Fig. 9-25.
Top view
dimension.*

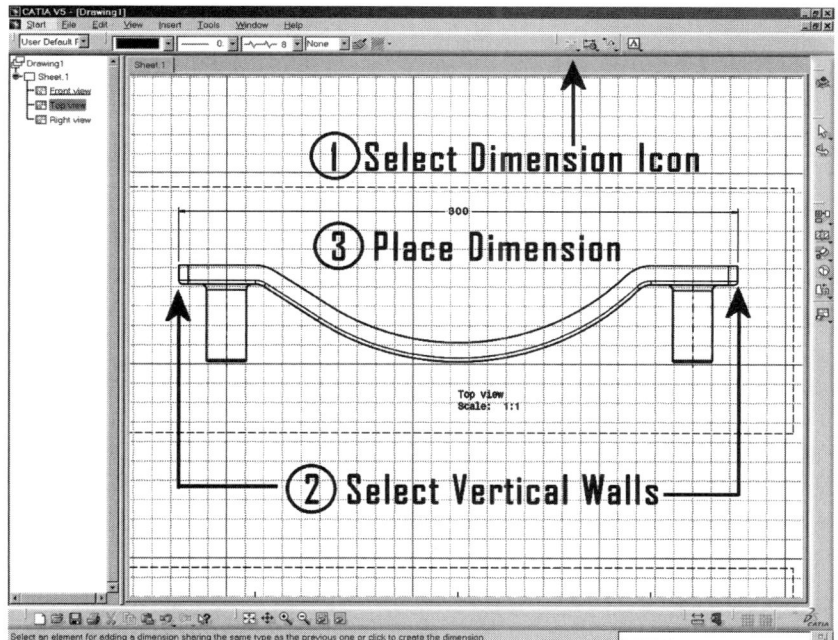

13 Using the same dimensioning techniques, create the dimensions shown in figure 9-26.

14 Click on the Text icon on the Annotations toolbar and create the note *Bracket Gear Select*. Locate the note at the bottom right-hand corner of the drawing, as shown in figure 9-27.

NOTE: *Text annotations will become attached to a particular view if that view is active upon creation of a note. Make sure all views are inactive if the note is to be independent of a view.*

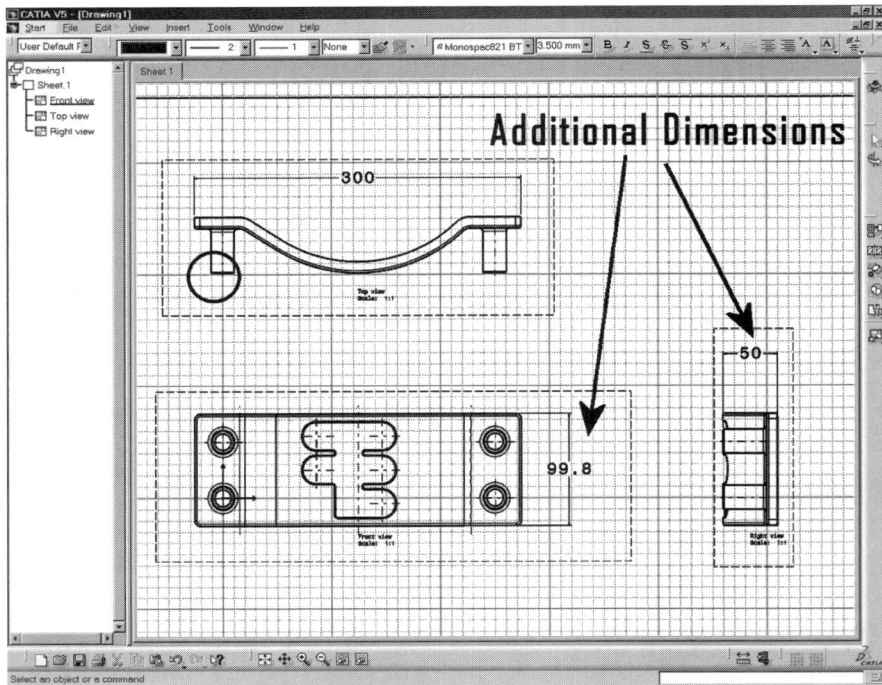

Fig. 9-26.
Additional
dimensions.

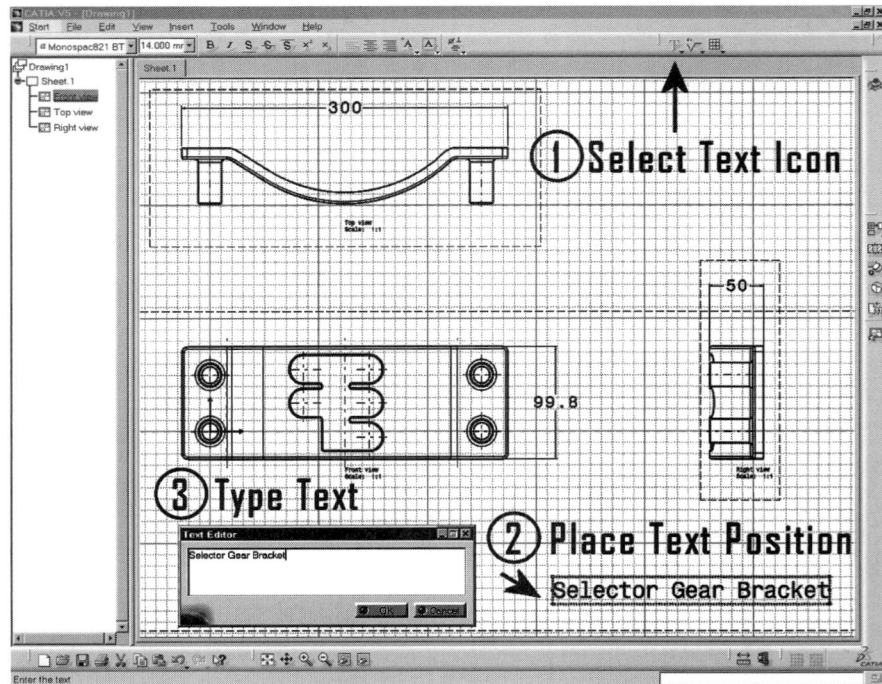

Fig. 9-27.
Text note.

15 Using the Text Properties toolbar, set the text font option to Arial and the letter height option to 10 (mm), as shown in figure 9-28.

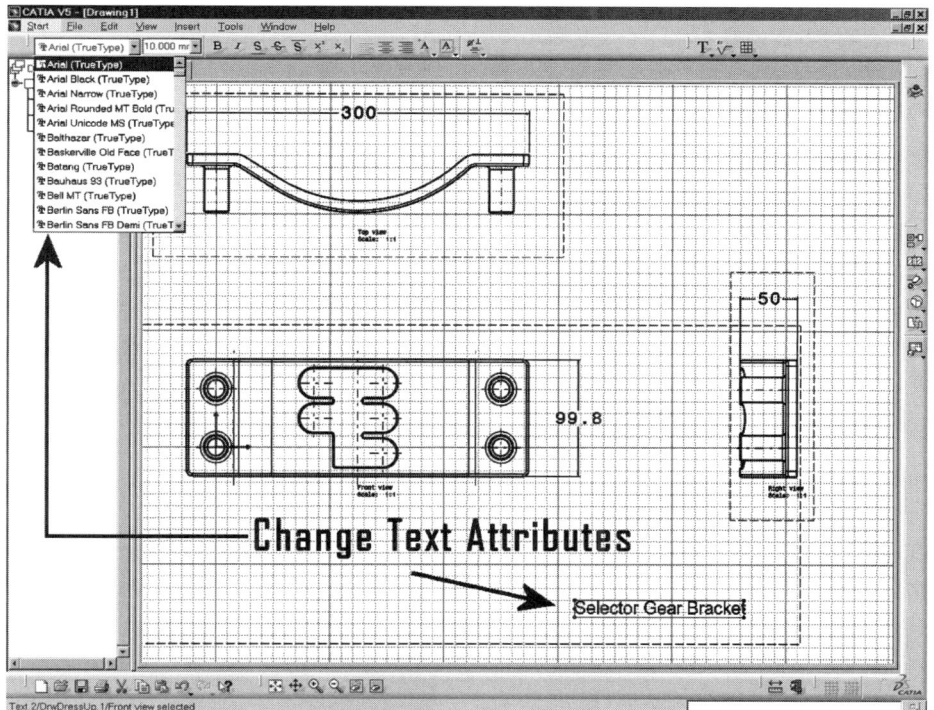

Fig. 9-28. Text modifications .

16 Minimize the drawing document and maximize the *Bracket Gear Select* part.

17 Expand the Constructions Elements Open Body and modify the Boss Constructions Plane, by double clicking from the specification tree. Change the offset distance from 50 (mm) to 35 (mm) and update the part to reflect the change, shown in figure 9-29.

18 Minimize the *Bracket Gear Select* part and maximize the drawing document.

Note that the drawing has not yet been updated to reflect the changes made to the 3D part. This is indicated by the small update icons attached to the drawing specification tree. The updated drawing is shown in figure 9-30.

19 Click on the Update icon to update the views of the drawing to reflect the current design.

Each view updates in the order of creation indicated in the specification tree. The basic drawing is now updated and complete, and is ready to be saved.

Fig. 9-29.
Boss updates.

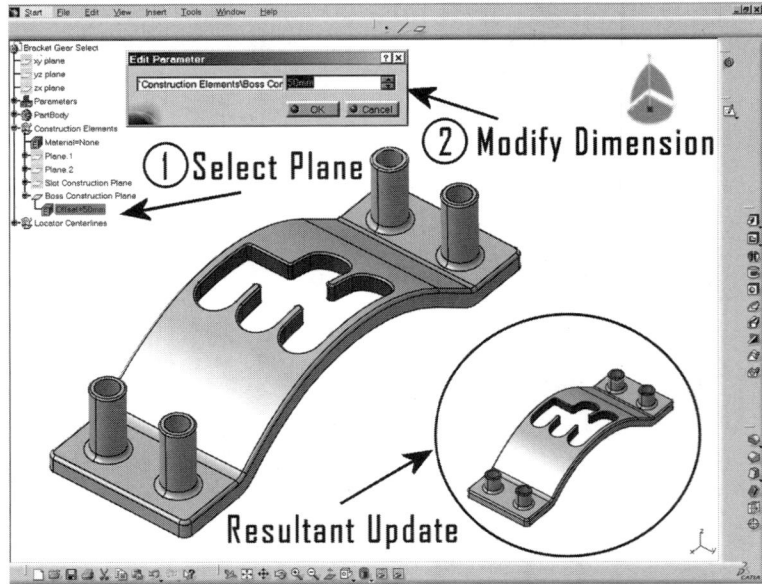

Fig. 9-30.
Updated
drawing.

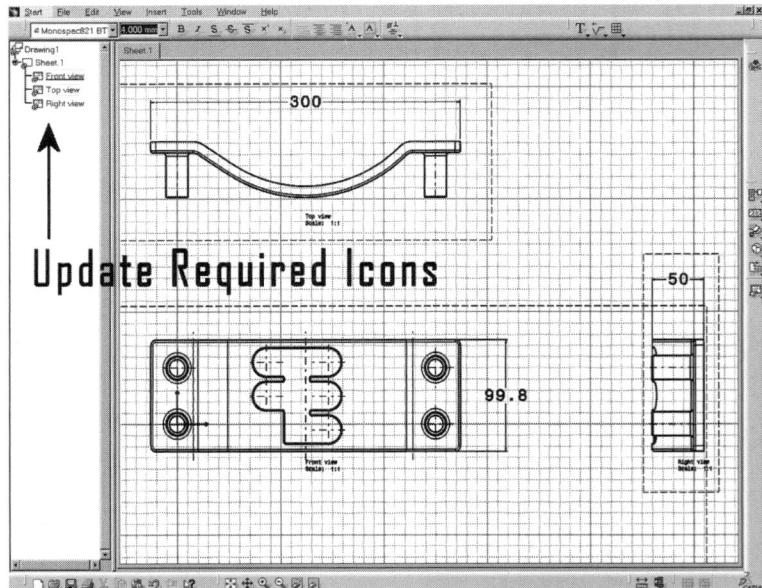

Summary

The Drafting workbench contains the tools used to create detailed drawings. You have the ability to create detailed views that incorporate dimensions, notes, symbols, and all elements necessary in capturing the manufacturing requirements of the design. The associative nature of drawing modules allows for the automatic update of a drawing based on changes made to 3D parts and assemblies. Using the functionality described in this chapter, you should be capable of quickly creating a drawing document for part or assembly design work. Users with 2D experience in other CAD systems should have a minimal learning curve regarding proficiency in working with the Drafting workbench.

Review Questions

1   Name two types of drafting methods (workbenches) available in CATIA V5.

2   In what instances would one method of question 1 be used over the other?

3   Name three types of views available for creation from the primary view.

4   When creating a new drafting document, is it advantageous to have the model already loaded into session?

5   What options does CATIA V5 offer for creating dimensions within the 2D drafting environment?

6   Why is one dimensioning option more critical than the other?

7   How are dimensions and text notes modified with CATIA V5?

8   Is it possible to modify 3D model dimensions within the 2D drafting environment?

Index